CHEMISTRY
INDUSTRY *AND THE*
ENVIRONMENT

CHEMISTRY
INDUSTRY *AND THE*
ENVIRONMENT

JAMES N. LOWE
University of the South

WCB
Wm. C. Brown Publishers
Dubuque, Iowa • Melbourne, Australia • Oxford, England

Book Team

Editor *Craig S. Marty*
Developmental Editor *Robert Fenchel*
Production Editor *Cathy Ford Smith*
Designer *K. Wayne Harms*
Art Editor *Mary E. Powers*
Photo Editor *Diane S. Saeugling*
Permissions Coordinator *Mavis M. Oeth*

Wm. C. Brown Publishers
A Division of Wm. C. Brown Communications, Inc.

Vice President and General Manager *Beverly Kolz*
Vice President Director of Sales and Marketing *Virginia S. Moffat*
Marketing Manager *Christopher T. Johnson*
Advertising Manager *Janelle Keeffer*
Director of Production *Colleen A. Yonda*
Publishing Services Manager *Karen J. Slaght*

Wm. C. Brown Communications, Inc.

President and Chief Executive Officer *G. Franklin Lewis*
Corporate Vice President, President of WCB Manufacturing *Roger Meyer*
Vice President and Chief Financial Officer *Robert Chesterman*

Cover photo © H.R. Bramaz/Peter Arnold, Inc.

A Times Miror Company

Library of Congress Catalog Card Number: 92–75073

ISBN 0–697–17087–X

Printed in the United States of America by Wm. C. Brown Communications, Inc.,
2460 Kerper Boulevard, Dubuque, IA 52001

10 9 8 7 6 5 4 3 2 1

This text was written for a one-semester course taken by non-science majors seeking to both satisfy a science requirement and gain some exposure to chemistry at a college level. Students reading this text will gain insight into the processes of science and an appreciation of the understandings that science offers.

The book was written with the conviction that chemistry offers important insights into both the benefits of technology and the environmental problems that accompany the use of this technology. It differs from the author's earlier text, *Worlds of Chemistry*, in its greater emphasis on the impact that science and technology have on our lives and on the quality of our environment. I hope that students using this text will increase their awareness and understanding of environmental issues so they might make more informed choices in their personal and public lives.

This text, in its early chapters, presents some of the history of chemistry. Many of the early experiments that enabled scientists to propose theories about atoms and molecules are simple in concept. I present some of the reasoning from experimental evidence that marked the development of ideas concerning atomic structure and chemical bonding. I also make some of the philosophy of science explicit. Students can better understand science by knowing some of its history, and they can better relate their study of science to other disciplines when its methodology is considered.

Applications are used to make concepts come alive for students and to arouse student interest in the underlying chemistry. In contrast with other texts for non-science students, I have presented fewer examples and sought to develop more complete connections between concepts and applications. For example, material on buffers is presented without using equilibrium calculations. The use of radioactive isotopes to estimate the age of the earth and the use of carbon-14 dating to trace human history illustrate nuclear decay reactions. A consideration of the rates of chemical reactions leads into a discussion of enzyme-catalyzed reactions and the mode of action of sulfa and penicillin. The control of automobile emissions is discussed when introducing the concepts of energy and entropy. The examples discussed in these chapters are ones that people encounter in reading newspapers and magazines; they are important for understanding the impact of science on modern thought and technology.

In the later chapters of the text, I emphasize the relationship between the structures of materials and their physical and chemical properties. The chemical structures of silicates and clays are related to the properties of glass and ceramics. The ion exchange properties of soils are presented along with the production of chemical fertilizers. The history of the introduction of copper, iron, and aluminum is related to differences in the chemistry of these metals and their compounds. Petroleum and gasoline, structure-property relationships for soaps, detergents and membranes, and environmental issues surrounding the use of halogenated hydrocarbons are used to illustrate some of the chemistry of organic compounds. Structure property relationships of synthetic polymers illustrate the way that scientists can design molecules with a wide variety of desired properties. A discussion of the important biochemical polymers provides a background for understanding the growing field of genetic engineering.

■ Special Chapters

Two chapters in the text focus on topics important for the consideration of technology and the environment and for science and society. In chapter 7, a consideration of the interaction of radiant energy with matter leads to a consideration of the environmental threats posed by the depletion of the ozone layer and by global warming. In chapter 12, problems associated with the growth of the use of energy and with extensive reliance on petroleum, coal, or nuclear energy are discussed.

■ Asides

Asides are an important feature of this text. They are used to enrich the reading, for subsequent material does not build on them. The asides go beyond the material under discussion in a variety of directions. They may present additional history, an aspect of the philosophy of science, an application to another scientific discipline, or a short exploration deeper into the topic at hand.

■ Reflections

Reflections are a novel feature of this text. In the reflections, I offer short essays that attempt to place concepts encountered in chemistry in a wider societal context. I may raise more questions than I offer answers. If the reflections provoke thought and promote a dialogue between student and student or student and instructor, they will have served their purpose well.

■ Questions

Questions are an important part of any chemistry text. Many of the questions in this text ask the student to think about an experiment or an application of chemistry. Students enjoy discovering that they too can understand experiments and can appreciate the scientific basis of some of their everyday experiences. Many of the questions are open ended and are addressed particularly to non-science majors.

■ Acknowledgments

I am grateful for the suggestions and support from students and colleagues at Sewanee. I appreciate the many students who have enrolled in Chemistry 100, a course taken to satisfy part of the math/science requirement. Their interest and occasional enthusiasm have been of great help in bringing this book to completion.

I appreciate the encouragement and contributions of people at William C. Brown. They have been supportive of the thrust of the book, and they have been committed to producing a quality textbook at an affordable price. Reviewers of this book have offered thoughtful criticisms and many detailed suggestions for improvements. They helped make this a better book. They are

Chester L. Campbell
Northeastern Oklahoma A & M College

Richard L. Snow
Brigham Young University

Karen C. Weaver
University of Central Arkansas

Clifford Meints
Simpson College

George Fleck
Smith College

Marvin J. Albinak
Essex Community College

Richard H. Eastman
Professor Emeritus, Stanford University

Allan K. Hovland
St. Mary's College of Maryland

My wife Martha, a special education teacher in a rural Tennessee school system, and my daughter Leah, a waitress and sometime director of plays in Minnesota, provided additional valuable criticisms from their non-science perspectives. I am especially grateful to the University of Kentucky Faculty Scholars Program and the University of the South for providing the opportunity and the time to complete this project and to the Pew Charitable Trusts for providing the much needed funding. Without their help this project would have been far more difficult.

1

From the Beginnings of Modern Chemistry to Dalton's Atomic Theory

Chemistry is a science that deals with everyday substances: the gases of the air, the rocks and soil of the earth, and the waters of the rivers and seas. Chemistry treats transformations of matter such as those occurring in the combustion of fuels, the smelting of metals, and the cooking of foods. The vocabulary of chemistry is familiar and pervasive. People eat fruits that are acidic and vegetables that are good sources of vitamins. They wash their hair with shampoos which may or may not be pH balanced. They wear rings of gold and clothing of natural fibers or they wear silver with polyester.

Chemistry played an essential role in the emergence of the modern world. The ability to make high quality steel made possible the machinery and transportation improvements that were the heart of the industrial revolution. The technology to produce strong, lightweight aluminum ushered in the age of flight. Fertilizers made agriculture more productive, and synthetic polymers replaced wool and cotton to help clothe growing populations. New drugs conquered many age-old diseases and ushered in the era of modern medicine. New explosives made war more terrible and yet helped to build dams that insured adequate water for cities and farms.

With the benefits of modern technologies come new and formidable problems. What do we do with our wastes? How do we provide clean water to drink and clean air to breathe? How long can we prolong life without answering questions about the quality of life? Chemistry offers insight into both the achievements of modern industry and the environmental issues that now confront society. It is a science that can be used to help provide enlightened responses to the problems we face.

Chemistry is the study of the composition and transformations of matter. Chemists seek to understand the structure of materials and the changes that occur when one material is converted to another. The development of chemistry as a science shows how people gained an understanding of the composition of matter and its changes. Even though chemical change has always been a part of human experience, the emergence of modern chemistry as a science is a recent development. During the period of 1750–1815, individuals performed experiments that led to discoveries of regular patterns in nature. Efforts to explain regularities in nature gave rise to the atomic theory that is basic to modern chemistry.

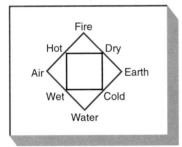

Figure 1.1
Early Greek ideas about matter emphasized properties of substances. For example, the element earth was thought to be dry and cold. The properties of real materials were thought to reflect the presence and properties of the elements earth, air, water, and fire.

■ Antecedents of Chemistry

The idea that all matter was composed of a small number of *elements* first arose in Greece. Greek philosophers considered air, earth, fire, and water to be the four elements from which all other substances were made. Elements endowed substances with properties: hot or cold, wet or dry, light or heavy (see Fig. 1.1). For example, a metal obtained by heating an ore was considered a compound of the elements earth and fire. It had substance due to the presence of earth, and it was shiny because it contained fire. Changes in materials were explained by changes in the quantities of the four elements within the compound. This viewpoint was a forerunner for the development of scientific ideas, as explanations were tested against observations.

The name, chemistry, is derived from *alchemy* which describes a wide variety of ideas and practices, over several centuries, and in many countries. Alchemy originated in Alexandrian Egypt. It flourished in the Arabic world and entered Europe through Moorish Spain. Its origins included both practical technology utilized by artisans and speculations about elemental nature as devised by the Greeks.

One aim of alchemy was *transmutation,* the effort to turn base metals such as iron into gold. From a modern chemical standpoint such an effort is doomed to failure, but to some alchemists it made sense. Gold was considered to be the noblest and most incorruptible of metals. According to Islamic records from the tenth century, the following doctrine formed a basis for attempting to transmute gold. Just as health is the state of perfection for the body, gold is the state of perfection for metals. Metals strive for the essence of goldness. In practice, alchemists sought to remove that which was imperfect and to add that which was lacking to attain perfection. At some times and in some places, alchemy also included recipes for making cheaper metals, such as silver, appear as gold.

Alchemists made important discoveries in the areas of early metallurgy and pharmacology. They extracted plant materials and they purified volatile materials by distillation. In the course of their work they developed recipes for the production of some acids and bases used in the investigations that established modern chemistry. Among the materials first described by alchemists are phosphorus, arsenic, antimony, bismuth, and zinc, all elements in the modern sense. Both apparatus and vocabulary in chemistry have been passed down from alchemy, and alchemy provided a crude framework of explanations for natural phenomena.

Unfortunately, competing and often confusing explanations for chemical changes abounded. Alchemy was, at times, forbidden by civil or religious authorities and practiced as a secret art. Its vocabulary was obscure, and its symbols carried both chemical and mystical meanings. For modern chemistry to develop, it was necessary to sort the natural explanations for chemical changes from the mystical interpretations of those changes.

■ Setting the Stage—Questions Facing Early Chemists

What are the properties of a substance? Is a given substance pure or is it a mixture? These are not trivial questions, and their answers were slow in coming. For example, even the description of something as familiar as water poses problems. Samples of water differ, for instance, sea water tastes salty, some spring water bubbles, and other spring water smells foul and tastes bad. How can one know if water is pure? Which properties of a sample of water are due to impurities and which are the properties of water itself?

If the analysis of water poses problems, consider the greater problems posed by a metal such as copper. Early samples of metals usually contained contaminants from the ore, and ores mined in different places often contained different impurities. Which sample of copper might be purer? How would one decide?

Greek philosophers, alchemists, and early chemists sought to explain chemical change in terms of elements and compounds. But what are elements? And what are compounds? Some substances had been observed in a wide variety of chemical reactions. For example, copper was obtained from a variety of ores, and sulfur (brimstone) reacted with most metals. One philosopher argued for a two element theory of matter based on copper and sulfur. Are copper and sulfur elements, or are fire and earth? The problem was how to decide whether a given substance was an element, a compound, or a mixture.

In asking these questions, a scientific attitude is implied, an attitude not found among educated people in the Middle Ages. Medieval Western attitudes were influenced by the Greek philosopher Plato. According to Plato's philosophy, real things were imperfect representations of the ideal just as circles drawn in the sand are imperfect representations of the ideal circle. When the goal of study is to understand the ideal or essence, experimental knowledge is of lesser value. The early chemist's decision to study the particular properties of particular samples of materials by observation represented an important change in people's thought.

■ The Beginnings of Modern Chemistry

The philosopher Francis Bacon (1561–1626) set forth a program for experimental science, calling for the direct observation of nature. In his preface to *True Directions Concerning the Interpretation of Nature,* he wrote:

> Those who have taken upon them to lay down the law of nature as a thing already searched out and understood . . . have therein done philosophy and the sciences great injury. For as they have been successful in inducing belief, so they have been effective in quenching and stopping inquiry; and have done more harm by spoiling other men's efforts than good by their own. .

In 1661, Robert Boyle (1627–1691) published *The Skeptical Chemist.* Boyle wrote that elements were:

> certain primitive and simple, or perfectly unmingled bodies; which not being made of any other bodies, or of one another, are the ingredients of which all those called perfectly mixt bodies are immediately compounded, and into which they are ultimately resolved.

Boyle criticized Greek ideas concerning a four-element explanation of matter, and he critically examined and rejected some of the claims of alchemy. Boyle's definition set the stage for the further evolution of ideas concerning elements and compounds, but it failed to indicate how one could decide if a particular substance was an element.

To find better explanations, additional and careful observations were required. Those qualities most useful for the characterization of substances had to be identified, and ways to prove or disprove the interpretations of chemical changes needed to be found. In this process, there was a shift from qualitative observations to quantitative measurements. The qualities "hot" and "cold" were replaced by thermometer readings; "light" and "heavy" by balance readings. Modern chemistry began with discoveries that disproved plausible explanations of natural phenomena based on

Figure 1.2
Early scientists collected gases by the displacement of water from an inverted container.

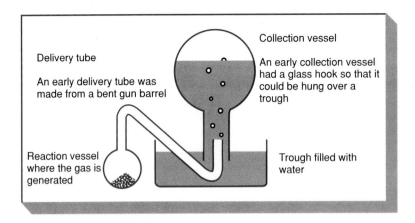

four elements. Air was shown to be a mixture, and water, a compound. We will examine the intimate relation of these experimental observations to the development of chemical explanations.

■ The Study of Gases

During the period from 1760 to 1780, scientists prepared, collected, and described a number of pure gases. This work contributed to the development of a new understanding of the nature of gases. Gases were collected by the displacement of water (Highly water soluble gases were collected over mercury.) from an inverted container as shown in figure 1.2. Properties of each gas were studied—Did it burn? Did it support the combustion of a candle? Did it dissolve in water? Similarities and differences of the collected gases were catalogued.

Because a gas separates from the liquid or solid from which it is generated, gases prepared by synthesis were generally very pure. This is important since one can conclude that the properties of these gases were not altered by the presence of impurities.

Acids and bases, substances that were known to the alchemists, were used to produce some gases. *Acids* taste sour, fizz with soda, and cause some vegetable dyes to change color. *Bases* taste bitter and reverse the color changes caused by acids. For the following studies concerning the generation and testing of gases we need to know two acids and one base. The common acids are *sulfuric acid,* prepared by heating an iron salt (hydrated ferrous sulfate), and *hydrochloric acid,* prepared by adding sulfuric acid to common salt and distilling a gas into water. The common base is *quicklime,* produced by heating limestone.

■ Air and Combustion

Early scientists were interested in combustion, for it was fire that brought about chemical change. Air would support the burning of a candle or the respiration of a mouse under a bell jar. When the flame went out or the mouse died, the remaining "air" had different properties. A part of this gas called "fixed air" readily dissolved in limewater (a solution of base made by dissolving quicklime in water) to produce a white precipitate (A *precipitate* is a solid formed by a reaction occurring in solution.). Another part, the "noxious air" remaining after the fixed air was dissolved in base, did not support combustion or respiration (see Fig. 1.3).

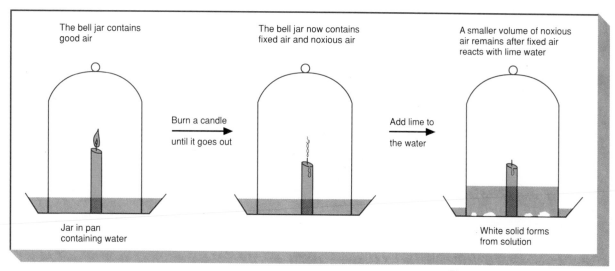

The bell jar contains good air

The bell jar now contains fixed air and noxious air

A smaller volume of noxious air remains after fixed air reacts with lime water

Burn a candle
until it goes out

Add lime to
the water

Jar in pan
containing water

White solid forms
from solution

Figure 1.3
In the eighteenth century, scientists studied the changes in air that accompanied the burning of a candle. In the course of their studies, they prepared fixed air now known as carbon dioxide and noxious air now known as nitrogen.

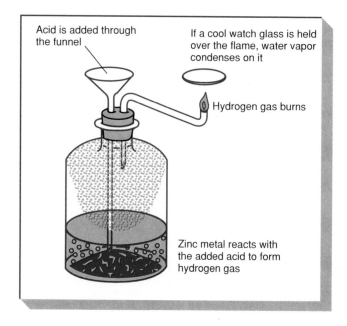

Acid is added through the funnel

If a cool watch glass is held over the flame, water vapor condenses on it

Hydrogen gas burns

Zinc metal reacts with the added acid to form hydrogen gas

Figure 1.4
Hydrogen gas is produced by the reaction of zinc metal with dilute sulfuric acid. Water is produced when the hydrogen gas burns in air.

The fixed air produced by combustion and the noxious air remaining after combustion were two of the first gases to be described and characterized. Fixed air, now known as *carbon dioxide*, was produced by respiration or combustion. It could be prepared in a pure form and studied by adding acid to the white precipitate formed by the reaction of carbon dioxide with limewater. Noxious air, now known as *nitrogen*, remained after the carbon dioxide was dissolved in base.

Early scientists also produced "inflammable air" or hydrogen, by the action of sulfuric acid or hydrochloric acid on zinc or iron. Hydrogen gas burns in air, and by condensing the vapor above the flame on a cold surface, it was seen that water is produced by the burning of hydrogen (see Fig. 1.4).

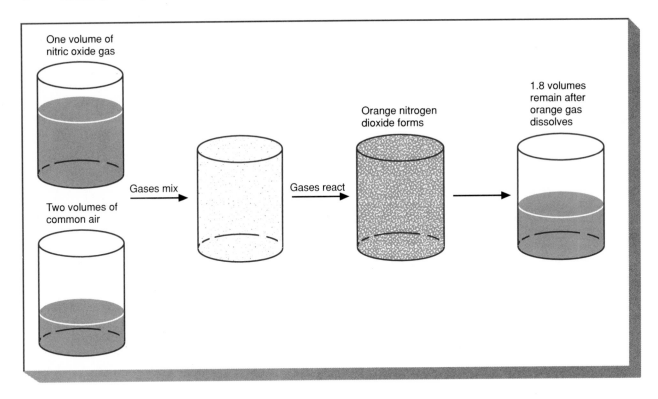

One volume of nitric oxide gas

Two volumes of common air

Gases mix

Orange nitrogen dioxide forms

Gases react

1.8 volumes remain after orange gas dissolves

Figure 1.5
Joseph Priestley devised a test for the goodness of air that involved the reaction of air with nitric oxide. A larger final volume of gas remained if the air had been spoiled by burning or breathing.

■ Joseph Priestley Discovers Oxygen

Joseph Priestley (1733–1804) prepared and described a number of previously unknown gases. In the course of his investigations, he developed a chemical test for the "goodness" of air, that is, its ability to support respiration. (Modern names rather than the names given by Priestley are used to describe his test.) When 2.0 L of common air was mixed with 1.0 L of nitric oxide over water, orange fumes formed and the volume decreased to 1.8 L as the orange gas dissolved (see Fig. 1.5). The remaining gas, being mostly nitrogen, would not support combustion. If the common air had been spoiled by breathing or burning, a larger final volume of gas remained.

Priestley used this test in experiments that led to the discovery of oxygen. He found that heating an orange solid known as mercury calx produced liquid mercury and a gas, which he collected and studied. Priestley found that glowing charcoal placed in the gas burst into flame. One day he tested the gas for its goodness as air by mixing two volumes of the gas with one volume of nitric oxide. Orange nitrogen dioxide formed and then dissolved in water. The next day, he happened to put a glowing candle in the gas remaining after the test. The candle burst into flame. Unlike common air, all of the gas obtained by heating the mercury calx would support combustion.

Priestley had discovered oxygen. The mercury calx that Priestley heated is now known to be mercury oxide. Priestley had decomposed mercury oxide into mercury and oxygen.

mercury oxide → mercury + oxygen

Priestley's test for the goodness of air can now be interpreted; nitric oxide reacts with oxygen in the air to form orange nitrogen dioxide.

nitric oxide + oxygen → nitrogen dioxide

When the reaction is run using air as the source of oxygen, the nitrogen in the air does not react. It is present both before and after the reaction with oxygen.

Nitric oxide + air → nitrogen dioxide + nitrogen
Nitric oxide + oxygen + nitrogen → nitrogen dioxide + nitrogen

When nitrogen dioxide dissolves in water, the total volume of gas decreases, and only nitrogen and excess nitric oxide remain. If the air had been used to support combustion, less oxygen would be present, less nitrogen dioxide would form and then dissolve, and a larger volume of gas would remain.

The discovery of oxygen was critical to the development of a clear distinction between the various pure gases and air. Because oxygen is responsible for the ability of air to support combustion, the identification and description of oxygen was also critical for unraveling the mystery of that most fundamental chemical process: fire.

■ Aside

Planning and Chance in Science

Joseph Priestley was a dissenting churchman in England who was persecuted for his religious and political views. For example, his sympathies toward the French Revolution led to his home and library being vandalized. He ended his career as an exile in America where he declined appointments to become a Congregational minister or a professor of chemistry at the University of Pennsylvania, founded by Benjamin Franklin, or at the University of Virginia being planned by Thomas Jefferson.

Was the discovery of oxygen an accident? Priestley explains:

I cannot, at this distance of time, recollect what it was that I had in view in making this experiment; but I know I had no expectation of the real issue of it. Having acquired a considerable degree of readiness in making experiments of this kind, a very slight and evanescent motive would be sufficient to induce me to do it. If, however, I had not happened for some purpose, to have a lighted candle before me, I should probably never have made the trial; and the whole train of my future experiments relating to this kind of air might have been prevented.

Seemingly chance discoveries frequently have played important roles in the development of science. Yet Joseph Priestley's curiosity and willingness to experiment illustrate important characteristics of many scientists. And, as Louis Pasteur, a nineteenth century scientist, said, "Chance favors only the mind that is prepared."

■ The Law of Conservation of Mass

Antoine Lavoisier (1743–1794), a Frenchman working around the time of the American Revolution, played a most important role in the birth of modern chemistry. Lavoisier systematically set out to bring order to the emerging science of chemistry.

Lavoisier studied the quantities of materials involved in chemical reactions. For example, he studied the formation and decomposition of mercury calx, the

Figure 1.6
Lavoisier heated mercury under air to form mercury calx. He measured the decrease in the volume of air. He also found that the final weight of the mercury plus the mercury calx plus the residual air is the same as the weight of the original mercury plus the original air.

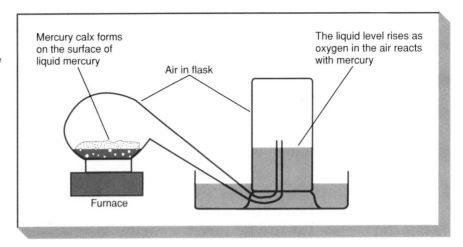

compound from which Priestley obtained oxygen. In one set of experiments, Lavoisier prepared mercury calx by heating mercury under a closed atmosphere of air in a furnace as shown in figure 1.6. He weighed the red powder that had formed, and he measured the decrease in the volume of air. When this sample of mercury calx was later heated more intensely, the solid lost weight to reform mercury and oxygen. The volume of oxygen was equal to the decrease in the volume of air during the formation of the solid.

Lavoisier carefully measured the mass of each material reacting and the mass of each product formed for a large number of reactions. He found that the total mass of the substances reacting was equal to the total mass of the products formed in each reaction, for example:

$$10 \text{ g sulfur} + 10 \text{ g oxygen} \rightarrow 20 \text{ g sulfur oxide}$$
$$7.75 \text{ g phosphorus} + 10.00 \text{ g oxygen} \rightarrow 17.75 \text{ g phosphorus oxide}$$
$$3.0 \text{ g carbon} + 8.0 \text{ g oxygen} \rightarrow 11.0 \text{ g carbon dioxide}$$

Lavoisier generalized his observations to formulate the *law of conservation of mass*. In a chemical reaction, the total mass of the reactants is always equal to the total mass of the products, so matter is neither created nor destroyed. *Mass is a fundamental measure of the quantity of matter.* The greater the mass of an object, the greater the force required to give it acceleration. (The weight of an object that we measure on earth is the gravitational force acting on its mass. Objects placed in orbit around the earth have the same mass as on earth even though they have less weight.)

Scientists, like Lavoisier, use laws to summarize the results of observations and to describe ways that the behavior of the physical world follows regular patterns. Scientific laws predict the results of new experiments and by generalizing from several experiments to all similar experiments in either formulating or using a law, scientists assert a belief that nature is predictable.

Sample Calculation 1

When strongly heated, limestone loses carbon dioxide to form lime. What mass of limestone would be needed to produce 100 g of lime if 78.5 g of carbon dioxide is also produced?

$$\text{limestone} \rightarrow \text{lime} + \text{carbon dioxide}$$
$$? \text{ g} \qquad 100 \text{ g} \qquad 78.5 \text{ g}$$

According to the law of conservation of mass, the mass of the reactant(s) must equal the mass of the products.

$$\text{Mass of limestone} = 100 \text{ g} + 78.5 \text{ g} = 178.5 \text{ g}$$

■ Lavoisier and Combustion

The examples given to illustrate the law of conservation of mass happen to involve combustion or burning. Lavoisier was the first to understand that when substances burn, they combine with oxygen. Laviosier's interpretation of combustion replaced an earlier explanation, the phlogiston theory that had been widely held during the previous hundred years.

According to the phlogiston theory, things rich in phlogiston burned. In combustion, phlogiston departed, and the remaining ash was "dephlogistonated." This theory had a certain success in explaining observations. For example, carbon burned because it was rich in phlogiston. Some metal calxes (metal oxides) could be converted to the uncombined metals by treating them with carbon. Because phlogiston from the carbon passed into the calx, these metals (such as iron) could burn.

It is instructive to consider why one "scientific" theory replaced another. The phlogiston theory was satisfactory as long as qualitative explanations were required, but it proved less attractive in a quantitative way. If a log burned because phlogiston left the wood, then the mass of the ashes is less than that of the wood because of the loss of phlogiston. However, iron and other metals gain mass when burning. To explain this observation, the departing phlogiston would have to have negative mass. The theory that phlogiston sometimes has positive mass and at other times has negative mass was a far more complicated and less attractive theory than the one proposed by Lavoisier.

Lavoisier extended his research into the field now called biochemistry. Carbon burns in air to give carbon dioxide. Since that same oxide is present in exhaled air but not in inhaled air, Lavoisier recognized that we obtain energy by the burning of foods to form carbon dioxide which we then exhale.

This promising line of inquiry was cut short by the beheading of Antoine Lavoisier in 1794. Although his services to the government of France ranged from promoting the metric system to improving gunpowder, his role as a bureaucrat and his association with the collection of taxes on tobacco led to his trial and execution during the French Revolution.

■ Operational Definitions Permit Science to Progress

Scientists use *operational definitions* to pose and resolve questions in science. To illustrate the use of operational definitions, consider a simple example. "Who is the smallest person in a class?" The question is not posed in terms of operations, so people may supply their own definitions and may disagree on the answer. Change the question to "Who is the shortest person in a class?" or "Who is the lightest person in a class?" Now measurements, height or weight, can be used to answer the questions posed.

Lavoisier introduced and used an operational definition to distinguish compounds from elements. Elements had been defined as those simple chemical substances from which all other substances were formed. But what does it mean to say

Figure 1.7

Lavoisier's table of the elements.

From Lavoisier, *Traite elementaire de chimie,* 1789.

	Noms nouveaux.	Noms anciens correfpondans.
Subftances fimples qui appartiennent aux trois règnes tqu'on peut regarder comme les élémensdescorps.	Lumière	Lumière,
	Calorique	Chaleur. / Principe de la chaleur. / Fluide igné. / Feu. / Matière du feu & de la chaleur.
	Oxygène	Air déphlogiftiqué. / Air empiréal. / Air vital. / Bafe de l'air vital.
	Azote	Gaz phlogiftiqué. / Mofète. / Bafe de la moféte.
	Hydrogène	Gaz inflammable. / Bafe du gaz inflammable.
Subftances fimples non metalliques oxidables & acidifiables.	Soufre	Soufre.
	Phofphore	Phofphore.
	Carbone	Charbon pur.
	Radical muriatique	Inconnu.
	Radical fluorique ...	Inconnu.
	Radical boracique .	Inconnu.
Subftances fimples metalliques oxidables & acidifiables.	Antimoine	Antimoine.
	Argent	Argent.
	Arfenic	Arfenic.
	Bifmuth	Bifmuth.
	Cobalt	Cobalt.
	Cuivre	Cuivre.
	Etain	Etain.
	Fer	Fer.
	Manganèfe	Manganèfe.
	Mercure	Mercure.
	Molybdène	Molybdène.
	Nickel	Nickel.
	Or	Or.
	Platine	Platine.
	Plomb	Plomb.
	Tungftène	Tungftène.
	Zinc	Zinc.
Subftances fimples falifiables terreufes.	Chaux	Terre calcaire, chaux.
	Magnéfie	Magnéfie, bafe du fel d'epfom.
	Baryte	Barote, terre pefante.
	Alumine	Argile, terre de l'alun, bafe de l'alun.
	Silice	Terre filiceufe, terre vitrifiable.

that a substance is simple? Lavoisier assumed that a substance in a chemical reaction is simpler if it has less mass than the substance from which it is formed. Mercury calx cannot be an element, for it was decomposed to mercury and oxygen, each of which has less mass than the original sample of the orange compound. *Chemicals are shown to be compounds when they can be formed from or decomposed into lighter materials.*

Lavoisier proposed that *a substance may be an element if it has not been shown to be decomposed by chemical reactions.* In 1789, Lavoisier published a treatise in which he named 33 substances that he considered to be elements. Many of these substances appear on current tables of elements (see Fig. 1.7).

The ability to use operational definitions to resolve questions speeds up the process of selection between competing explanations. By choosing to use measurement in deciding questions, scientists restrict the domain of scientific inquiry. In particular, many questions of value lie outside the domain of science.

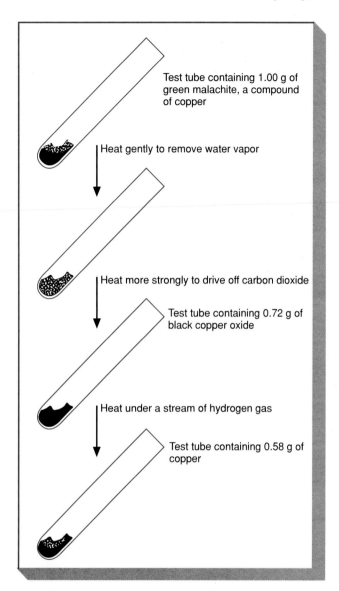

Test tube containing 1.00 g of green malachite, a compound of copper

Heat gently to remove water vapor

Heat more strongly to drive off carbon dioxide

Test tube containing 0.72 g of black copper oxide

Heat under a stream of hydrogen gas

Test tube containing 0.58 g of copper

Figure 1.8
Joseph Proust carefully analyzed samples of malachite mined in different places as well as a sample of the same compound prepared in the laboratory. He found that each sample of malachite contained 58% copper and that each sample of copper oxide contained 80% copper. From these and other analyses he formulated the law of constant composition.

■ The Law of Constant Composition

A period of increased activity in analytical chemistry followed Lavoisier's demonstration of the importance of measuring weights in chemical reactions. A large number of compounds were analyzed by decomposing them to their constituents. About 1800, Joseph Proust (1754–1826) analyzed samples of uncontaminated malachite, a dark-green copper ore, obtained from several different sources. From other chemicals, Proust prepared the same compound present in the malachite. Gentle heating of each sample of malachite produced water. Upon stronger heating, carbon dioxide evolved. The product of these operations was a black oxide of copper. When heated, the black oxide reacted with hydrogen gas to produce water vapor and copper metal (see Fig. 1.8).

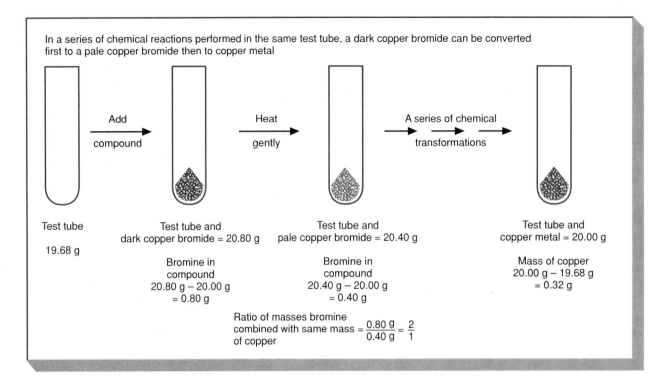

In a series of chemical reactions performed in the same test tube, a dark copper bromide can be converted first to a pale copper bromide then to copper metal

Add compound → Heat gently → A series of chemical transformations → →

Test tube
19.68 g

Test tube and
dark copper bromide = 20.80 g

Bromine in
compound
20.80 g − 20.00 g
= 0.80 g

Test tube and
pale copper bromide = 20.40 g

Bromine in
compound
20.40 g − 20.00 g
= 0.40 g

Test tube and
copper metal = 20.00 g

Mass of copper
20.00 g − 19.68 g
= 0.32 g

Ratio of masses bromine
combined with same mass $= \dfrac{0.80 \text{ g}}{0.40 \text{ g}} = \dfrac{2}{1}$
of copper

Figure 1.9
The law of multiple proportions can be illustrated using the analyses of two compounds composed of the elements copper and bromine.

When the mass driven off in each transformation was expressed as a percentage of the original mass, Proust found that each sample gave the same analysis. Every sample of each intermediate compound had the same percentage copper as all other samples of that compound.

Proust formulated *the law of constant composition*. He generalized that *every compound contains a fixed ratio of elements by mass.* For example, in any sample of pure water, the ratio of the mass of oxygen to the mass of hydrogen is 8:1. The fraction of the total mass contributed by each element is also constant for every compound. For example, carbon dioxide is always three-elevenths carbon by mass.

■ The Law of Multiple Proportions

John Dalton (1766–1844) developed the *law of multiple proportions.* This law was first proposed in 1803, about the same time Dalton formulated an atomic theory. It is the most striking evidence for the atomic theory. To illustrate the law of multiple proportions consider this experiment: two compounds are formed between the pair of elements, copper and bromine (see Fig. 1.9). When one of these copper bromides, a dark shiny compound, is heated, orange bromine gas is evolved and the second compound, a pale copper bromide, remains. This second copper bromide can be converted to copper metal (and other products containing the bromine) in a series of chemical operations. At successive times, there is present a dark shiny copper bromide, a pale copper bromide, and finally, copper metal. Both of the copper bromides contain the same quantity of copper. By subtracting the mass of the copper isolated at the end of the experiment from the mass of each compound, the mass of bromine in each compound can be determined. There is a striking outcome when these masses are compared. The mass of bromine in the dark solid is exactly twice the mass of bromine in the pale copper bromide. (Note that this is a statement about masses. Neither the names nor the formulas of the copper bromides were necessary to show this mass relation.)

The law of multiple proportions generalizes the results of this and similar experiments. *When two compounds are formed between a pair of elements, with a fixed mass of one element, the masses of the second element in the compounds are expressed in small whole number ratios.* Examples of such simple integer ratios are 2:1, 3:1, and 3:2.

Sample Calculation 2

Two gaseous compounds of carbon and oxygen are known. From 6.0 g of carbon, 14.0 g of carbon monoxide or 22.0 g of carbon dioxide can be prepared. Show that the composition of these gases is in accord with the law of multiple proportions.

$$6.0 \text{ g carbon} + \text{oxygen} \rightarrow 14.0 \text{ g carbon monoxide}$$
$$\text{oxygen in } 14.0 \text{ g carbon monoxide} = 14.0 \text{ g} - 6.0 \text{ g carbon} = 8.0 \text{ g}$$
$$6.0 \text{ g carbon} + \text{oxygen} \rightarrow 22.0 \text{ g carbon dioxide}$$
$$\text{oxygen in } 22.0 \text{ g carbon dioxide} = 22.0 \text{ g} - 6.0 \text{ g carbon} = 16.0 \text{ g}$$

The ratio of the masses of oxygen combined with a fixed mass of carbon is 8.0:16.0 or 1:2, a ratio of small whole integers.

■ Aside

Experimental Results Vary from the Ideal

What does it mean to say the results are in a simple integer ratio? This is an idealization of experimental results because measured ratios from individual experiments vary. Consider the experiment performed with compounds of copper and bromine. When a class of students performs the experiment converting samples of the dark copper bromide first to the pale solid then to copper, results vary. Most students find that their measured ratio of masses of bromine in the two compounds lie between 1.90 to 1.00 and 2.10 to 1.00 rather than being exactly two to one. One student may fail to drive all the readily removable bromine from the first compound. Another may expose hot copper to the air and have some of it react with oxygen, giving too great a mass of copper in each of the compounds. These and other experimental errors would cause the measured ratio of masses of bromine to vary. With small errors, the measured ratios are close to the integer ratio, 2:1. Variation is assumed to be due to experimental errors and uncertainties. An assumed, precise regularity of nature is expressed in the scientific law.

■ Dalton's Atomic Theory

Three important generalizations about the composition of pure substances have been discussed. First, mass is always conserved when substances react. Second, a pure compound always contains a fixed ratio of elements by mass. Third, if two compounds form between a pair of elements, the masses of the one element found combined with a fixed mass of the other element in the two compounds are in a simple integer ratio. To the scientist, these regularities in nature invite explanation.

Figure 1.10
In 1803, John Dalton formulated an atomic theory that incorporated these assumptions.

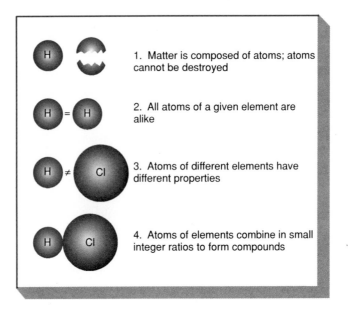

1. Matter is composed of atoms; atoms cannot be destroyed

2. All atoms of a given element are alike

3. Atoms of different elements have different properties

4. Atoms of elements combine in small integer ratios to form compounds

In 1803, Dalton postulated an *atomic theory* to account for the observed mass relations in chemical combination. Dalton made the following assumptions:

1. All matter is composed of tiny indivisible, indestructible particles called atoms.
2. All atoms of a given element are alike. In particular, they have the same mass.
3. Atoms of different elements have different properties. Most importantly, they have different masses.
4. Atoms of different elements combine in small integer ratios to form compounds (see Fig. 1.10).

Because atoms can neither be created nor destroyed, masses are conserved in chemical reactions. All of the atoms that start out in the reactants end up in the products. Hence the mass of the products is equal to the mass of the reactants. In a compound, the atoms of different elements are present in the ratio of small whole numbers, and the masses of the atoms are fixed. Therefore, the fraction of the total mass of each element in a compound is fixed. Finally, since chemical combinations of atoms are in simple integer ratios, the law of multiple proportions follows. For example, if the copper bromides in the previous experiment had the formulas $CuBr_2$ and $CuBr$, then with a given number of atoms of copper, one compound would have twice the number of atoms of bromine and twice the mass of bromine as the other. (The recognition of the mass relations summarized by the law of multiple proportions can either be considered powerful evidence for postulating chemical atoms, or it can be seen as evidence confirming a prediction of atomic theory.)

Dalton's atomic theory was a major milestone in the development of the science of chemistry. His theory successfully explained observed mass relations in chemical combination and provided the framework for the subsequent exploration of chemical structures and transformations. The refinement of atomic theory to generate rules for chemical structure and reactions was to be a major activity in chemistry after Dalton.

■ Aside

Dalton, the Weather, and Atoms

John Dalton, a Quaker, was born into a poor English family. Although he had little formal schooling, a wealthy patron helped him acquire some instruction in science and mathematics. He studied Newtonian science and began to teach school at an early age. For most of his life, he modestly supported himself by tutoring.

For more than fifty years, Dalton regularly made and recorded weather observations. Among other things, Dalton studied humidity, the contribution of water vapor to the atmosphere. (Those of you who have experienced hot days on the Gulf Coast and in the desert Southwest know that humidity can make a difference.) Dalton measured the relative humidity by measuring the number of degrees the air must be cooled to reach the dew point.

His atmospheric studies convinced him that the gases in air acted independently and were not chemically combined, as others thought. Dalton thought the gases in air consisted of tiny solid particles or atoms that repelled one another. He believed that the properties of the mixture of gases were due to the numbers and properties of each kind of atom present.

Dalton applied the early atomic ideas from his weather studies to the consideration of masses in chemical compounds. He pursued these ideas to develop an atomic theory for chemical combination. Dalton is an example of a scientist who brought insights gained in one scientific area to the solution of problems and advancement of understanding in another.

■ *Reflection*

Atoms and Inherent Limits

In the absence of modern technology, whether in an earlier time or in an undeveloped country, the world is often perceived as a dangerous place. In some years agricultural harvests are bountiful, in others lean. Storms that cause crops to fail bring the threat of starvation. Wild animals may inhabit the forest or plain. Sicknesses, particularly of children, are life-threatening, and in past centuries, plagues have caused drastic population reductions throughout the world.

Though many of the dangers of an earlier time have been tamed, the world is still a dangerous place, even with modern technology. Problems of food production and storage have been alleviated by the introduction of new insecticides, herbicides, and preservatives, but health and environmental concerns accompany the use of these chemicals. Short supplies of natural materials for clothing and shelter have been augmented by the introduction of synthetic substitutes, but new problems of waste disposal become apparent. The industry that raises living standards also generates mountains of garbage.

The chemist's description of matter has important implications for those thinking about technology and the environment. Matter is transformed in chemical reactions. Atoms are rearranged and put into new combinations and are neither created nor destroyed. Fossil fuels (oil, coal, and natural gas) were deposited over millions of years. When they have been extracted and burned, they are gone. Ores rich in iron and other scarcer metals also exist in limited quantities. When the ores have been mined and smelted, they are gone. Yet the atoms of carbon in coal and the atoms of iron in steel remain forever, sometimes forming a part of the waste that we call pollution. This result may not have been intended or planned; at first, it may simply have been viewed as of no consequence.

The production of chemical wastes necessarily accompanies virtually all chemical transformations. Consider the chemistry in the production of iron from its ore. (Ignore for the moment the environmental problems that accompany the mining of the iron ore and of the coal needed to make coke to refine the ore.) Iron ores contain iron bonded to oxygen as in ferric oxide Fe_2O_3. The iron compounds occur with silicate rocks and often with phosphate-containing rocks as well. Simply put, the atoms of iron must be separated from those of oxygen in the ore and from those of silicon and phosphorous in the rock.

The environmental costs of extracting resources from the earth will grow. Again, consider the case of iron. The high grade iron ores of the Mesabi Range of Minnesota are long gone. We now depend on lower grade ores containing less iron. It is apparent that to move more earth in mining and to remove more impurities in the ore requires greater expenditures of energy. The disposal of waste by-products is one part of the environmental costs of obtaining iron from ore. The environmental cost of supplying the required energy is a second and growing expense.

Chemistry is a science that offers insights into the dilemmas faced by society as we seek to reconcile acceptable levels of technology with the need to preserve and protect the quality of our environment. To live in a chemical world is to pollute. All options are limited, for the same technologies that deliver benefits in one area necessarily pose dangers in other ways. Chemistry describes a world in which all resources are limited, and it can be used to estimate how close we are to reaching some of those limits. In this post-modern world, problems are inherent and solutions are constrained. So let us explore a chemical world in which matter can be transformed but in which constraints are inherent.

■ Questions—*Chapter 1*

1. A philosophy student argues that it should be no harder to distinguish an element from a compound than it is to distinguish a planet from a star. Why were ancient astronomers able to distinguish planets from stars at a time that alchemists were unable to distinguish most elements and compounds?
2. Hydrogen, oxygen, and nitrogen were among the first gaseous elements to be studied by early chemists. How was each gas prepared? Which of these gases do you think was least pure when it was first studied? Why?
3. Collecting gases by the displacement of mercury, Priestley was able to identify and describe some previously undiscovered gases. What property of these gases prevented their collection by the displacement of water?

4. In Priestley's test for the goodness of air using nitric oxide, what gas remains if the sample tested is air? If the sample tested is pure oxygen?

5. How does the burning of hydrogen gas to form water contradict an explanation of matter based on the existence of the elements air, earth, water, and fire?

6. How could the reaction of hydrogen gas with oxygen to form water be used to show that water is a compound under Lavoisier's operational definition of elements?

7. What mass of carbon dioxide is produced in the following reaction? What law is invoked to answer this question?

$$100 \text{ g calcium carbonate} \rightarrow 56 \text{ g lime} + ? \text{ g carbon dioxide}$$

8. What mass of oxygen is required to burn 1000 g of isooctane, a component of gasoline?

$$1000 \text{ g isooctane} + ? \text{ g oxygen} \rightarrow 3082 \text{ g } CO_2 + 1419 \text{ g } H_2O$$

9. How does the formation of lime and carbon dioxide from limestone prove that limestone is not an element?

10. How can it be shown that carbon dioxide is not an element?

11. Which of the following analyses were obtained from samples of the same compound?

Sample	Mass of Sample, g	Mass iron, g	Mass Chlorine, g
A	0.600	0.206	0.394
B	0.750	0.258	0.492
C	0.500	0.220	0.280

12. The composition of air is about 20% oxygen and 80% nitrogen. Why do you believe air to be a mixture rather than a compound of oxygen and nitrogen?

13. A student wakes up from a nightmare in which rivers and streams were aflame. He knows that mixtures of hydrogen gas and oxygen gas burn. How can he reassure himself that water does not burn even though it is composed of hydrogen and oxygen? (Note: the meanings in chemistry, as in all other areas, depend upon context.)

14. Lavoisier considered lime to be an element. Why might his operational definition of an element have led to the inclusion of the compound lime in a list of elements?

15. From 100 g of a red oxide of iron, 70 g of iron can be obtained by reaction with air and charcoal. The same mass of iron can be obtained from 90 g of a black oxide of iron. Show that these results are consistent with the law of multiple proportions.

16. An oxide of sulfur is composed of equal masses of oxygen and sulfur. According to Dalton's postulates, this compound could not have the formula SO. Why?

17. Acetylene and ethane are both compounds of carbon and hydrogen. When acetylene is burned, 4.89 g of carbon dioxide are formed for each gram of water. When ethane is burned, 1.63 g of carbon dioxide is formed per gram of water. Show that these compounds illustrate the law of multiple proportions.

18. In some texts, an atom is defined as the smallest part of an element that retains the physical properties of the element. Is this an operational definition of an atom? Criticize this definition using the fact that diamonds and graphite used in pencil leads are two forms of the element carbon.

19. Does the fact that three forms of the element carbon exist contradict Dalton's postulate that all atoms of an element are alike? That they have the same mass?

20. Does Dalton's atomic theory lead to a new operational definition of an element? If so, what would the operation be?

21. If Dalton's atomic theory had been proposed on the basis of the law of conservation of mass and the law of constant composition, the law of multiple proportions would have been a consequence of atomic theory. Describe how the law of multiple proportions could be predicted from atomic theory.

22. Lavoisier proposed his theory of combustion in 1783, only a few years before the convening of the constitutional convention in the United States. Which has undergone the greater amount of change in the subsequent two centuries—chemistry or political science? Why?

2

From Molecules to the Periodic Table

Dalton's atomic theory provided a framework for the exploration of the chemistry of elements and compounds. However, two major questions needed to be addressed. First, in what ratio do atoms of elements combine in a compound? For example, is water HO_2 or HO or H_2O? Second, how are the properties of compounds related to their elemental constituents? For example, oxygen combines with carbon and nitrogen to form gases, but with iron and aluminum it forms solids that melt at high temperatures.

To answer these difficult questions, chemists needed to find independent ways to determine formulas and they needed to accumulate a body of knowledge describing the chemistry of elements and their compounds. This chapter opens with the development of a successful approach to determining chemical formulas, and it closes with the revelation of patterns of chemical behavior as summarized in Mendeleev's periodic table.

■ Dalton's Atomic Theory Is Incomplete

In assigning formulas to compounds, Dalton made an incorrect assumption when formulating his atomic theory. He assumed that if only one compound is known to occur between two elements, the elements combine in a 1:1 ratio of atoms. This assumption led him to assign an incorrect formula (HO) for water.

Consider the consequences of Dalton's assumption for the relative weights of oxygen and hydrogen atoms. It can be demonstrated by experiment that 8.0 g of oxygen combines with 1.0 g of hydrogen to form water. If water has the formula HO, then in every sample of water, no matter how large or small, there would be equal numbers of hydrogen and oxygen atoms. In 8.0 g of oxygen there would be the same number of atoms as in 1.0 g of hydrogen. Each oxygen atom would weigh 8.0 times as much as each hydrogen atom.

The assumption that atoms combine 1:1 when there is a single compound between a pair of elements led Dalton to calculate incorrect relative atomic weights for carbon, oxygen, and some other elements. Confusion about relative atomic weights slowed the acceptance of Dalton's atomic theory. To construct a consistent and correct table of relative atomic weights, scientists needed to determine formulas of compounds by experiment.

Does this incorrect assumption mean that Dalton's atomic theory is wrong? With hindsight we see that some of his other assumptions are also faulty. As you

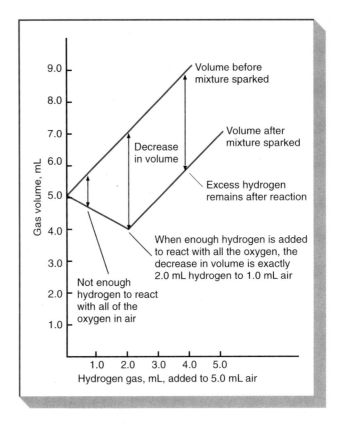

Figure 2.1
By adding small amounts of hydrogen gas to air, then sparking the mixture and measuring the remaining gas, Gay-Lussac found that in the reaction of hydrogen gas with oxygen gas to form water, the ratio of reacting gases was exactly 2:1.

explore chemistry, you will learn of the existence of isotopes (atoms of an element that differ in weight) and that atoms are divisible and composed of subatomic particles. Finally, some compounds are known in which atoms do not combine in simple integer ratios. For each of Dalton's assumptions there are exceptions, yet his atomic theory continues to shape our perception of the physical world.

Theories such as Dalton's, which chart new ground in science are often simple. When nature is found to be more complex than predicted by the simple theory, the theory is modified. Dalton's theory has been modified as scientists learned more about atoms. But if the complexity of modern atomic theory had been presented in 1810, it would have been unbelievable. It is only after a theory has been found to explain phenomena that it is accepted. Later, the inadequacy of the theory in explaining additional phenomena can be seen as a problem that requires the theory's modification.

■ The Law of Gay-Lussac

Measurements of the combining gas volumes in chemical reactions led to a method for determining chemical formulas. Joseph Louis Gay-Lussac (1778–1850), a Frenchman, measured the oxygen content of air by adding hydrogen and then sparking the mixture. Oxygen and hydrogen combined to form liquid water and a smaller volume of gas remained. When excess hydrogen is present, the decrease in volume was equal to three times the volume of oxygen present in the air. Gay-Lussac concluded that two volumes of hydrogen gas react with one volume of oxygen gas to make water (see Fig. 2.1).

Gay-Lussac then investigated volume relationships in the reactions of other gases. Quantities of gases in chemical reactions were studied by measuring volumes of both the reactants and the products at the same temperature and pressure. For

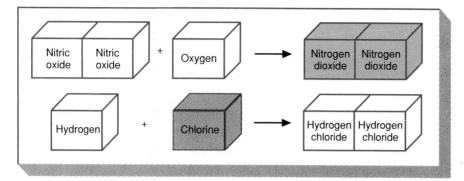

Figure 2.2
In reactions of gases, Gay-Lussac found that volumes of reacting gases and of product gases (measured at the same temperature and pressure) were in small whole number ratios. For example, 2.0 L of nitric oxide combine with 1.0 L of oxygen to form 2.0 L of orange nitrogen dioxide. In a second example, 1.0 L of hydrogen reacts with 1.0 L of chlorine to form 2.0 L of hydrogen chloride gas.

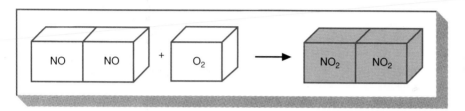

Figure 2.3
The formation of 2.0 L of NO_2 from 1.0 L of oxygen is evidence for the existence of oxygen molecules with the formula O_2.

example, in Priestley's chemical test for the goodness of air, nitric oxide reacted with oxygen to give nitrogen dioxide. When volumes of the reacting gases and of the product gases are measured, it is found that 2.0 L of nitric oxide reacts with 1.0 L of oxygen to give 2.0 L of nitrogen dioxide. In another example, 1.0 L of hydrogen gas reacts with 1.0 L of chlorine gas to form 2.0 L of hydrogen chloride gas (see Fig. 2.2).

Gay-Lussac discovered that a very simple relationship described the proportion of volumes observed for reacting gases. The *law of Gay-Lussac*, formulated in 1806, states that *volumes of reacting gases measured at the same temperature and pressure can be expressed as small whole number ratios.*

■ Gases Are Composed of Molecules

The simplicity of the relationship as formulated by Gay-Lussac invites explanation. In 1811, Amedeo Avogadro (1776–1856), an Italian scientist, offered one. Avogadro hypothesized that *equal volumes of gases measured at the same pressure and temperature contain equal numbers of particles,* known as *molecules.* Avogadro's hypothesis, along with measurements of volume in the reactions of gases, provides the information needed to determine formulas for gaseous elements such as hydrogen and oxygen, and gaseous compounds such as hydrogen chloride.

The relation of atoms to molecules can best be shown by considering as an example, the reaction of nitric oxide with oxygen to form nitrogen dioxide as shown in figure 2.3.

$$\text{Nitric oxide} + \text{oxygen} \rightarrow \text{nitrogen dioxide}$$
$$\text{2.0 L} \qquad \text{1.0 L} \qquad \text{2.0 L}$$

According to Avogadro's hypothesis, since the volumes of these two gases are equal there are equal numbers of nitric oxide and nitrogen dioxide molecules. The number of oxygen molecules is only half the number of either the nitric oxide or nitrogen dioxide molecules since the oxygen volume is only half that of each compound. Each molecule of the reactant nitric oxide must gain at least one oxygen atom when

Figure 2.4
The formation of 2.0 L of hydrogen chloride gas from 1.0 L of hydrogen and 1.0 L of chlorine is evidence for the diatomic molecules H_2 and Cl_2. Each molecule of hydrogen and chlorine furnishes one atom to a molecule of HCl. Because hydrogen chloride occupies double the volume of hydrogen and chlorine, there are twice as many molecules of the compound as there are of either element.

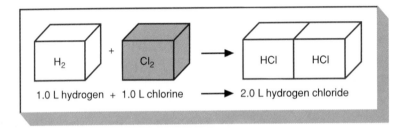

forming a product. The atoms contained in one volume of oxygen are divided equally between two volumes of the product. Therefore, each oxygen molecule must consist of an even number of atoms (at least two).

The reaction of hydrogen and chlorine to form hydrogen chloride provides evidence about molecules with the formulas H_2 and Cl_2. Each molecule of hydrogen chloride must have at least one atom of hydrogen and one atom of chlorine. The hydrogen atoms (combined with chlorine atoms) that go into two liters of hydrogen chloride are contained in just one liter of hydrogen gas. This requires that the molecule of hydrogen contains an even number of hydrogen atoms. A similar argument indicates that a chlorine molecule contains an even number of chlorine atoms (see Fig. 2.4). The assumption that the smallest even integer 2 is correct in the formulas for molecules of oxygen, hydrogen, and chlorine has been proven. These elements are composed of *diatomic* (di-, meaning two) molecules.

The law of Gay-Lussac, together with Avogadro's hypothesis is applied to determine the formula of water despite the practical difficulties present in the experiment. (Mixtures of oxygen and hydrogen may explode—to measure the volume of water as a gas requires a temperature above 100° C, the boiling point of water.) Experimentally, it is found that 2 L of hydrogen react with 1 L of oxygen to form 2 L of water vapor. The 2 L of hydrogen has the same number of hydrogen atoms as 2 L of water vapor. Each hydrogen molecule contributes its two hydrogen atoms to one molecule of water. Because 2 L of water vapor are formed from just 1 L of oxygen, one molecule of water has one-half the number of oxygen atoms as one molecule of oxygen. Since oxygen is diatomic, water must have the formula, H_2O (see Fig. 2.5).

Sample Calculation 1

When 1.00 L of the gas methane, composed of the elements carbon and hydrogen, is burned, 1.00 L of carbon dioxide (CO_2), and 2.00 L of water vapor are formed. What is the formula for methane?

One volume of methane contains the same number of carbon atoms as are found in one volume of carbon dioxide. Each molecule of methane contains the same number of hydrogen atoms as are found in two volumes of water and therefore each methane molecule must contain four atoms of hydrogen. The formula of methane is CH_4.

■ Chemical Equations Describe Reactions

Chemists write *chemical equations* to summarize chemical reactions. Chemical equations can be interpreted on a macroscopic, observational level. In a reaction of gases, coefficients appearing before formulas may represent relative volumes (if a coefficient in an equation is not given, it is assumed to be 1). For example, 2 L of nitric oxide gas combines with 1 L of oxygen gas to form 2 L of nitrogen dioxide gas:

$$2\,NO + O_2 \rightarrow 2\,NO_2$$

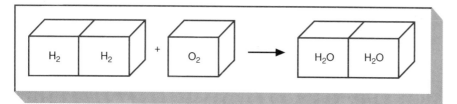

Figure 2.5
Since 2.0 L of H$_2$ combine with 1.0 L of O$_2$ to form 2.0 L of water vapor, water must have the formula H$_2$O. Each molecule of water has the same number of atoms as one molecule of hydrogen, and it has one-half the number of oxygen atoms as one molecule of oxygen.

On the microscopic level, equations represent molecular descriptions of reactions. For example, two molecules of nitric oxide react with one molecule of oxygen to form two molecules of nitrogen dioxide. Coefficients appearing before formulas represent relative numbers of molecules on a microscopic scale. Material is balanced, as there are equal numbers of nitrogen atoms and oxygen atoms on each side of the arrow in this equation and mass is conserved.

The reactions presented in figures 2.2 and 2.3 provide additional examples of chemical equations:

$$H_2 + Cl_2 \rightarrow 2\,HCl$$
$$2\,H_2 + O_2 \rightarrow 2\,H_2O$$

The use of chemical equations to calculate the masses of reactants and products involved in chemical reactions is elaborated in an appendix.

■ Relative Atomic Weights

Using the combining weights of elements and the formulas of compounds, *relative atomic weights* can be calculated. With the correct formula (H$_2$O) and the weight ratio of oxygen to hydrogen in water (8:1), it is possible to calculate the atomic weight of oxygen relative to hydrogen. Since one oxygen atom weighs eight times as much as two hydrogen atoms, an oxygen atom must be sixteen times as heavy as a hydrogen atom.

Chemists needed to use a standard element for expressing relative atomic weights. Atomic weights could be expressed relative to bromine, or to sulfur, or to any other element. A standard element had to be selected and a value had to be assigned to that element.

Hydrogen was one possible candidate for that standard since it is the lightest element. If the relative atomic weight of hydrogen were defined as exactly 1 (1.000), then all other elements would have atomic weights greater than one. However, hydrogen proved to be a poor candidate in other ways. Few compounds containing only hydrogen and a metal were known, and some of these were reactive and difficult to analyze. Since weighing chemical samples on a balance was the principal procedure used in determining chemical atomic weights, these shortcomings were major obstacles.

Chemists initially adopted a scale that compared the atomic weight of other elements to that of oxygen because most elements do form compounds with oxygen. The relative atomic weight of oxygen was defined to be exactly 16. The value 16.000 for oxygen had the advantage of assigning a value of 1.008, a number very close to 1, to hydrogen.

This oxygen-based scale was used by chemists, but physicists generated a different scale of relative weights using charge-to-mass measurements similar to those described in chapter 3. In 1961, a new unified scale was chosen (the new standard is presented in chapter 3). On this scale, the relative atomic weight of oxygen is 15.9994 and that of hydrogen is 1.0079.

Figure 2.6
An ice calorimeter can be used to determine the heat capacity of metals. To melt 1 g of ice, 80 cal are required.

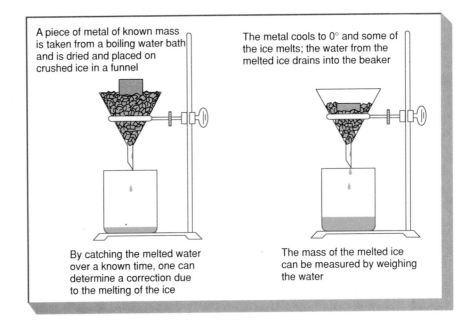

A piece of metal of known mass is taken from a boiling water bath and is dried and placed on crushed ice in a funnel

The metal cools to 0° and some of the ice melts; the water from the melted ice drains into the beaker

By catching the melted water over a known time, one can determine a correction due to the melting of the ice

The mass of the melted ice can be measured by weighing the water

■ Determining Atomic Weights for Metallic Elements

Chemists applied the law of Gay-Lussac and Avogadro's hypothesis to determine atomic weights of gaseous elements. Elements such as bromine and iodine that readily vaporize and elements such as carbon that form a number of gaseous compounds could be measured. However, these methods could not be used to determine the atomic weights for many metals. To determine atomic weights of these metallic elements measurements of specific heats were used.

The *specific heat* of a substance is the quantity of heat required to increase the temperature of 1 g of the material 1° C. Specific heats can be measured using an ice calorimeter as shown in figure 2.6. When a piece of metal of known weight is taken from a boiling water bath and dropped into a container of cracked ice, some ice melts and the metal cools to the temperature of the ice water. The melted ice water can be separated and weighed, and the specific heat of water is exactly 1 cal/(g-° C) when measured at 14° C. The greater the amount of ice that is melted by a fixed weight of a metal, the greater the metal's specific heat.

Two French scientists, Pierre DuLong (1785–1838) and Alexis Petit (1791–1820), observed that the product of the specific heat times the atomic weight is a constant for many metallic elements. The *law of DuLong and Petit*, formulated in 1819, states that for a given metal the specific heat of an element times its atomic weight (in grams) is approximately 6 cal/° C (see Tab. 2.1 for representative values).

The relationship described by DuLong and Petit is approximate and does not apply to all solid elements. However, it enabled early chemists to correct atomic weights calculated for copper, zinc, nickel, and iron. Early atomic weights for these elements had been calculated from analyses of their oxides, assuming a formula of MO_2 for each oxide. When the relative atomic weights were recalculated assuming that the correct formula was MO, the calculated atomic weights of the metals were in accord with the law of DuLong and Petit.

Table 2.1

Data Illustrating the Law of DuLong and Petit

Metal	Atomic Weight (AW)	Specific Heat (SH) cal/°C	AW × SH cal/°C
Aluminum	27.0	0.215	5.81
Iron	55.8	0.1075	6.00
Copper	63.5	0.0924	5.87
Silver	107.9	0.0562	6.06
Gold	197.0	0.0308	6.07
Lead	207.2	0.0305	6.32

Sample Calculation 2

In a determination of the atomic weight for copper reported in an 1826 table, it was found that 63.31 g of copper combined with 16 g of oxygen in copper oxide. What was the atomic weight of copper?

Assumed Formula	Atomic Weight of Copper
CuO	63.31
CuO_2	126.62

The approximate atomic weight of copper, applying the law of DuLong and Petit is:

$$AW \times \text{specific heat} = 6.0$$
$$AW = 6.0 / \text{sp. heat} = 6.0/0.0924$$
$$= 65$$

The approximate weight permits the formula CuO to be assigned and the atomic weight of copper was determined to be 63.31. The accepted value today is 63.546.

■ Measuring Quantities in Chemical Reactions: The Mole

Chemists use a chemical counting unit, the *mole*, that can be measured with a balance. One mole of an element's atoms is the mass in grams equal to the relative atomic weight of the element. One mole of hydrogen atoms is 1.0079 grams. A mole of oxygen atoms is 15.9994 g.

A mole of a particular compound is the quantity having a weight in grams equal to the formula weight. The formula weight of a compound is numerically equal to the sum of relative atomic weights of the atoms in the formula unit. (If the compound is known to consist of molecules, the formula weight is usually referred to as the molecular weight.)

The weight of 1 mole of hydrogen gas, H_2, is:

$$\text{mole } H_2 \times \frac{2 \text{ mole H}}{\text{mole } H_2} \times \frac{1.008\text{g}}{\text{mole H}} = 2.016 \text{ g}$$

The molecular weight of water, H_2O, is:

$$\frac{2 \text{ moles H}}{\text{mole } H_2O} \times \frac{1.008 \text{ g}}{\text{mole H}} + \frac{1 \text{ mole O}}{\text{mole } H_2O} \times \frac{16.00 \text{ g}}{\text{mole O}} = 18.02 \text{ g}$$

Sample Calculation 3

Calculate the mass of 1 mole of calcium carbonate $CaCO_3$.

$$1 \text{ mole Ca} \times \frac{40\,g}{\text{mole Ca}} + 1 \text{ mole C} \times \frac{12\,g}{\text{mole C}} + 3 \text{ moles O} \times \frac{16\,g}{\text{mole O}} = 40\,g + 12\,g + 48\,g$$

$$\text{One mole of } CaCO_3 = 100\,g$$

A balanced chemical equation can be interpreted either as a statement about atoms and molecules or as a statement about moles in chemical reactions. In the balanced equation for the formation of carbon dioxide from carbon monoxide and oxygen, 2 moles of carbon monoxide react with 1 mole of oxygen to give 2 moles of carbon dioxide.

$$2\,CO + O_2 \rightarrow 2\,CO_2$$

■ Groups of Elements with Similar Chemistry

As observations accumulated from various laboratories, the body of the descriptive chemistry of elements and their compounds grew. Chemical and physical similarities were noted for groups of elements. For example, chemists classified most elements as either *metals* or *non-metals.* Most metals are shiny solids that conduct heat and electricity well. When combined with one another, metals form alloys that may vary widely in composition. Most nonmetals, including all the gaseous elements, do not conduct electricity. Nonmetals form compounds, having fixed composition, with both metallic and non-metallic elements.

As the catalogue of elements and compounds grew larger, the search for underlying principles became more important. Was the chemical world to be as diverse as that of biology? Would explorers of new lands bring home a collection of new elements and compounds? If so, were there underlying principles as in biology? Gregor Mendel had found simple statistical ratios (such as three to one) in the distribution of physical properties in a generation of peas, and he had formulated laws of genetic inheritance. What were the laws of chemical combination?

There were tantalizing hints of order in chemistry. Triads of elements with similar properties were noted. The earliest chemical family to be recognized may be the *coinage metals:* copper, silver, and gold (see Fig. 2.7). Copper, silver, and gold are lustrous metals that are resistant to oxidation. The coinage metals are among the very few metals to be found uncombined in nature. From the time of the pharaohs, people have valued these metals. Artisans have shaped these metals into jewelry and vessels for religious celebration. Explorers crossed oceans and continents in search of gold and silver.

The *halogens* are nonmetals with strikingly similar chemistry. All are diatomic molecules in the elemental state. The heavier the halogen, the deeper its color and the greater its density. Fluorine is a pale yellow gas, chlorine is a pale green gas, bromine is a red-brown liquid that boils at a low temperature, and iodine is an almost black crystalline solid, which readily sublimes to a violet vapor. The hydrogen halides, HF, HCl, HBr, and HI, are gases that dissolve in water to form acidic solutions.

Yet another triad of elements consists of the *alkali metals*, lithium, sodium, and potassium. Early alchemists knew the salts of sodium and potassium and gave names to the elements. But it was not until Humphry Davy (1778–1829) passed

Figure 2.7
Photographs of gold, silver, and copper coins.

electric currents through their molten compounds that tiny globules of the shiny, reactive metals were isolated. In contrast to silver and gold, sodium and potassium are soft, low-melting solids that react violently with water to form hydrogen gas and basic (alkaline) solutions. Compounds of the alkali metals are crystalline solids that dissolve readily in water. The chlorides of these elements have the formulas LiCl, NaCl and KCl.

The elements carbon, silicon, germanium, tin, and lead form another chemical family. The lower atomic weight elements, carbon and silicon, are nonmetals. The higher atomic weight elements, tin and lead, are metals. There are similarities, however, carbon monoxide (CO) and carbon dioxide (CO_2) and the two oxides of lead $(PbO$ and $PbO_2)$ have similar formulas. The halides of elements in this family are volatile liquids. Carbon tetrachloride (CCl_4), boils at $77°C$ and tin tetrachloride $(SnCl_4)$, boils at $114°C$. This group of elements shows greater differences and fewer similarities than the coinage metals, the alkali metals, or the halogens.

■ Mendeleev's Periodic Table

Did the existence of chemical triads, like that of musical chords, indicate that there was harmony in nature? (In Greek philosophy, the Pythagorean school believed that harmonies of nature are displayed in arithmetic and geometric relationships.) An affirmative answer to this question was proposed by a Russian chemist, Dmitri Mendeleev (1834–1907), in 1869 in the form of a Periodic Table. Mendeleev ordered known elements on the basis of increasing atomic weight and assigned an atomic number to each element based on rank. He then displayed their symbols in rows of increasing atomic number across a table and started new rows so that chemically similar elements fell in columns.

The elements of two rows of Mendeleev's table are shown below with the relative atomic weights given to the nearest integer value.

| Li=7 | Be=9 | B=11 | C=12 | N=14 | O=16 | F=19 |
| Na=23 | Mg=24 | Al=27 | Si=28 | P=31 | S=32 | Cl=35 |

Tabelle II.

Reihen	Gruppe I. $\overline{R^2O}$	Gruppe II. $\overline{RO}$	Gruppe III. $\overline{R^2O^3}$	Gruppe IV. RH^4 RO^2	Gruppe V. RH^3 R^2O^5	Gruppe VI. RH^2 RO^3	Gruppe VII. RH R^2O^7	Gruppe VIII. $\overline{RO^4}$
1	H = 1							
2	Li = 7	Be = 9,4	B = 11	C = 12	N = 14	O = 16	F = 19	
3	Na = 23	Mg = 24	Al = 27,3	Si = 28	P = 31	S = 32	Cl = 35,5	
4	K = 39	Ca = 40	– = 44	Ti = 48	V = 51	Cr = 52	Mn = 55	Fe = 56, Co = 59, Ni = 59, Cu = 63.
5	(Cu = 63)	Zn = 65	– = 68	– = 72	As = 75	Se = 78	Br = 80	
6	Rb = 85	Sr = 87	?Yt = 88	Zr = 90	Nb = 94	Mo = 96	– = 100	Ru = 104, Rh = 104, Pd = 106, Ag = 108.
7	(Ag = 108)	Cd = 112	In = 113	Sn = 118	Sb = 122	Te = 125	J = 127	
8	Cs = 133	Ba = 137	?Di = 138	?Ce = 140	–	–	–	– – – –
9	(–)		–		–		–	
10	–	–	?Er = 178	?La = 180	Ta = 182	W = 184	–	Os = 195, Ir = 197, Pt = 198, Au = 199.
11	(Au = 199)	Hg = 200	Tl = 204	Pb = 207	Bi = 208	–	–	– – – –
12	–	–	–	Tb = 231	–	U = 240	–	

(From Annalen der Chemie, supplemental vol. 8, 1872.)

Figure 2.8
Mendeleev's horizontal periodic table of 1871.

This arrangement of elements is repeated below showing the formulas of compounds of these elements with chlorine and with hydrogen. Note that the formulas of the chlorides and of the hydrides are characteristic of families and that nonmetals are found to the right in these rows and metals are to the left.

LiCl   BeCl$_2$ BCl$_3$ CCl$_4$

NaCl   MgCl$_2$ AlCl$_3$ SiCl$_4$

CH$_4$ NH$_3$ H$_2$O   HF

SiH$_4$ PH$_3$ H$_2$S   HCl

One version of Mendeleev's periodic table is shown in figure 2.8. Symbols of metallic elements are found at the left (at the start of rows), and those of nonmetallic elements are at the right (at the end of rows). Close chemical relationships among elements are indicated using an eight column table. The chemical symbols of copper and gold are written above and below that of silver, and the triads of coinage metals, halogens and alkali metals become parts of larger chemical families. Of the families discussed in this chapter, the coinage metals and the alkali metals are in column I, the family containing carbon and lead is in column IV, and the halogens are found in column VII.

If the relationships were as apparent as those shown thus far, one might wonder why the periodic law was so long in coming. Proceeding by trial and error with elements ranked in order of increasing atomic weight and with their chemistry described on 3 × 5 index cards, one would soon discover a periodic arrangement. In part, the answer is that in the rows starting with Li and Na, the periodic similarities and differences within and between families seem most dramatic. In later rows, Mendeleev sometimes placed three similar elements, for example, the metals iron, cobalt, and nickel, in an eighth column. The chemistry of some elements in the

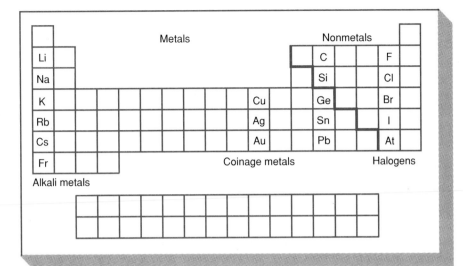

Figure 2.9
The outline of the modern long version of the periodic table is shown with the symbols of the alkali metals, the coinage metals, the halogens, and the family that runs from carbon to lead. The heavy jagged diagonal line separates the metals on the left from the nonmetals on the right. A full version of the periodic table is shown inside the front cover.

same column also differed in important ways indicating that two families may be in each column. Mendeleev indicated that the chemistry of the metal manganese was both similar and different from that of the halogens by placing Mn with a predicted heavier metal on the left side of column VII and placing Cl, Br, and I on the right.

Current versions of the periodic table as outlined in figure 2.9 avoid these difficulties by having longer horizontal rows and rows of different lengths. In these versions of the periodic table, metals are found to the left and nonmetals on the right. Until very recently, the numbering of columns in the long table reflected that of Mendeleev. For example, Group VII of Mendeleev's table was split into two columns: VIIA containing Mn, and VIIB containing the halogens.

A greater understanding of the achievement represented by the development of a periodic law can be gained by looking at some problems Mendeleev faced. Not all the naturally occurring elements had been discovered and for some known elements, the reported atomic weights were in disagreement or error. Mendeleev applied the law of DuLong and Petit to correct the reported atomic weights of some elements.

Mendeleev did not always order elements according to increasing atomic weights. For some pairs of adjacent elements, he reversed their order, asserting that the reported atomic weights must be wrong. The pair of elements tellurium and iodine are an example of an inversion cited by Mendeleev. Although iodine was reported to have a lower atomic weight than tellurium, Mendeleev placed iodine in the halogen column and he put tellurium, which forms H_2Te and Na_2Te, under oxygen. (Mendeleev's assignment of family relationships for iodine and tellurium was correct although the atomic weight of iodine is, indeed, less than that of tellurium.)

More dramatically, Mendeleev proposed the existence of new elements and predicted their properties. To obtain correct family placement of heavier elements, Mendeleev left four gaps in his periodic tables. Three of these gaps were filled by elements discovered during the following fifteen years. An illustration of the power of the periodic table to correlate and forecast chemical behavior is in the prediction and later, observation of the element germanium. Mendeleev asserted the existence of an element, eka silicon, to lie below silicon and above tin in column IV of his periodic table. Its properties and those of its compounds would be intermediate between those of silicon and tin and their compounds. When this element was discovered, it was named germanium.

Property	Eka Silicon	Germanium
Atomic Weight	72	72.3
Density, g/cm^3	5.5	5.47
Specific heat cal/g • °C	0.073	0.076
Boiling Point of GeCl$_4$	Under 100°	86

Mendeleev's periodic table was provocative. It correlated a great deal of chemical information, although the underlying basis of the correlation was not readily apparent. Why should the ranking of elements in order of increasing atomic weight lead to an arrangement in which elements of similar chemistry are listed in columns in a table? As scientists learned more about the nature of atoms, they developed a more satisfying rationale for the periodic table. An explanation for chemical periodicity based on atomic structure is developed in chapter 4.

■ Questions—*Chapter 2*

1. Dalton assumed that water had the formula HO. What formula would he have attributed to hydrogen peroxide, H_2O_2, that would be consistent with his assumption?

2. A 2 L amount of the gas ammonia is produced by the reaction of 1 L of nitrogen and 3 L of hydrogen, H_2. What is the simplest formula for nitrogen? For ammonia?

3. Isotopes of uranium have been separated by the diffusion of a gaseous uranium fluoride, UF_x. A 1 L amount of UF_x can be prepared by reacting 3 L of fluorine gas with solid uranium. What is the simplest formula for this uranium fluoride? If fluorine is diatomic, what is the formula of uranium fluoride?

4. A 1 L amount of a hydrocarbon gas C_xH_y burns completely with oxygen to form 2 L of carbon dioxide and 2 L of water vapor. What is the formula for the hydrocarbon?

5. Balance the following equations by supplying the missing coefficient:
   (a) $C_3H_8 + ? O_2 \rightarrow 3 CO_2 + 4 H_2O$
   (b) $C_6H_6 + 7/2 O_2 \rightarrow ? CO_2 + ? H_2O$
   (c) $H_2 + Br_2 \rightarrow ? HBr$

6. When 1.00 L of ammonia gas, NH_3, is mixed with 1.00 L of gaseous hydrogen chloride, HCl, all the gas reacts and a white solid is produced. What is the formula of the white solid?

7. When a direct electric current is passed through a dilute aqueous solution of sodium sulfate, the water reacts to form hydrogen gas at one electrode and oxygen gas at the other. How many milliliters of hydrogen form for every 25 mL of O_2?

8. A confused chemistry student asks you the difference between an atom and a molecule. How would you explain the difference? What experimental evidence would you cite for atoms? For molecules?

9. Why would confusion between atoms and molecules make it difficult to determine accurate atomic weights?

10. The heat capacity of the element lead is one-fourth that of iron. What is the approximate atomic weight of lead if the atomic weight of iron is about 56?

11. A metal oxide with the formula MO contains 4.09 g of the metal for every gram of oxygen. Calculate the relative atomic weight of the metal on a scale with O = 16. Use the periodic table to identify the metal.

12. Use the information in table 2.1 to estimate approximate values for the heat capacity of lithium (AW = 6.94) and tin (AW = 118.7).

13. Calculate the mass of one mole of each of the following compounds.
    (a) SO_2
    (b) SO_3
    (c) CaO
    (d) $CaSO_4$

14. How many moles of oxygen are required to burn 1 mole of each of the following hydrocarbons?
    (a) $CH_4 + ? \, O_2 \rightarrow CO_2 + 2 \, H_2O$
    (b) $C_7H_8 + ? \, O_2 \rightarrow 7 \, CO_2 + 4 \, H_2O$

15. On what basis did Mendeleev group elements into families? On what basis did he order elements across the periodic table? Give some specific examples of periodic similarities and differences in your answer.

16. Use the periodic table to write formulas for the compounds formed between the following pairs of elements.
    (a) Potassium (K) and bromine (Br)
    (b) Barium (Ba) and chlorine (Cl)
    (c) Gallium (Ga) and fluorine (F)
    (d) Silicon (Si) and iodine (I)

17. Use the periodic table to predict the formulas of the compound formed between hydrogen and each of the following elements.
    (a) Arsenic (As)
    (b) Germanium (Ge)
    (c) Selenium (Se)
    (d) Iodine (I)

18. Selenium is in the same family of the periodic table as oxygen. Which of the following properties of selenium can be predicted on the basis of family relationships? If a property can be predicted, give a reason why.
    (a) Small amounts of selenium are essential to a good diet.
    (b) Hydrogen and selenium combine to form a compound H_2Se.
    (c) Selenium is more dense than sulfur (S) and less dense than tellurium (Te).

19. Mendeleev predicted the existence of eka boron and eka aluminum, elements to fill empty positions below B and Al in the Periodic Table. What are the symbols and names of these elements?

20. Which of the following elements has chemistry most similar to antimony (Sb)? Briefly justify your answer.
    (a) Tin (Sn)
    (b) Bismuth (Bi)
    (c) Nitrogen (N)
    (d) Tellurium (Te)

21. What does the work of Mendeleev suggest about the role of the scientist's belief concerning order in nature?

3

Radioactivity and the Structure of Atoms

T he pace of discovery in physics and chemistry quickened as the nineteenth century drew to a close. Growing knowledge of the electrical nature of matter together with new discoveries in physics provided a background for inquiry into the structure of the atom.

In 1896, Henry Becquerel (1852–1908) discovered radioactivity. Following his discovery, scientists found that atoms of radioactive elements eject charged particles that become atoms of different elements. The exploration of the properties of charged particles and of the interaction of radiation with matter soon led to the discovery of subatomic particles and to the formulation of a nuclear model for atomic structure.

■ The Discovery of Radioactivity

The discovery of radioactivity began with a false hypothesis. When fluorescent minerals are illuminated with ultraviolet light, they emit visible light. Since X rays also cause some compounds to glow or fluoresce, Becquerel thought that fluorescence might be related to the newly discovered X rays; in particular, he believed the X-ray emission might accompany fluorescence.

Becquerel used the observations that X rays penetrate matter and cause film to become fogged to test his hypothesis. Becquerel placed fluorescent materials over photographic plates wrapped in black paper as shown in figure 3.1. He exposed the materials to light so that they fluoresced and then he developed the plates to look for evidence of X rays. Potassium uranyl sulfate, a salt of uranium, caused the plates to appear exposed, but most of the other fluorescent substances did not.

A chance observation produced evidence that led to the discovery of radioactivity. During a period of cloudy weather, some wrapped photographic plates and the fluorescent salt of uranium lay together in a drawer. This time the compound had not been exposed to light, but Becquerel developed the plates just the same. He found a characteristic fogged halo below where the compound had been. When materials were placed between the compound and plates, they shielded the plates from exposure and left shadows. To account for these observations, Becquerel postulated that radiation from uranium caused the plates to fog.

Samples of pitchblende, an ore of uranium, emit more radiation than do pure uranium compounds. Marie Curie (1867–1934) attributed the radiation in pitchblende to the presence of one or more new radioactive elements, and, with her

Figure 3.1
Henri Becquerel discovered that radiation from compounds of uranium could pass through paper and leave an image on photographic plates. These rays did not penetrate a coin so that a "shadow" could be cast by blocking some of the radiation from the uranium.

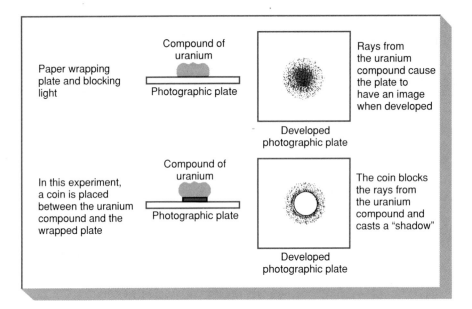

Figure 3.2
Marie Curie separated traces of highly radioactive radium from pitchblende by precipitating it with a salt of barium, an element in the same family of the periodic table. From several tons of the ore, she obtained less than 0.1 g of radium chloride following repeated recrystallizations.

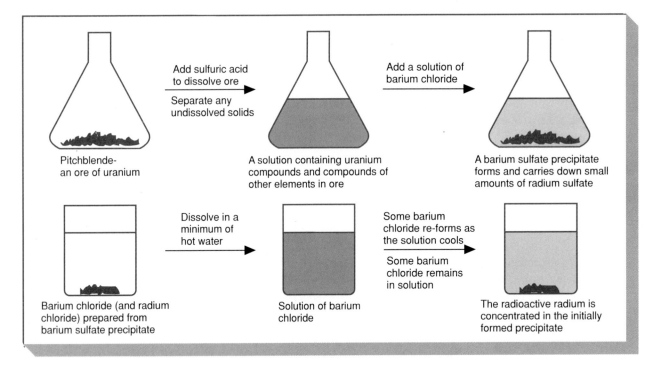

husband Pierre (1859–1906), she sought to isolate them. The amounts of these radioactive elements were too small to separate by themselves. Her plan was to precipitate insoluble compounds of known elements that might carry the new elements with them.

Curie dissolved the pitchblende in acid, then added compounds of known elements to the solution. When she precipitated these "carrier" elements, some samples contained increased portions of radioactivity. (Radioactive elements in the pitchblende separated with carrier elements of the same family in the periodic table, for they have similar chemistry.) By redissolving radioactive samples and reisolating portions by precipitation, she obtained portions richer in radioactivity (see Fig. 3.2). In

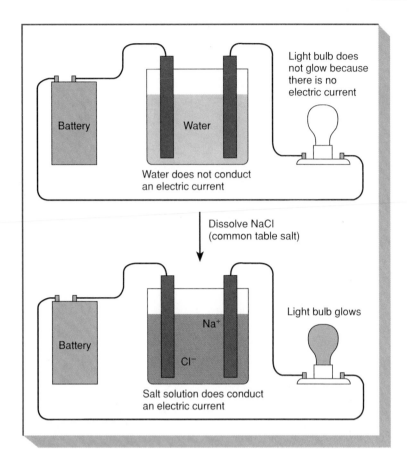

1898, the Curies isolated both the radioactive element polonium (named for Marie Curie's native Poland) and a salt of radium, an element so radioactive that it glows in the dark. Becquerel and the Curies shared a Nobel Prize in 1903.

■ The Electrical Nature of Matter

Studies of the electrical nature of matter provided additional background knowledge for gaining insight into atomic structure. Solutions of salt in water conduct electricity as demonstrated using a light bulb in an electric circuit that has two leads immersed in water. When the circuit is connected to a power source, no current flows if the water is pure, for water acts with large resistance to the flow of electricity. But when common table salt, sodium chloride, is dissolved in the water, current flows and the bulb lights up (see Fig. 3.3)

In solutions that conduct electricity, there are carriers of electric charge called *ions.* Positive ions are called *cations,* and negative ions are called *anions.* In a sodium chloride solution the carriers of electric charge are the sodium cations and the chloride anions. When an external voltage is applied, cations move toward the negative electrode, and anions move toward the positive electrode.

When a direct current is passed through molten sodium chloride as shown in figure 3.4, sodium metal is produced at one electrode and chlorine gas is formed at the other. Positive sodium ions gain negative charge to form the metal at one electrode, and chloride ions lose negative charge to form the gas at the other electrode. (In 1807, Humphry Davy had first isolated the alkali metals, sodium and potassium, by passing an electric current through melted samples of compounds of these metals.)

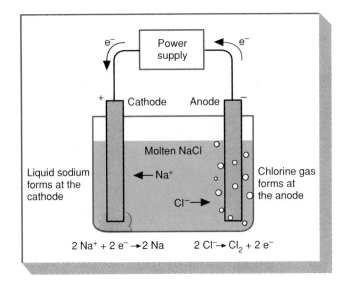

These experiments demonstrate the electrical properties of some chemical compounds and the electrical nature of some chemical reactions. However, to understand the charged particles found in some chemical compounds and the electrical nature of some chemical reactions, it is first necessary to consider the structure of the atom.

■ Subatomic Particles and the Nuclear Atom

Experiments in physics at the beginning of the twentieth century dramatically changed scientists' model of the atom. From the indivisible atom postulated by Dalton, scientists moved to a description of an atom as consisting of charged subatomic particles. This model of the atom is so important for an understanding of bonding and reactions in chemistry that it is desirable to understand its experimental basis.

In the following sections, we will consider the design of three basic experiments. In each of these experiments, scientists interpreted observations to infer properties of subatomic particles. These experiments are simple in concept, but rely on a knowledge of physics that is beyond the background of most students taking an introductory chemistry course. We are able, however, to connect each concept to its experimental basis without going into the physics used in the calculations. In a qualitative way, we can understand the model of the atom developed by physicists near the turn of the century.

■ Electrons

In 1897, J. J. Thomson (1856–1940) discovered the electron in the course of his investigation of cathode rays. (Cathode ray tubes are the ancestors and cousins of the picture tubes used in television sets.) When a high voltage is imposed across a vacuum tube containing electrodes, the end of the tube farthest away from the cathode glows. If this end is then coated with a phosphorescent material, as it is in picture tubes, the glow is greatly enhanced. Objects placed between the cathode and the screen cast shadows indicating that the glow is caused by rays emanating from the cathode.

Thomson studied the nature of cathode rays. He could block most of the cathode rays by using an anode with a slit in it, and then observe the position of a

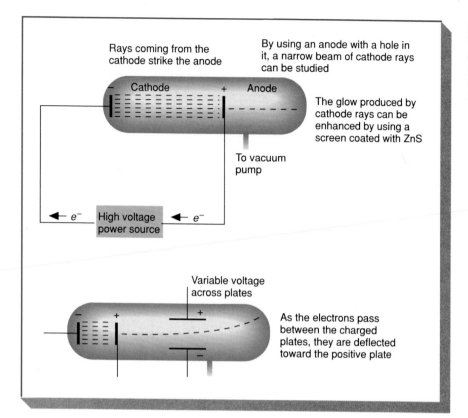

Figure 3.5
When a high voltage is applied to
an evacuated tube, cathode rays
are produced. Scientists showed that
cathode rays were composed of
negative particles called electrons.

glowing spot at the end of the tube (see Fig. 3.5). When the narrow beam was passed through an electric field imposed between two plates, the spot was deflected. This indicated that the rays were charged particles rather than a form of light. Because the direction of the deflection was toward the positive plate, Thomson knew he was studying negatively charged particles. These negative particles are called *electrons*.

Thomson determined the ratio of the charge of the electron (e) to the mass of the electron (m) or the e/m for the electron. He measured the deflection of a beam of electrons by an electric field of known strength. The amount of deflection depended on the charge, the mass, and the velocity of the electron since, for a given force, an object is deflected less if it is heavier or if it is moving faster. Thomson then used a magnet to deflect the electrons in the direction opposite to that caused by the electric field (see Fig. 3.6). When the electric and magnetic forces acting on the electrons were made equal in strength, no deflection of the beam occurred. With this additional measurement, Thomson could calculate the value of e/m for the electron.

■ Determining the Charge of the Electron

In 1908, Robert Millikan (1868–1953) measured the charge on the electron by studying the movement of charged oil droplets in an electric field. In the absence of a field, the tiny droplets fell at a constant speed. The downward gravitational force was offset by a frictional force, just as is the case of a falling person suspended from a parachute. Millikan used a radioactive source to generate ions, and negatively charged some of the oil drops. In an electric field, the charged drops rose at a constant speed with an electric force, an altered frictional force, and a gravitational force acting on them (see Fig. 3.7).

Figure 3.6

J. J. Thomson measured the charge-to-mass ratio, e/m, for the electron. A moving charge is deflected in a magnetic field. Thomson adjusted the voltage across the plates so that the electric force and the magnetic force on the electrons were equal and there was no net deflection. He also measured the deflection in the electric field alone. From these two measurements, he could calculate the value of e/m.

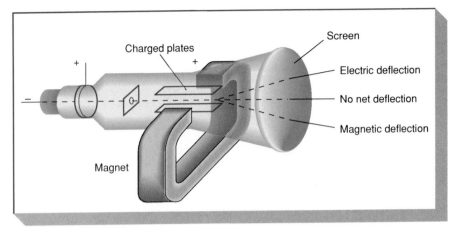

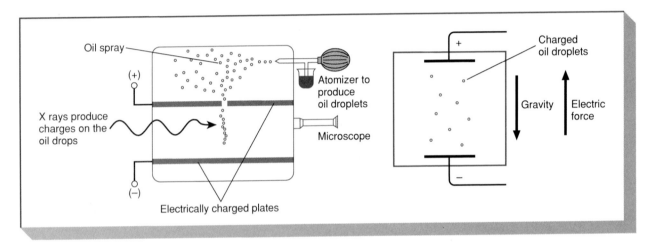

Figure 3.7

Robert Millikan determined the charge on a single electron by studying the motion of oil droplets in an electric field. From the rate of fall in the absence of the field, he could calculate the size of an individual drop. From the rate of its rise in an electric field, he could then calculate the charge on a drop. All of the measured charges on individual drops could be expressed as an integer times the smallest unit of charge, which is the charge on a single electron.

By repeatedly turning the voltage on and off, Millikan and his students could observe individual drops rise and fall. The size of any drop could be calculated from the rate at which it fell. The charge on any drop could be calculated from its size and the rate at which it rose. By determining the charges on many drops, Millikan accumulated a table of charges.

Millikan found that the measured charges could all be expressed as multiples of a single charge. The largest common divisor of the charges was the smallest unit of charge, the charge on a single electron. That charge, e, is 1.602×10^{-19} coulomb. Since the ratio of charge-to-mass, e/m, is known from Thomson's experiment, it is also possible to calculate the mass of an electron. The mass of an electron is 1/1840 times the mass of the lightest atom, hydrogen.

■ The Nuclear Atom

Are electrons found in atoms? Scientists explored the implications of this question. If atoms do contain negative electrons, they must also contain positive particles since atoms are neutral. Because an atom is much heavier than its electrons, these positive particles must be much heavier than electrons.

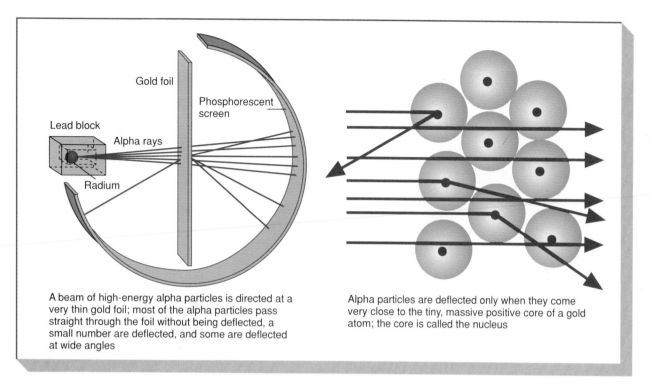

A beam of high-energy alpha particles is directed at a very thin gold foil; most of the alpha particles pass straight through the foil without being deflected, a small number are deflected, and some are deflected at wide angles

Alpha particles are deflected only when they come very close to the tiny, massive positive core of a gold atom; the core is called the nucleus

How do the positively charged particles and the negatively charged particles fit together in an atom? Are negative and positive particles distributed randomly, are they spread out evenly throughout an atom, or are they clustered? Ernest Rutherford (1871–1937) provided the answer to these questions by studying the deflection of radioactive decay products, called *alpha particles*, by a very thin gold foil. In 1911 he proposed the nuclear atom to account for his observations.

Rutherford studied the interactions of the most massive known radiation, alpha particles, with gold foil, the thinnest possible solid. Alpha particles are high-speed positive particles. Each is about four times as heavy as a hydrogen atom and about 7000 times as heavy as an electron. Since a collision of an alpha particle with an electron would resemble that of a high-speed bowling ball with a ping pong ball, the path of an alpha particle would scarcely be perturbed by interactions with electrons in the foil. Alpha particles served as probes to measure the distribution of positive charge and mass.

Rutherford measured the number of alpha particles deflected by the foil as a function of the angle of deflection as shown in figure 3.8. Most alpha particles passed through the foil undeflected. A small fraction were deflected, some at large angles. Amazingly, some came back toward the alpha emitter source. Rutherford later commented, "It was about as credible as if you had fired a 15-inch shell at a piece of tissue paper and it came back and hit you."

To account for the large deflections, Rutherford proposed a model for the atom which consists of a tiny, massive, highly charged positive core called the *nucleus* that is surrounded by electrons. Since most alpha particles were not deflected, Rutherford concluded that most of the gold foil consisted of empty space or space occupied only by electrons. When infrequent collisions did occur, rebounding alpha

Figure 3.8
Ernest Rutherford proposed a nuclear model of atomic structure to account for the deflection of alpha particles by a thin gold foil.

Figure 3.9
In a neutral atom the positive charge on the nucleus is balanced by the negative charge of electrons. A hydrogen atom has a +1 nucleus and one electron. A gold atom has a +79 nucleus and 79 electrons.

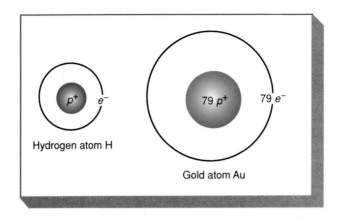

particles were deflected at large angles because there were large forces in the collisions due to the repulsion of the positive alpha particles by the large positive charge concentrated in the very small nucleus. The other reason for the large deflect was that the massive gold nuclei recoil far less than the lighter alpha particles.

■ Atomic Structure

After Rutherford's formulation of a nuclear model for the structure of atoms, scientists made additional discoveries that enabled them to relate differences between elements to their different nuclear structures. Nuclei are composed of subatomic building blocks called *protons* and *neutrons*. A proton is a positive particle with a charge equal in magnitude to the negative charge on an electron. A neutron is an uncharged particle with a mass close to that of a proton. Both the proton and the neutron are about 2000 times as heavy as an electron, and both are called *nucleons*.

All atoms of a given element have the same number of protons in their nuclei. The number of protons in an atom of an element, known as the *atomic number* of that element, is identical to the ranking of that element in the periodic table. (The atomic weight of elements, the quantity used by Mendeleev to rank them, usually increases in the same order as the numbers of protons in the nucleus increases.)

In a neutral atom, the positive charge on the nucleus is balanced by the negative charge of electrons about the nucleus. An atom of hydrogen, the lightest element, has a single proton and a single electron. An atom of gold has 79 electrons encircling a nucleus that contains 79 protons (see Fig. 3.9).

The description of an atom consisting of a positive nucleus surrounded by electrons helps explain the formation of ions. Positive cations are formed from atoms by the loss of electrons while negative anions are formed by the gain of electrons. An atom of sodium can lose an electron to form a cation. The resulting ion Na^+ has a nucleus with 11 protons surrounded by 10 electrons. An atom of chlorine can gain an electron to form an anion. The Cl^- anion has 18 electrons surrounding a nucleus with 17 protons (see Fig. 3.10).

■ Aside

Counting the Atoms in a Mole

How many atoms are in a mole? The determination of the charge on the electron enabled chemists to answer this question. When a direct current of electricity passes through a dilute solution of a silver compound, silver metal is deposited at one electrode. Scientists

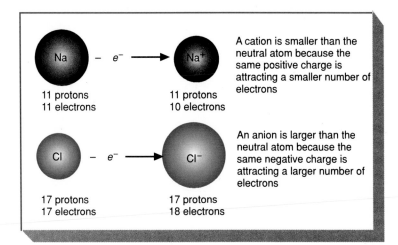

Figure 3.10
Cations are formed from atoms by the loss of one or more electrons. Anions are formed from atoms by the gain of one or more electrons.

measured the quantity of electric charge (current × time) required to deposit 1 mole of silver (107.9 g). This quantity of charge, 9.649×10^4 coulombs, is called a faraday (F). Because it takes one electron to convert each Ag^+ to a silver atom, the charge required to yield 1 mole of silver metal is the charge on 1 mole of electrons. By dividing 1 F of an electrical charge by the charge on one electron, scientists calculated Avogadro's number, that is, the number of electrons in a faraday and the number of silver atoms in a mole of silver metal. Avogadro's number N has the value 6.022×10^{23}.

$$N = \frac{F}{e} = \frac{9.649 \times 10^4}{1.602 \times 10^{-19}} = 6.022 \times 10^{23}$$

■ Isotopes

Isotopes are atoms of an element containing the same number of protons but a different number of neutrons in their nuclei. The most abundant isotope of hydrogen has a nucleus that consists of a single proton. A small fraction of hydrogen atoms are those of a heavier isotope called deuterium that has both a proton and a neutron in its nucleus. The least abundant isotope of hydrogen, tritium, has one proton and two neutrons in its nucleus. Isotopes of an element have the same chemical properties, but they differ in their mass and nuclear chemistry. For example, tritium is the only radioactive isotope of hydrogen.

Naturally occurring fluorine consists of a single isotope with 9 protons and 10 neutrons in its nucleus. The mass number for an isotope is the sum of the number of protons and the number of neutrons in the nucleus. Fluorine has a mass number of 19. In the symbol of an isotope, the mass number is given in the superscript and the atomic number in the subscript. The symbol of the naturally occurring isotope of fluorine is $^{19}_{9}F$.

An isotope of carbon, $^{12}_{6}C$ was chosen as the standard for measuring isotopic masses and atomic weights. The atomic mass of the $^{12}_{6}C$ isotope is defined to be exactly $12.\overline{000}$ atomic mass units (amu). For every isotope other than $^{12}_{6}C$, the mass is close but not identical to the mass number. For example, the atomic mass of naturally occurring fluorine is 18.998403 amu.

Figure 3.11
Isotopes are atoms of the same element that differ in the number of neutrons in their nuclei. These two isotopes of chlorine both have 17 protons and 17 electrons, but one has 18 neutrons and the other 20 neutrons.

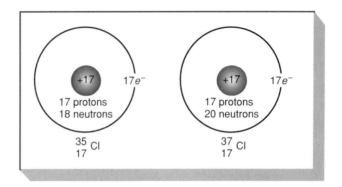

Many elements are found in nature as mixtures of isotopes. For example, there are two naturally occurring isotopes of chlorine, $^{35}_{17}\text{Cl}$ and $^{37}_{17}\text{Cl}$ (see Fig. 3.11). Each isotope of chlorine has 17 protons in its nucleus. The more abundant isotope of chlorine has 18 neutrons in its nucleus. The mass number of this isotope is 35 (17 protons + 18 neutrons = 35 nucleons). The isotope with mass number 37 has 37 nucleons (17 protons + 20 neutrons). Atoms of each isotope of chlorine have 17 electrons to balance the +17 charge of the nucleus. The atomic weight of chlorine, 35.453, is the weighted average of the masses of the two isotopes.

A graph showing the trend in the number of protons versus the number of neutrons in stable nuclei is shown in figure 3.12. For light elements other than hydrogen, nearly equal numbers of protons and neutrons are found in their most abundant isotopic nuclei. For heavier elements, the number of neutrons in the isotope's nucleus exceeds the number of protons. Heavy elements have relative atomic weights that are more than double their atomic numbers since their nuclei contain more neutrons than protons.

■ Nuclear Reactions

Scientists write equations for nuclear reactions using symbols that identify both the atomic number and the mass number of each reactant and product. In nuclear reactions, both electrical charge and the number of nucleons are conserved. The sum of the charges on the products of a nuclear reaction must equal the sum of the charges on the reacting nuclei, and the total number of nucleons in the products must equal the total number of nucleons in the reactants. An equation for a nuclear reaction is balanced when the charges and the mass numbers are balanced.

The following equations illustrate characteristic modes of nuclear decay. (High-energy electromagnetic radiation called *gamma rays* are produced in many nuclear reactions, but they do not appear in the balance of charge or mass number.) All nuclei containing more than 83 protons are radioactive. Many of them undergo alpha decay (An alpha particle is a helium nucleus emitted from a radioactive nucleus.):

$$^{238}_{92}\text{U} \rightarrow \, ^{234}_{90}\text{Th} + \, ^{4}_{2}\text{He}$$

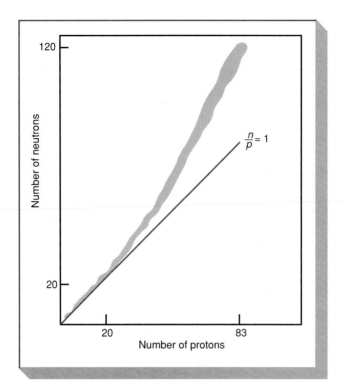

Figure 3.12
Stable nuclei are found in the band shown in the graph of the number of neutrons verses the number of protons in a stable isotope. For elements with an atomic number greater than 20, there are more neutrons than protons in their stable nuclei. No stable isotope has more than 83 protons in its nucleus.

Nuclei with a high neutron-to-proton ratio often undergo beta decay (A beta particle is an electron emitted from a radioactive nucleus.):

$$^{14}_{6}\text{C} \rightarrow\ ^{14}_{7}\text{N} +\ ^{0}_{-1}\text{e}$$

In the case of beta decay, a neutron in the reacting nucleus emits an electron to become a proton in the product nucleus. Some nuclei with low neutron-to-proton ratios decay by positron emission, and others decay by electron capture.

$$^{30}_{16}\text{S} \rightarrow\ ^{30}_{15}\text{P} +\ ^{0}_{+1}\text{e} \text{ (positron emission)}$$

$$^{7}_{4}\text{Be} +\ ^{0}_{-1}\text{e} \rightarrow\ ^{7}_{3}\text{Li} \text{ (electron capture)}$$

A *positron* has the same mass as an electron but it carries a +1 charge. Positrons are short-lived; upon collision with an electron, both particles are annihilated and gamma radiation is produced. When a positron is emitted, a proton is converted to a neutron. In the case of *electron capture,* an electron is captured by the nucleus to convert a proton to a neutron. Both positron emission and electron capture raise the neutron-to-proton ratio (see Fig. 3.13).

The mass number and the atomic number of the products of nuclear reactions can be readily calculated. By balancing the charge numbers and the mass numbers, one can readily complete nuclear reactions. However, it is necessary to know either the starting isotope and one of the products or the two products of a nuclear decay reaction.

$$^{226}_{88}\text{Ra} \rightarrow\ ^{4}_{2}\text{He} + ? \text{ (alpha decay)}$$

Figure 3.13
For radioactive elements, the mode of decay is in the direction of the zone of stability.

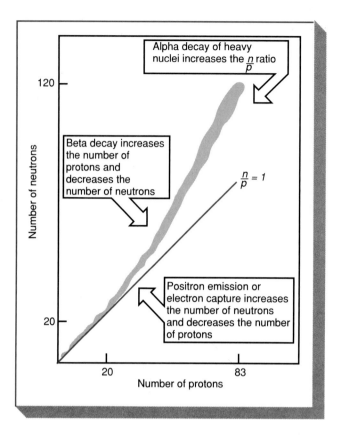

Mass number of product = 226 − 4 = 222

Atomic number of product = 88 − 2 = 86

The symbol of element 86 is Rn, so the product is $^{222}_{86}$Rn.

$$^{214}_{83}\text{Bi} \rightarrow {}^{0}_{-1}\text{e} + ? \text{ (beta decay)}$$

Mass number of product = 214 − 0 = 214

Atomic number of product = 83 − (−1) = 84

The product of the beta decay reaction is $^{214}_{84}$Po.

Identify which of the following nuclear reactions of thorium isotopes produces an alpha particle and which produces a beta particle.

$$^{230}_{90}\text{Th} \rightarrow {}^{226}_{88}\text{Ra} + ?$$

$$^{234}_{90}\text{Th} \rightarrow {}^{234}_{91}\text{Pa} + ?$$

Products of nuclear reactions may be radioactive themselves. For example, a chain of nuclear decay reactions that starts with $^{238}_{92}$U does not stop until it reaches the stable isotope $^{206}_{82}$Pb. Both polonium (Po) and radium (Ra), the radioactive elements first isolated by Marie Curie from a uranium ore, are "daughters" of uranium 238. Intermediate radioactive isotopes in this decay series undergo either alpha decay or beta decay. (Note that considerations about the safe use of a radioisotope need to take into account the possible production and safety of radioactive daughters.)

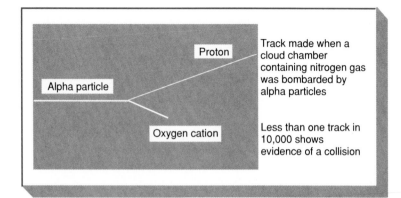

Proton

Alpha particle

Oxygen cation

Track made when a cloud chamber containing nitrogen gas was bombarded by alpha particles

Less than one track in 10,000 shows evidence of a collision

Figure 3.14
When an alpha particle strikes the nucleus of a nitrogen atom, a nuclear reaction produces a proton and an oxygen ion. In a cloud chamber, moving charges give rise to a trail of water droplets. These droplets are illuminated and photographed against a black background. The slower a charged particle moves, the more ions it produces and the thicker its track.

■ Aside

Observing Protons and Neutrons

On the basis of his gold foil experiment, Rutherford formulated the model of an atom in which electrons surrounded a small, massive positive core. The gold foil experiment does not require that nuclei consist of protons and neutrons. In fact, the existence of protons was known before the Rutherford experiment, and neutrons were not observed directly until two decades later.

Positive particles, both protons and ions of elements other than hydrogen, were observed in experiments similar to that in which J. J. Thomson measured e/m for the electron. Positive particles were produced by electrical discharges through evacuated glass tubes containing small residues of gases. These positive particles could be deflected by electric or magnetic fields. Scientists measured the charge-to-mass ratio (e/m) for these ions. The positive particle with the largest value of e/m was produced using tubes containing hydrogen gas. This was the proton, $_1^1 H$, produced by the loss of an electron from a hydrogen atom. (When the gas was oxygen, the positive particle produced was O^+ with 16 times the mass of H^+.)

There remained the task of demonstrating the existence of protons and neutrons in nuclei. Rutherford and his associates bombarded a wide variety of targets with alpha particles. Using nitrogen gas as the target, they produced new positive particles. These particles passed through foils that would stop alpha particles. By deflecting the new particles with a magnetic field, they showed them to be hydrogen nuclei.

Charged particles moving through the air knock electrons from some atoms to form ions. These ions initiate the formation of water droplets from water vapor. Scientists use trails of water droplets to track the motion of charged particles. Measurements using a cloud chamber show that protons are produced by the reaction of an alpha particle with a nitrogen nucleus (see Fig. 3.14). From the thousands of pictures taken in cloud chambers, scientists found a small number in which the trail of an alpha particle ended abruptly and a new forked trail began. An alpha particle had been absorbed, and a proton and an oxygen ion had been produced.

$$^{14}_{7}N + ^{4}_{2}He \rightarrow ^{17}_{8}O + ^{1}_{1}H$$

When a beryllium target was bombarded with alpha particles, a new intense radiation was produced. This radiation penetrated materials that stopped charged particles and interacted with paraffin, a substance rich in hydrogen, to knock out protons. (Gamma radiation does not do this.) In 1932, James Chadwick showed that this penetrating radiation consisted of neutrons. These long-sought particles had been produced by the reaction of alpha particles with beryllium nuclei.

$$^{9}_{4}Be + ^{4}_{2}He \rightarrow ^{12}_{6}C + ^{1}_{0}n$$

■ Effects of Radiation

Particles ejected in nuclear reactions interact strongly with matter. Alpha and beta particles produce ions by knocking electrons from atoms along their path. Gamma radiation can also cause ionization. In addition, radiation can interact with molecules to break them into highly reactive fragments.

Although the more massive, higher-charged alpha particles cause greater damage along their paths, they are more easily stopped than beta particles. Alpha particles can be stopped by a thin piece of paper while the lighter beta particles penetrate further but are still readily contained by thin shields. Thick concrete or heavy lead shields are required to contain high-energy gamma radiation.

Radiation can cause mutations in living organisms because both the ionization caused by radiation and the production of highly reactive fragments of molecules serve to bring about changes in DNA, the genetic material of cells. Because the body has some ability to repair radiation damage, the biological effects of long-term exposure to very low levels of radiation are controversial. There is some "natural radiation" that is called background radiation. Some scientists believe that any increase in radiation levels above background levels constitutes a health hazard; others do not.

■ Isotopes Are Tools in Biology, Geology, and Medicine

Scientists use both man-made and naturally occurring isotopes as research tools. Most elements occur as a mixture of isotopes, while nuclear reactions are used to produce radioactive isotopes. All of these isotopes can be incorporated into compounds and used to follow underwater drainage or the fate of a compound in metabolism. The following examples describe some of the range of applications of isotopes in research and medicine. Other examples will be presented later in the text when additional features of radioactive decay have been discussed.

In 1952, Alfred Hersey and Martha Chase used radioactive isotopic tracers to demonstrate that the DNA of viruses, rather than their protein, is responsible for

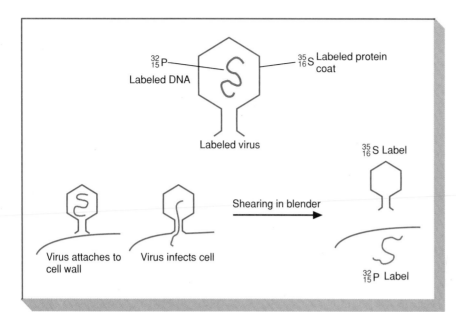

Figure 3.15
Hersey and Chase used isotopic markers to show that DNA, rather than protein, carried the genetic information of viruses.

infection. Many viruses that infect normal cells and cause them to make new viral particles consist only of protein and a small amount of DNA. Proteins contain a small amount of sulfur but no phosphorus. Phosphorus, and not sulfur, is present in DNA. Hersey and Chase grew viruses in a medium that contained $^{32}_{15}P$ and $^{35}_{16}S$, man-made radioactive isotopes. Then they used the isotopically labeled viruses to infect normal cells. After a short time the mixture was placed in a blender to shear away viral material adhering to the outside of cell surfaces and then Hersey and Chase tested the infected cells for the presence of $^{32}_{15}P$ and $^{35}_{16}S$. They found the phosphorus isotope but not the radioactive sulfur, showing that viral DNA had entered the cells while viral protein remained outside. They had conclusively demonstrated that DNA carried the instructions for the synthesis of new viral particles (see Fig. 3.15).

Radiologists use radioactive isotopes to help treat cancers. High levels of radiation are directed at tumors to kill cancerous cells. For example, needles of radioactive cobalt are placed near a tumor to irradiate the tumor for a controlled amount of time. Since they multiply more rapidly, cancer cells are more vulnerable to radiation than normal cells, although both suffer damage. For this reason, radiation therapy can only be used intermittently. Doctors often use radiology in combination with chemotherapy to treat tumors inaccessible to surgery.

■ Mass and Energy in Nuclear Reactions

Albert Einstein (1879–1955) postulated that matter and energy are related. Einstein's equation is one of the more widely known equations of physics:

$$E = mc^2$$

Energy is equal to mass times the square of the speed of light.

Figure 3.16
The binding energy per nucleon as a function of mass number. The most stable nuclei are at the top of the curve. The most stable nucleus is $^{56}_{26}$Fe.

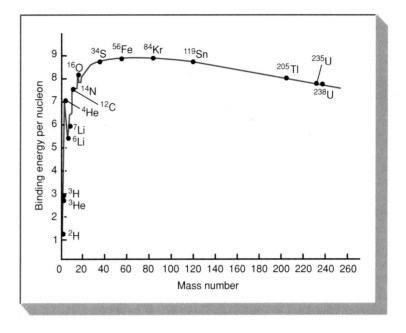

Physicists apply Einstein's equation to estimate the energy holding nuclei together. The mass of a nucleus is less than the sum of the masses of protons and neutrons present in the nucleus. For example, the mass of a helium nucleus is slightly less than the sum of the masses of two protons and two neutrons.

$$2\,^{1}_{1}\text{H} \quad + \quad 2\,^{1}_{0}\text{n} \quad \rightarrow \quad ^{4}_{2}\text{He}$$

$$1.0078 \text{ amu} \qquad 1.0087 \text{ amu} \qquad 4.0026 \text{ amu}$$

$$\Delta m = 4.0026 \text{ amu} - 2(1.0078 \text{ amu}) - 2(1.0087 \text{ amu}) = -0.0304 \text{ amu}$$

The difference between the mass of a nucleus and the sum of the masses of its nucleons is called the *mass defect*.

The energy calculated from the mass defect is called the *binding energy*, because that much energy would be required to break the nucleus into its separate nucleons. (The mass defect of nuclei are very small, but because c^2 in the Einstein equation is a very large number, the energies involved in nuclear binding are far greater than those involved in chemical binding.) The binding energy per nucleon first increases as the number of nucleons increases, and then it decreases as shown in figure 3.16. The most stable nuclei, those of iron and elements near iron in atomic number, have a greater binding energy per nucleon than the nuclei of lighter or heavier elements.

In many nuclear reactions, the mass of the products is measurably less than the mass of the reactants. So the *m* in the Einstein equation is the mass converted into energy—the mass of reactants minus the mass of the products.

■ Transmutation and Nuclear Fission

When alpha particles strike some nuclei, transmutations occur. One element is converted into another by a nuclear reaction. For example, the nuclear reaction of an alpha particle with nitrogen produces a proton and an isotope of oxygen, and the nuclear reaction of an alpha particle with beryllium produces a neutron and an isotope of carbon. Scientists looked for other nuclear reactions by using protons and neutrons to bombard atomic targets.

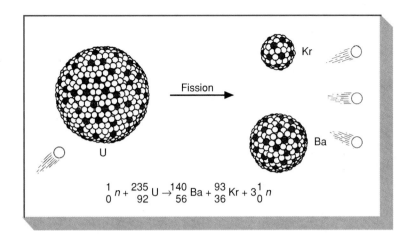

Figure 3.17
Fission occurs when a uranium-235 nucleus absorbs a neutron. The uranium nucleus splits into two product nuclei and produces additional neutrons. Large amounts of energy are released because the product nuclei have a greater binding energy per nucleon than does the uranium nucleus.

Otto Hahn (1879–1968), a German chemist, separated the elements barium, lanthanum, and cerium from a uranium source that had been bombarded by neutrons. He concluded that these lighter elements were produced by *fission*. When a nucleus of uranium-235 absorbs a neutron, it spontaneously splits to form two lighter nuclei and additional neutrons. (For example, see Fig. 3.17.)

$$_{0}^{1}n + _{92}^{235}U \rightarrow _{56}^{140}Ba + _{36}^{93}Kr + 3\,_{0}^{1}n$$

Scientists immediately recognized that fission produces energy far in excess of any chemical reaction. In the fission of uranium-235, the mass of the products is less than the mass of the reactants. The missing mass is converted into energy. The energy appears as the recoil energy of the particles produced by fission and as gamma rays.

Hahn communicated his discovery to Lisa Meitner, a former coworker, who had fled fascist Germany in 1938. Through refugee scientists from Europe, the news reached the scientific community in America. In a letter dated August 2, 1939, Albert Einstein, at the urging of fellow physicists, wrote to President Roosevelt calling attention to the use of fission as a source of energy and advising further investigation. This letter paved the way for the Manhattan Project, the effort that culminated in the production of the first atomic bombs dropped on Japanese cities at the close of World War II (The peaceful use of atomic energy for the production of power is discussed in Chapter 12.).

■ Questions—*Chapter 3*

1. Humphry Davy was the first person to isolate the metals magnesium, calcium, strontium, and barium, as well as sodium and potassium. Write an electrode reaction for the formation of barium from the Ba^{+2} ion found in compounds.
2. Molten calcium bromide conducts electricity. What can you infer about the particles present in the melt?
3. In his discovery of radioactivity, did Becquerel actually observe the alpha decay of $_{92}^{238}U$? (Consider the penetrating power of different types of radiation.) What radiation might have caused the wrapped photographic plates to fog?
4. Why are alpha and beta particles deflected in opposite directions by an electric field? Why are beta particles deflected more than alpha particles?

5. What modification of Dalton's atomic theory is required by the discovery of radioactivity? (Which of his postulates must be modified?)

6. When did the statement "Uranium is radioactive" become scientific fact? Would it have been meaningful to Dalton? Are scientific facts separable from theory?

7. How did the isolation of a radioactive element by the Curies contribute to the understanding of radioactivity? (Without this discovery, could it have been known that radioactive decay is a nuclear reaction?)

8. In the Millikan oil drop experiment, drops occasionally changed charge. How would a change in charge affect the rate of fall? Of rise? What size charge might a drop gain or lose? Why?

9. Why would it be necessary to measure the charge on many drops to obtain the charge on an electron using Millikan's method?

10. The charge on the electron is 1.602×10^{-19} C. The ratio e/m for the electron is close to 1.76×10^8 C/g. What is the mass of an electron?

11. If Rutherford had used a foil of a metal with a much lower atomic weight than gold, some of his observations would have been different. Would the number of alpha particles passing through the foil have increased greatly? Would the number of alpha particles deflected through wide angles have decreased significantly? Why or why not?

12. In Rutherford's gold foil experiment, what is the significance of each of the following observations?
    (a) Most alpha particles are not deflected when they pass through the foil.
    (b) Some alpha particles are deflected through wide angles.

13. Between 1800 and 1900, the formal educational background of people entering science changed greatly. In an encyclopedia, locate information on the educations of Joseph Priestley, Antoine Lavoisier, and John Dalton, and compare it to the corresponding information on Marie Curie, J. J. Thomson, and Ernest Rutherford.

14. Mass spectrometers are instruments used to separate positive ions. Positive ions can be accelerated, passed through a slit, and deflected by either a magnetic field or an electric field. Using a mass spectrometer, ions of isotopes of an element differing in the number of neutrons in their nucleus can be separated. Why is this possible? (*Hint:* Consider the value of e/m for each isotope.)

15. Complete the following table.

Element	Protons	Neutrons	Mass Number	Isotopic Symbol
U	92	238		$^{238}_{92}U$
C			14	$^{14}_{6}C$
	17	18		

16. In figure 4.9, the drawing indicates that the Na^+ ion is smaller than an Na atom. Consider the attraction of the nucleus for electrons. Why should the removal of one electron produce a smaller species?

17. Why is the chloride ion, Cl^-, larger than the chlorine atom?

18. Complete the following table.

Species	Protons	Electrons	Net Charge
S^{-2}	18		-2
Cl^-	17		-1
	18	18	
K^+		18	
	20		+2

19. Complete the following nuclear reactions found in the U-238 decay series.

    (a) $^{226}_{88}\text{Ra} \rightarrow ? + {}^{4}_{2}\text{He}$

    (b) $^{222}_{86}\text{Rn} \rightarrow {}^{218}_{84}\text{Po} + ?$

    (c) $? \rightarrow {}^{214}_{82}\text{Pb} + {}^{4}_{2}\text{He}$

    (d) $? \rightarrow {}^{214}_{83}\text{Bi} + {}^{0}_{-1}e$

20. Complete the following nuclear reactions. Identify the reactions as examples of alpha decay, beta decay, positron emission, or nuclear fission.

    (a) $^{14}_{6}\text{C} \rightarrow {}^{14}_{7}\text{N} + ?$

    (b) $^{38}_{19}\text{K} \rightarrow {}^{38}_{18}\text{Ar} + ?$

    (c) $^{210}_{84}\text{Po} \rightarrow {}^{206}_{82}\text{Pb} + ?$

    (d) $^{1}_{0}n + {}^{235}_{92}\text{U} \rightarrow {}^{90}_{38}\text{Sr} + {}^{144}_{54}\text{Xe} + ?$

21. Frederic Joliot-Curie and his wife, Irene, daughter of Marie Curie, showed that the bombardment of aluminum by alpha particles produced both neutrons and positrons. When the alpha particle source was removed, only the positron emission continued. They concluded that they had produced a radioactive isotope of phosphorus, and that the newly formed isotope of phosphorus was a positron emitter. For their demonstration of artificially induced radioactivity, they shared the Nobel prize in chemistry in 1935. Complete and balance the equations for the nuclear reactions discovered by the Joliot-Curies.

$$^{4}_{2}\text{He} + {}^{27}_{13}\text{Al} \rightarrow {}^{?}_{?}\text{P} + {}^{1}_{0}n$$

$$^{?}_{?}\text{P} \rightarrow {}^{?}_{?}? + {}^{0}_{+1}e$$

22. What was the significance of Otto Hahn's discovery that the element strontium had been produced in a sample of uranium that had undergone neutron bombardment?

4

The Modern Atom and the Periodic Table

A revolution in the ways to view the nature of both light and matter occurred in physics early in the twentieth century. ("Classical physics" and "modern physics" are sometimes used to designate the prerevolutionary and postrevolutionary eras in the field.) In classical physics, light and matter were considered to be separate and distinct entities. After the revolution in physics, light was thought to have particle-like properties and electrons in atoms were thought to resemble waves. This revolution in physics led to a model of the atom that helps account for the chemical similarities and differences summarized in the periodic table. This model can be extended to explain the bonding of elements to form compounds and the properties of the compounds formed by chemical combination.

■ Light, Waves, and Particles

Light can be considered as consisting of electromagnetic waves (see Fig. 4.1). For each wave, the distance from crest to crest is called the *wavelength*, λ. Units used for reporting wavelengths are frequently meters (m) or centimeters (cm). The number of waves passing a stationary point in a unit of time is called the *frequency, v*. Units of frequency are cycles per second or sec^{-1}. Note that one cycle per second is also called one hertz (Hz). *The speed of light (c) is the product of the wavelength of the radiation times its frequency.*

$$c = \lambda v = 3.00 \times 10^8 \text{ m/sec}$$

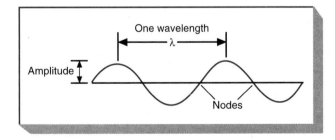

Figure 4.1
A wave is characterized by a wavelength, an amplitude, and a frequency. The frequency of a wave is the number of waves passing a fixed point per second.

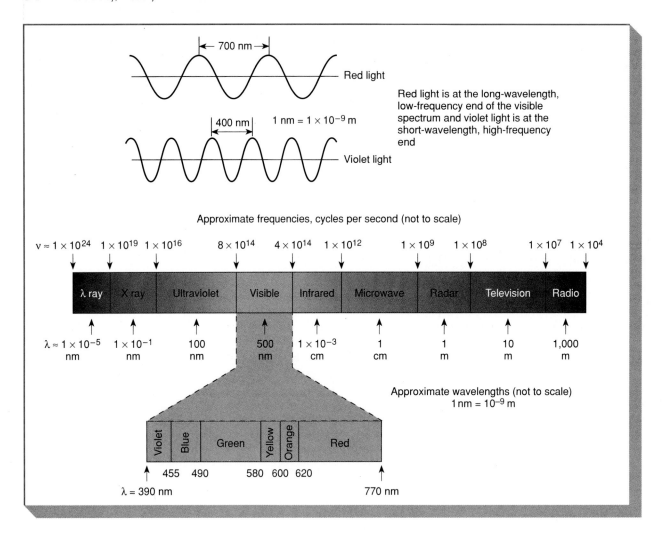

Figure 4.2
Visible light is a small portion of the electromagnetic spectrum.

Visible light comprises only a small portion of the electromagnetic spectrum as shown in figure 4.2. In the visible spectrum, red light has the longest wavelength and lowest frequency, and violet light has the shortest wavelength and highest frequency. Lower-frequency, longer-wavelength radiation includes infrared rays, microwaves, and radio waves. Shorter-wavelength, higher-frequency radiation includes ultraviolet light, X rays, and gamma rays. All these forms of electromagnetic radiation travel at the speed of light.

A *wave model of light* is proposed to explain the bending of light by a prism or the diffraction of light by a grating (see Fig. 4.3). A prism separates a beam of white light into its component colors because the bending of light at the surfaces of the prism depends on the wavelength. Longer-wavelength red light is bent less than shorter-wavelength blue. A diffraction grating is another device that separates white light into its colored components. The closely spaced rulings on a grating scatter light into its colored components because the direction of maximum reinforcement depends on the wavelength of light. Waves in phase reinforce one another and waves out of phase cancel one another.

However, the wave model of light fails to explain the *photoelectric effect*. Light with sufficiently high frequency causes electrons to be emitted from some surfaces as shown in figure 4.4. Light of lower frequency, no matter how bright, has no effect. For example, yellow light causes electrons to be emitted from potassium, but red light does not. A *particle model of light* is used to interpret the photoelectric

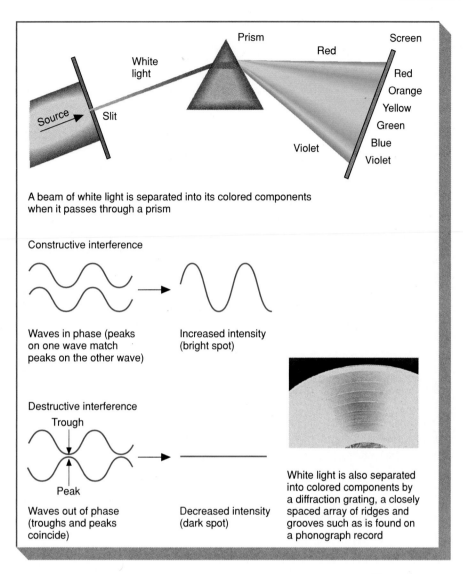

A beam of white light is separated into its colored components when it passes through a prism

Constructive interference

Waves in phase (peaks on one wave match peaks on the other wave)

Increased intensity (bright spot)

Destructive interference

Trough

Peak

Waves out of phase (troughs and peaks coincide)

Decreased intensity (dark spot)

White light is also separated into colored components by a diffraction grating, a closely spaced array of ridges and grooves such as is found on a phonograph record

Figure 4.3
A wave model of light is used to explain the bending of light by a prism or the diffraction of light by a grating.

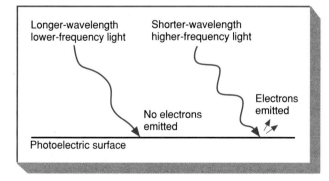

Longer-wavelength lower-frequency light

Shorter-wavelength higher-frequency light

No electrons emitted

Electrons emitted

Photoelectric surface

Figure 4.4
Light of high frequency causes electrons to be emitted from some surfaces. A particle model of light in which energy of the light depends on frequency, $E = h\upsilon$, is used to explain this phenomenon.

Figure 4.5
In a flame test, a drop of a metal halide solution is heated by an intense flame. If the light from the flame is passed first through a slit and then through a prism, the resulting spectrum consists of a series of bright lines that are characteristic for that metallic element. A similar line spectrum can be obtained if the vapor of the element is excited by an electric discharge.

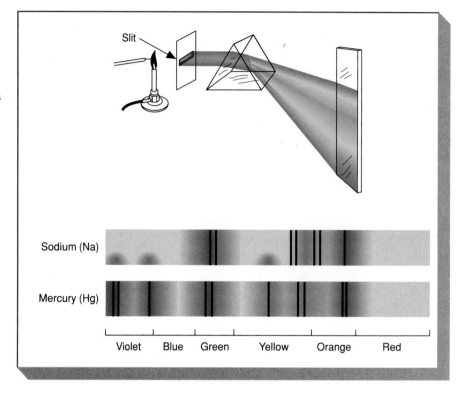

effect. Albert Einstein postulated that *light consists of particles called photons,* and that a photon with sufficiently high energy can cause an electron to be ejected while one or more lower-energy photons cannot.

In 1900, Max Planck (1858–1947) had first proposed that light energy is emitted in discrete units called photons to account for the distribution of "black body" radiation, which is emitted from an idealized hot object. According to Planck, the energy of a photon is proportional to its frequency. In *Planck's equation, E* is the energy of the photon, *h* is a constant known as Planck's constant, and υ is the frequency of the light.

$$E = h\upsilon$$

where *h* is Planck's constant and equals 6.63×10^{-34} J-sec.

In some experiments light behaves like a wave and in other experiments light acts as if it were a particle. The paradox that one experiment may be interpreted by a wave model of light and another experiment may be interpreted by a particle model is sometimes referred to as the *wave-particle duality of light.*

■ The Interaction of Atoms and Light

The study of the colors in flame (and in gas discharge tubes) test results played a critical role in the development of modern ideas concerning atomic structure. When a drop of a solution of a salt suspended on a wire loop is placed in the flame of a Bunsen burner, the solution vaporizes and then a flash of a characteristic bright color appears in the flame. Sodium compounds give intense yellow; lithium compounds, red; barium compounds, green; and copper compounds, blue. (Some of these same compounds are used in fireworks to generate the bright reds, greens, and blues that can be seen in a fireworks display on the Fourth of July.) When the light from a flame test is passed through a slit and then through a prism, an emission spectrum consisting of a few bright lines can be observed (see Fig. 4.5). For

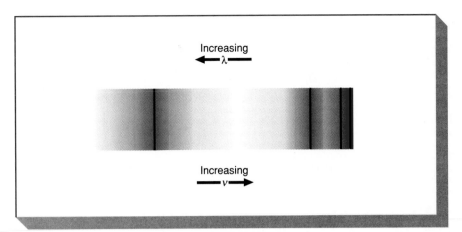

Figure 4.6
The emission spectrum of the
hydrogen atom consists of a small
number of lines increasingly close
together at shorter wavelengths.

example, the emission spectrum obtained from a sodium flame test features bright yellow lines. These same bright yellow lines can be seen in the spectrum of light emitted by sodium vapor lamps used as street lights.

The study of the interaction of matter and light is known as spectroscopy. In a flame or in an electric discharge, some atoms are excited to a high-energy state. These excited atoms can lose energy by emitting light. The shorter the wavelength of light, the greater the energy of a photon. (A photon of shorter-wavelength blue light carries almost twice the energy as a photon of longer-wavelength red light.)

■ A Simple Equation Describes Hydrogen Spectra

Hydrogen's emission spectrum is relatively simple. Its visible spectrum, known as the Balmer series, consists of a few lines that are increasingly close together toward the violet end of the spectrum as shown in figure 4.6. A simple mathematical expression describes an empirical relationship between the wavelengths of lines in the spectrum of hydrogen:

$$1/\lambda = R_H(1/2^2 - 1/n^2); \ n = 3, 4, 5 \ldots$$

where R_H is a constant equal to 109,677.6 cm^{-1}.

For the first three lines in the Balmer series, the expression in the parentheses $(1/2^2 - 1/n^2)$ has the values $(1/4 - 1/9)$, $(1/4 - 1/16)$, and $(1/4 - 1/25)$.

Scientists looked for, and found, additional series of emission lines at wavelengths predicted by substituting the integers 1 (the Lyman series), and then 3 (the Paschen series), for the integer 2 in the equation for the Balmer series. Lines of the Lyman series have shorter wavelengths than visible light and are found in the ultraviolet portion of the spectrum. (The expression in parentheses has the value $(1/1 - 1/4)$ for the first line in the Lyman series.) Lines of the Paschen series have longer wavelengths than visible light and are found in the infrared portion of the spectrum. (The expression in parentheses has the value $(1/9 - 1/16)$ for the first line in the Paschen series.)

Notice again the appearance of integers in an empirical relationship—the equation for the emission spectrum of hydrogen. Recall that hydrogen is the simplest of atoms, for it consists of a nucleus and a single electron. The negative charge of the electron is balanced by the positive charge of the nucleus. What is the connection between the structure of the hydrogen atom and the existence of integers in the equation for its emission spectrum?

Figure 4.7
In the Bohr atom, an electron circles the nucleus in an orbit. The radius of the orbit depends on the value of the quantum number n.

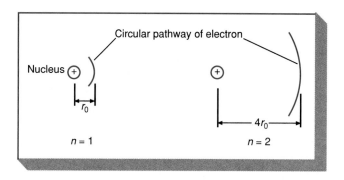

■ The Bohr Atom

In 1913, Neils Bohr (1885–1926), a young Danish physicist, formulated a model of the hydrogen atom to explain its spectrum. Bohr proposed that in this simple atom, an electron circles the nucleus as the earth circles the sun. The circular path of the electron was called an *orbit*, and the energy of an electron in an orbit was fixed. To account for the emission lines in the spectrum of hydrogen, Bohr postulated that the radius of each orbit is restricted; each radius is an integer squared times the radius of the lowest energy orbit:

$$r = n^2 \, r_0; \; n = 1, 2, 3, 4, \ldots$$

n is a positive integer and r_0 is the radius of the lowest-energy orbit. The integer n is called a *quantum number*. This quantum number also appears in the calculated value of the energy of each orbit in the Bohr atom (see Fig. 4.7).

Because the radius of the orbit is restricted, the energy of an electron in the orbit is restricted. Bohr expressed the energy of the electron in an orbit as the sum of an energy term involving the attraction of the negative electron to the positive nucleus and of an energy term involving the movement of the electron. Bohr then solved the equation for the energy of the hydrogen atom. The calculated energy of an electron in the Bohr atom depends upon the quantum number n, Planck's constant h, the mass m and charge e of the electron, and the speed of light c.

$$\text{Energy} = - (2\pi^2 m e^4 c^4 \times 10^{-14}/h^2) \, (1/n^2); \; n = 1, 2, 3, 4, \ldots$$

The complex term in parentheses is equal to R_H, the empirical constant in the equation for the emission spectrum of hydrogen.

A totally separated electron and proton are considered the starting point for measuring energy. The closer the electron is to the nucleus, the lower its energy. The lowest-energy state of the hydrogen atom is associated with the quantum number $n = 1$; larger values of n correspond to higher-energy states.

A photon of light is emitted when an electron drops from a higher-energy orbit, one further from the nucleus, to a lower-energy orbit, one nearer the nucleus. The energy of the emitted photon accompanying a change from one orbit to another can be readily calculated as the difference between the energies of the two orbits (see Fig. 4.8).

$$E_{\text{upper}} - E_{\text{lower}} = \Delta E = h\upsilon = hc/\lambda$$

The changes in energies predicted by the Bohr model correspond to the energies of the emission lines in the hydrogen spectrum. These results imply a dramatic connection between line spectra and the electronic structure of atoms.

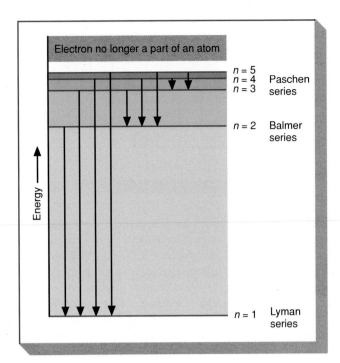

Figure 4.8
According to the Bohr model of the atom, an electron falls from a higher-energy orbit (larger value of n) to a lower-energy orbit (smaller value of n) by emitting a photon with energy equal to the energy difference between the two orbits.

■ Problems with the Bohr Atom

Although the Bohr atom accounted for the line spectrum of hydrogen, the appearance of integers, or quantum numbers, was arbitrary. This model had other deficiencies. According to classical physics, electrons could not circle the nucleus in stable orbits; instead, they would radiate energy and spiral into the nucleus. Because the introduction of quantum numbers was arbitrary, and physical theory failed to account for the path of the electron, the Bohr model was recognized to be artificial even as it was proposed.

Why was the Bohr model of the atom important, even though it was recognized as deficient? The Bohr model was the first to describe the electronic structure of atoms, and it successfully accounted for the emission spectrum of hydrogen. It was even extended to account for some features of the periodic table. Although inadequate in some ways, the Bohr model of the atom pointed to new physical phenomena and aided scientists during a period of rapid change.

■ Aside

Spectroscopy Leads to the Discovery of the Noble Gases

Astronomers were the first to find evidence of a new element in the sun's corona during a solar eclipse in 1868. A dark line was present in the spectrum of the corona that could not be associated with the spectrum of any known element or compound. The sun's light, at the proper wavelength, had been absorbed by atoms of an unknown element. The element was named helium after the Greek "helios" meaning sun.

The dark line present in the spectrum of the sun was later shown to be an absorption line of a gaseous element that was first isolated from an

Figure 4.9
An entire family of elements, the noble gases, was discovered after Mendeleev had formulated the periodic table.

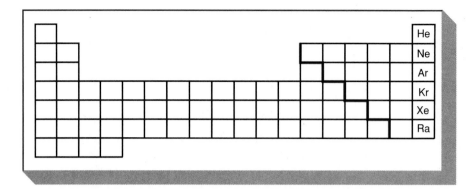

uranium ore in 1888. The helium trapped in the ore was a product of radioactive decay. When alpha particles are slowed by their interactions with matter, they eventually pick up electrons to form helium atoms. When the ore was dissolved in acid, the helium gas was released.

A search that led to the discovery of other gaseous elements began when discrepancies were found in the density of nitrogen. Nitrogen gas prepared from compounds of nitrogen was less dense than nitrogen from air. William Ramsey and his associates sought to remove nitrogen and oxygen from air and examine the residue. In one set of experiments, oxygen-enriched air was repeatedly sparked to form nitrogen oxides, which dissolved in sodium hydroxide solution. The remaining gas was denser than nitrogen, and its spectrum contained new lines. Ramsey had isolated the element argon. In the following years, Ramsey also isolated neon, krypton, and xenon from liquefied air (see Fig. 4.9).

At various times these unreactive, monatomic elements have been called noble gases, inert gases, or rare gases. Not only are these elements abundant (argon is one percent of air and helium is one of the most abundant elements in the universe), but some, like xenon, undergo chemical reactions. Although no compounds of any of these elements were known before 1956, a few compounds of the heavier elements of the family have been prepared since then.

■ Electrons Can Behave as Waves

If light could behave as a particle, could an electron behave as a wave? The answer to this question is yes. In 1923, Louis de Broglie (1892–1987) postulated that a particle could behave as a wave with a characteristic wavelength, $\lambda = h/mv$, where λ is the wavelength, h is Planck's constant, m is the mass of the particle, and v is its velocity or speed.

The *de Broglie hypothesis* was tested experimentally using electrons. It was known that atoms in a crystal diffract X rays. Electrons were accelerated to a velocity such that their wavelength, calculated using the de Broglie relationship, was equal to the wavelength of diffracted X rays. When this beam of electrons was aimed at a nickel crystal, diffraction was observed as shown in figure 4.10. The de Broglie hypothesis was correct; electrons can behave as waves.

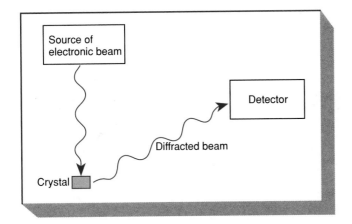

Figure 4.10
To demonstrate the wave nature of electrons, scientists studied their diffraction. In this experiment, electrons are diffracted by a crystal of nickel. The diffracted beam of electrons is observed at the same position as diffracted X rays of the same wavelength.

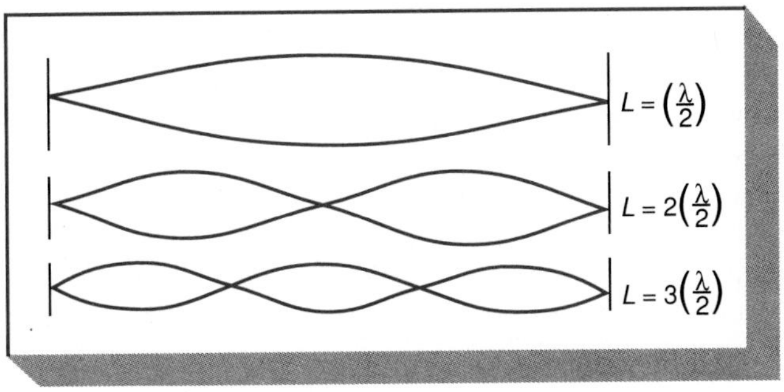

$$L = \left(\frac{\lambda}{2}\right)$$

$$L = 2\left(\frac{\lambda}{2}\right)$$

$$L = 3\left(\frac{\lambda}{2}\right)$$

Figure 4.11
For a string anchored at both ends, standing waves are restricted to a small number of wavelengths. Each possible standing wave consists of an integral number of half-waves.

■ The Wave Mechanical Atom

In 1926, Erwin Schrödinger (1887–1961) proposed a *wave mechanical model of atomic structure.* He postulated that electrons in atoms behave like waves, rather than like particles. If electrons behave as waves in atoms, their behavior must also be similar to familiar examples of waves such as vibrating strings, waves on water, and the sound of music. The mathematics necessary to describe wave behavior is well-known but beyond the scope of an introductory chemistry course. We can, however, reason by analogy. If we consider the vibrations of a guitar string as a guide to the behavior of electrons in atoms, then we might better understand the description of the atom developed in the late 1920s.

Solutions of the wave equation for a vibrating string are quantized. Only certain wavelengths are observed. Since a guitar string is anchored at each end, sustained vibrations or standing waves must consist of an integral number of half-waves. To put this relationship in mathematical form, only wavelengths satisfying the following equation are allowed:

$$n\lambda/2 = L; \; n = 1, 2, 3 \ldots$$

where L is the length of the string. The integer n in the equation is called a *quantum number.* Each solution to the wave equation for a vibrating string has a different value of n (see Fig. 4.11).

For a vibrating string, the wavelength and energy of a wave are independent. A musical note may be either high or low, loud or soft. The wavelength and the

Figure 4.12

Atomic orbitals differ in their size and shape. The darker the shading in a diagram, the higher the probability of finding an electron in a small region of space. Note that some orbitals have nodal planes.

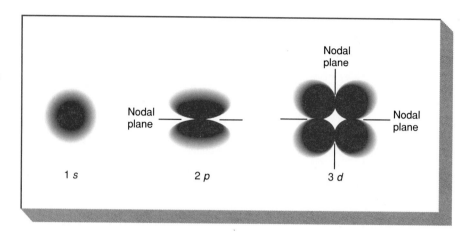

frequency of vibration are related to pitch—the shorter the wavelength is, the higher the frequency and the pitch are. When shortening the string by pressing on a fret, the guitarist creates a different string length, a new solution for the wave equation, and a higher note. In contrast, the energy of the wave is related to its amplitude. The greater the amplitude of the vibration, the louder the sound produced.

In what ways is the treatment of electrons in atoms similar to the treatment of a vibrating guitar string? For the electron in the atom, there are boundary conditions analogous to the anchoring of the guitar string at its ends. Similarities are, therefore, found in the mathematical form of their wave equation solutions. Solutions to the wave equation for the atom are also characterized by integers called quantum numbers. For an electron in an atom, there are three quantum numbers *n*, *l*, and *m* corresponding to three dimensions in space, and a fourth quantum number *s* related to the magnetic properties of the electron.

Differences between wave descriptions of vibrating strings and of electrons in atoms lie in their physical interpretations. For the guitar string, amplitude is related to energy, and wavelength is related to pitch. For the electron, amplitude is related to the electron's preference for various positions in the atom, and wavelength is related to the electron's energy.

■ Atomic Orbitals

Each solution to the wave equation for an electron in an atom is called an *orbital*. An orbital is a region in space where an electron can be found. Associated with each orbital is a unique set of quantum numbers, *n*, *l*, and *m*. *Each orbital can contain a maximum of two electrons,* for two electrons having different values of the fourth quantum number *s* can share an orbital. Such electrons are said to be "paired."

Orbitals having the same value for *n*, the principal quantum number, belong to the same *shell*. The quantum number *n* can have the integer values 1, 2, 3, ... *The number of orbitals in a shell is equal to n^2.* The maximum number of electrons in a shell is $2n^2$. For n = 1, 2, 3, or 4, this limits the first four shells to 2, 8, 18, and 32 electrons, respectively.

Representations of three different orbitals are shown in figure 4.12. One cannot speak of the pathway of an electron in an orbital but only of the probability of finding it in a region of space. The darker the shading in a particular place in the diagram of an orbital, the greater the probability of finding an electron. Note that the orbitals labeled 2*p* and 3*d* have nodal planes. Nodes are a feature of the wave nature of electrons in atoms. An electron in a 2*p* orbital may be found above or below the nodal plane but never in the plane. The existence of nodal planes in the description of orbitals illustrates the dramatic difference between a wave description and a particle description of an electron in an atom.

In the quantum mechanical description of the atom, electrons occupy orbitals, probability descriptions of the space in which electrons might be found, rather than the fixed planetary orbits described by the Bohr model. Orbitals in the hydrogen atom's wave description have the same energy as orbits in the Bohr atom. Either model of atomic structure can be used to account for the observed emission spectrum of hydrogen. However, the appearance of integers that appeared arbitrary in the Bohr model is shown to be an outcome of the electron's wave nature in the quantum mechanical model.

■ **Aside**

Modern Physics Changes Our World View

Since the time of the Greeks, physics was considered to be an extension of common experience. It is said that Archimedes claimed that, given a long enough lever and a fulcrum, he could move the earth. Newton connected the movements of the planets to the fall of apples. After Newton, people knew that the same laws of physics apply on earth and in the heavens. Using Newtonian mechanics, people calculated the orbits of planets and the trajectories of artillery shells.

In the twentieth century, a new generation of young physicists developed a radically different world view. They used *gedanken,* or thought experiments to change our view of nature and to call into question the separation of mass and energy, and of time and space. Werner Heisenberg (1901–1976), who helped develop a wave description of electrons in atoms, argued that there is a limit to which we can simultaneously measure position and movement. He said that we change a system when we measure it. We can measure the position of an electron by bouncing a photon of light off of it. The light is then collected by a lens and returned to the observer. There is uncertainty to the path of the light through the lens; so there is some uncertainty in the position of the electron. With shorter-wavelength light, that uncertainty is reduced. However, since the light transfers some energy to the object being measured, the movement or momentum of the electron is changed. The shorter the wavelength of the light, the higher its energy; consequently, more energy will be transferred when a photon bounces off of an electron. There is an absolute limit to which we can know both position and path at the same instant. The uncertainties are small, but their effects become important for the small particles found in atoms (see Fig. 4.13).

This change in world view from the certainties of Archimedes and Newton to the uncertainties symbolized by wave-particle duality was wrenching, even for the physicists involved. Heisenberg wrote:

I remember discussions . . . which went through many hours till very late at night and ended almost in despair; and when at the end of the discussion I went alone for a walk in the neighboring park I repeated to myself again and again the question: Can nature possibly be as absurd as it seemed to us in these atomic experiments?

Figure 4.13
Werner Heisenberg used a thought experiment to demonstrate that there is a limit to which we can know both the position and path of an elementary particle at the same instant.

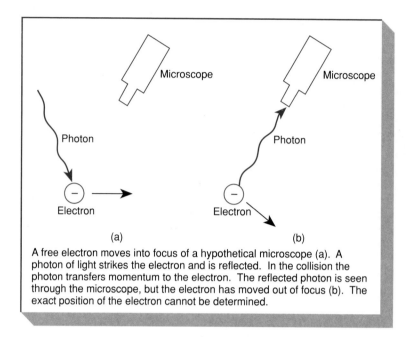

A free electron moves into focus of a hypothetical microscope (a). A photon of light strikes the electron and is reflected. In the collision the photon transfers momentum to the electron. The reflected photon is seen through the microscope, but the electron has moved out of focus (b). The exact position of the electron cannot be determined.

■ Atomic Structure and the Periodic Table

To extend the wave model of the atom to elements other than hydrogen, approximate methods are used. Orbitals are assumed to be like those of hydrogen, so with each increase in atomic number, one proton is added to the nucleus and one additional electron is placed in an orbital. Orbitals of lower energy (closer to the nucleus) are filled before those of higher energy. In this manner, electron configurations are built up within the atom.

Electron shells contain *subshells* that differ in energy. Orbitals in a subshell share the same value of l, the second quantum number. Since the value of l goes from 0 to $(n - 1)$, the number of subshells in a shell is equal to n. Subshells with $l = 0$ are designated s subshells, those with $l = 1$ are p, and those with l values of 2 and 3 are labeled d and f. Within each subshell, there are $(2l + 1)$ orbitals. Each s subshell has one orbital, each p subshell has three orbitals, each d subshell has five orbitals, and each f subshell has seven orbitals. Finally, each orbital can hold a maximum of two electrons. Thus, an s subshell can hold 2 electrons; a p subshell, 6; a d subshell, 10; and an f subshell, 14.

n	l	Designation of Subshell	Orbitals in Subshell	Orbitals in Shell	Maximum number of Electrons in Shell
1	0	$1s$	1	1	2
2	0	$2s$	1		
2	1	$2p$	3	4	8
3	0	$3s$	1		
3	1	$3p$	3		
3	2	$3d$	5	9	18
4	0	$4s$	1		
4	1	$4p$	3		
4	2	$4d$	5		
4	3	$4f$	7	16	32

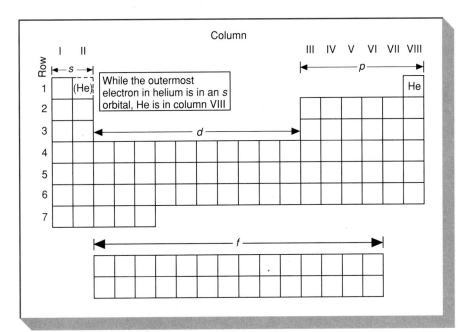

Figure 4.14
In modern versions of the periodic table, elements are grouped according to subshells being filled.

The design of a modern periodic table reflects this buildup of the electronic configuration of atoms. Rows in the periodic table correspond to the filling of subshells and shells. Just as in the Bohr model, electrons with larger values of n are farther from the nucleus. Atoms in the same column of the periodic table have the same number of outer electrons in subshells of the same type (see Fig. 4.14). The lower the atom in a column, the more filled shells beneath the outer electrons. Note the focus on *outer electrons*, those in orbitals furthest from the nucleus and having the highest value of the quantum number n. Electrons having lower values of n are *inner electrons* in orbitals closer to the nucleus.

Not all subshells within a shell are filled before the next shell is begun. Note, for example, that the 4s subshell is filled before the 3d. The filling of subshells can be recognized by the position of an element in the periodic table or by applying the following scheme. When the subshell notations are listed in indented fashion as shown, the columns then give the sequence of filling.

$1s$

$2s$ $2p$

$3s$ $3p$ $3d$

$4s$ $4p$ $4d$ $4f$

$5s$ $5p$ $5d$ . . .

$6s$ $6p$ . . .

. . .

■ Elements in a Chemical Family Have Similar Electronic Structures

The chemistry of an element is closely related to its outer electron configuration. Atoms of the elements in the same family in the periodic table have the same number of outer electrons in the same type of subshell. Atoms of the alkali metals, for instance, all have only one outer electron, which is in an s orbital. The halogens have seven outer electrons, two in an s subshell, and five in a p subshell.

Figure 4.15
Within a column of the periodic table, atoms of different elements have the same kernel charge and the same number of outer electrons. For the alkali metals, lithium and sodium, one outer electron is attracted to a +1 kernel.

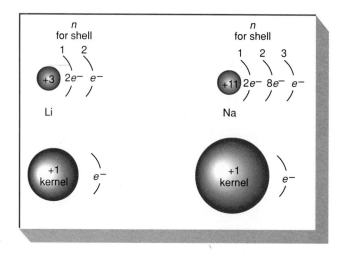

Why do elements with the same outer electron configuration have similar chemical behavior? We can best understand the reason if we develop one more concept, that of the kernel. The *kernel* of an atom consists of the nucleus and the inner electrons. Inner electrons are closer to the nucleus than the outer electrons, and the repulsion of an outer electron by an inner electron cancels the attraction of one proton. The net charge attracting the outer electrons is the kernel charge, the charge on the nucleus minus the charge on the inner electrons. Members of a chemical family have the same kernel charge acting on the same number of outer electrons (see Fig. 4.15).

Atomic Number	Element	Protons in Nucleus	Total Inner Electrons	Kernel Charge	Total Outer Electrons
3	Li	3	2	+1	1
11	Na	11	10	+1	1
19	K	19	18	+1	1
37	Rb	37	36	+1	1
55	Cs	55	54	+1	1
9	F	9	2	+7	7
17	Cl	17	10	+7	7
35	Br	35	28	+7	7
53	I	53	46	+7	7

For the main group elements, the kernel charge and number of outer electrons can readily be calculated from the number heading the column in the periodic table. For the elements found in columns I and II and in columns III through VIII, the number of outer electrons is equal to the column number.

We have used the wave mechanical model of the electron to develop a picture of the atom in accordance with the periodic table. Having developed this picture, we can now simplify our approach to chemistry. Atoms of elements in a family have similar chemistry because they have the same kernel charge and the same number of outer electrons. Those in different families have different kernel charges and different numbers of outer electrons. In chapter 5, we will develop ideas about chemical bonding and properties of compounds by considering only the outer electrons of atoms and the positive kernels attracting those electrons.

1. Which of the following colors of visible light has the longest wavelength? The shortest?
    (a) Blue
    (b) Red
    (c) Yellow

2. In which of the following regions of the radiation spectrum are photons with the lowest energy found? With the highest?
    (a) Infrared
    (b) Visible
    (c) Ultraviolet

3. Which element in the second row of the periodic table is not known to form compounds? How many outer shell electrons do atoms of this element have?

4. If scientists had not measured the spectral lines of all known elements, would the appearance of a dark line in the spectrum of the sun have been evidence for the existence of helium? (How did the routine measurement of the position of lines in spectra contribute to the discovery of new elements?)

5. Why were the noble gases discovered later than the elements in other families of the periodic table?

6. Why does the emission spectrum of hydrogen have fewer lines than those of other elements?

7. In the Balmer series for the hydrogen atom, the spacing of lines becomes closer toward the violet end of the spectrum. Why do they converge toward the high-energy side of the range?

8. The lines in the Balmer series of the hydrogen spectrum obey the following equation:

    $$1/\lambda = R_H(1/2^2 - 1/n^2); \; n = 3, 4, 5 \ldots$$

    Calculate the value of the fraction that is equal to $(1/2^2) - (1/n^2)$ for $n = 3,4,5,$ and 6.

9. Justify the statement that for all lines in the Balmer series of the hydrogen spectrum $5/36 \, R_H < 1/\lambda < 1/4 \, R_H$.

10. What is the value of n for the orbit in the Bohr atom with $r = 25 \, r_o$?

11. What is the maximum number of electrons that can be accommodated in the shell with $n = 5$?

12. Sketch a wave for a vibrating string with ends fixed and a length L equal to $5\lambda/2$. How many nodes does this wave have?

13. Sketch a d orbital and identify its nodal planes.

14. Is an electron a wave or a particle in the Bohr model of the atom?

15. Does an electron behave as a wave or as a particle in an atom?

16. Henry Moseley (1887–1915) found that the absorption of X rays, high-energy electromagnetic radiation, provided an independent way to measure the number of protons in the nucleus. The frequency of X rays emitted from target metals in X-ray tubes increases with increasing atomic number. This radiation arises when an electron in an outer orbital of an excited atom falls to a vacant position in an innermost orbital. Why should the energy emitted when an electron falls to an inner orbital increase with increasing atomic number? Why is the energy emitted not a periodic function of the atomic number?

17. How many orbitals are there in each of the following subshells?
    (a) 3s
    (b) 3p
    (c) 3d
    (d) 4f
    (e) 5g

18. The nonmetals, carbon and silicon, and the metals, tin and lead, are in Column IV of the periodic table. What is the kernel charge and the number of outer electrons for each of these elements? How does the size of atoms change in this family going down the periodic table? Why is lead a metal and carbon is not?

19. Complete the following table.

Atomic Number	Element	Protons	Inner Electrons	Kernel Charge	Outer Electrons
4		4	2		
	Mg	12		+2	
	Ca				2
	O	8		+6	
16				10	
28					6

20. Write a letter to Mendeleev telling him the modern basis for grouping elements into families in the periodic table.

21. For each of the following elements, list the subshell in which their outermost electrons are found.

Symbol	Element	Subshell Being Filled
O	Oxygen	
Ca	Calcium	
Fe	Iron	
Li	Lithium	
Ag	Silver	
Hg	Mercury	
U	Uranium	

5

Bonding in Elements and in Compounds

How do atoms combine in elements and in compounds? The formula of an element or compound alone tells us little about how the atoms are bonded or what properties a substance might have. Chemists seek to connect the observed physical and chemical properties of elements and compounds to their underlying atomic architecture. They have identified three major ways that atoms join together—through metallic, ionic, or covalent bonding.

Consider the following reaction in which the reactants and product can serve to illustrate the major types of chemical bonding. When a small piece of sodium metal is placed in a container of chlorine gas, a violent reaction occurs that forms salt (sodium chloride). Sodium is a metal; metallic bonding enables it to conduct heat and electricity well. Chlorine demonstrates covalent bonding since two atoms of chlorine share electrons to form a molecule of chlorine with only weak attraction for other chlorine molecules. Sodium chloride features ionic bonding. Its lattice of positive and negative ions causes it to be a rigid, high-melting solid that does not conduct electricity. When salt is dissolved in water or is molten, however, it can conduct an electric current by the movement of its ions (see Fig. 5.1).

■ Bonding in Metals

Metals conduct electricity and heat well. Their shiny appearance and rapid conduction of electricity and heat are associated with mobile electrons. Ductile metals can be drawn to form wires while malleable metals can be hammered into different shapes. Metals are ductile and malleable because bonding is retained even as atoms are pulled or pushed past one another, changing their atomic neighbors.

In a simple model for *metallic bonding*, we can consider a metal to be a lattice of positive ions surrounded by a mobile electron gas as shown in figure 5.2. (A *lattice* is a regular, repeating array of particles in space.) In this model, the charged ions are atomic kernels, and the electron gas consists of the loosely held outer-shell electrons. Atoms of metallic elements have, at most, only a few outer electrons, and they all have empty orbitals in their outer shell. In a solid or liquid, where atoms are close to one another, atomic orbitals on one atom overlap those on neighboring atoms. Electrons can flow from one atom to another along paths of overlapping atomic orbitals. This easy movement of electrons accounts for the conductivity of the metal.

Figure 5.1
Sodium metal (a) and chlorine gas (b) react (c) to form crystalline sodium chloride (d).

(a)

(b)

(c)

(d)

Figure 5.2
In one description of bonding in metals, a lattice of positive kernels is surrounded by a cloud of mobile outer electrons. The easy movement of electrons accounts for the conductivity of metals.

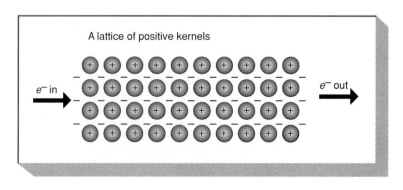

A lattice of positive kernels

e^- in

e^- out

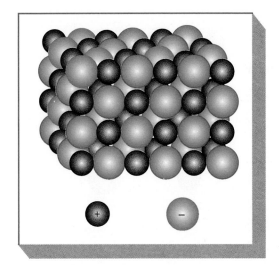

Figure 5.3
An ionic lattice is a regular array of positive and negative ions. Strong forces between nearby oppositely charged ions account for the high-melting point and hardness of ionic solids.

■ Bonding in Ionic Compounds

Sodium chloride is a crystalline solid that melts at a high temperature. Its crystals are hard, yet they shatter when struck. Sodium chloride conducts electricity when molten or when dissolved in water, but the solid itself does not conduct.

A lattice of positive and negative ions accounts for the hardness and high melting point of sodium chloride as well as for its brittleness (see Fig. 5.3). Forces of attraction between positive and negative ions are great, so considerable energy needs to be supplied in order to melt the solid. The brittle nature of the solid results from the strong repulsive forces that develop when an external force causes adjacent layers of ions to slide, bringing positive ions near positive ions and negative ions near negative ions (Fig. 5.4).

In the solid's lattice, the ions are not free to move. In solution or in the melt, positive and negative ions can slide past one another; this movement of positive and negative ions in opposite directions is why a current of electricity can be conducted.

■ Ionic Bonding and the Periodic Table

Compounds formed between metals and nonmetals usually exhibit ionic bonding. Atoms of metals lose electrons to form cations, and atoms of nonmetals gain electrons to form anions. To understand why the formation of ionic compounds may be favorable, we need to consider the processes of cation and anion formation as well as the forces between the ions in a lattice.

The charges found on many cations in compounds are related to the group numbers of the elements in the periodic table (see Fig. 5.5). Sodium and other alkali metals in column I form +1 ions in compounds. Calcium and other alkaline earth elements in column II form +2 ions. Aluminum and other metals in column III can form +3 ions in order to bond ionically (some compounds of aluminum are ionic and others are covalent).

Why do cations found in compounds have a charge related to their position in the periodic table? Because the energy required to form a cation from an atom depends on both the charge of the ion and the position of the element in the periodic table. For example, sodium in column I is found as Na^+ in compounds but never as Na^{2+}. The energy required to remove one electron from a sodium atom is relatively small, for the outer electron is attracted only to a +1 kernel. It costs much more energy to remove an electron from Na^+ to form Na^{2+}. Not only are the

Figure 5.4
Although they are hard and high-melting, ionic solids also tend to be brittle. When like charges are brought near one another, there are strong forces of repulsion between them.

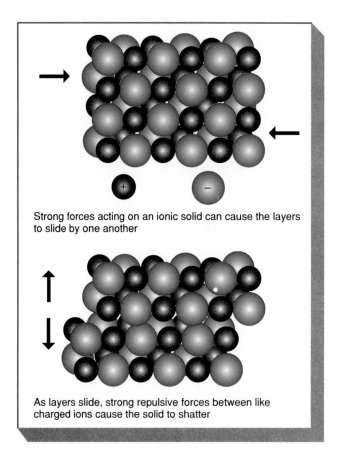

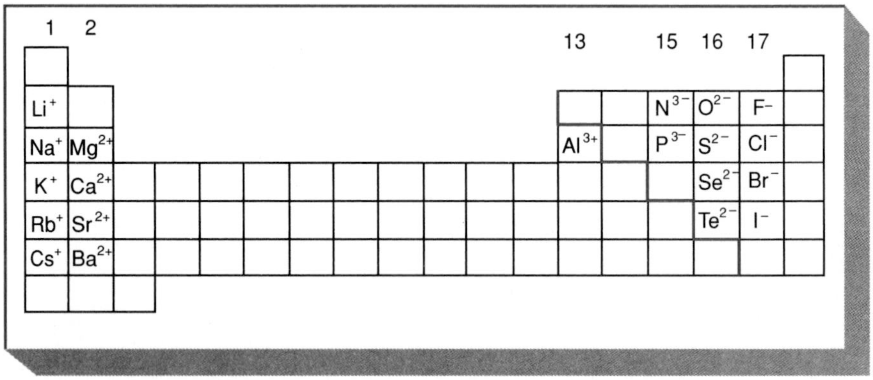

Strong forces acting on an ionic solid can cause the layers to slide by one another

As layers slide, strong repulsive forces between like charged ions cause the solid to shatter

Figure 5.5
Ionic bonding is favored in compounds formed between the metals in columns 1 and 2 of the periodic table and the nonmetals in columns 16 and 17. The charge on these simple ions is directly related to their column number in the periodic table.

1	2												13		15	16	17	
Li^+															N^{3-}	O^{2-}	$F-$	
Na^+	Mg^{2+}												Al^{3+}		P^{3-}	S^{2-}	Cl^-	
K^+	Ca^{2+}															Se^{2-}	Br^-	
Rb^+	Sr^{2+}															Te^{2-}	I^-	
Cs^+	Ba^{2+}																	

electrons in Na^+ closer to the nucleus than in the larger sodium atom, but they are in an inner shell that experiences very strong attraction to the nucleus. In contrast, it is energetically much less expensive to form Mg^{2+} than to form Na^{2+}. Since magnesium is in column II of the periodic table, both of the electrons removed from an atom of Mg in order to form Mg^{2+} come from the outer shell (see Fig. 5.6).

As we go down a column in the periodic table, it becomes easier to form a cation. Although the kernel charge remains the same, the atoms are larger and the outer electrons are farther from the nucleus and more easily removed. For example, it is easier to form K^+ from an atom of potassium than Na^+ from an atom of sodium.

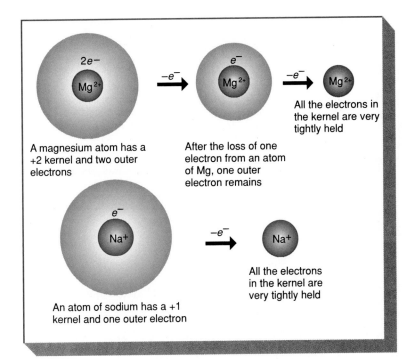

Figure 5.6
Less energy is required to form a +1 sodium ion than a +1 magnesium ion, but it is far easier to form a +2 magnesium ion than it is to form a +2 sodium ion.

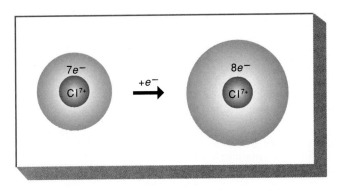

Figure 5.7
The strong attraction between the +7 kernel of a chlorine atom and its outer electrons permits the chlorine atom to gain an electron to form a -1 chloride ion.

There is a corresponding relationship for the charges on anions in compounds and the positions of nonmetals in the periodic table. Chlorine and the other halogens form −1 ions; members of the oxygen family form −2 ions. These anions have an octet of outer electrons consisting of filled s and p subshells.

The addition of one electron to a halogen atom, such as chlorine, is favored. The added electron is attracted more to the +7 kernel than it is repelled by the other seven outer electrons. The chloride anion is larger than the chlorine atom since the +7 charge on the kernel cannot hold eight electrons in the ion as closely as it holds seven (see Fig. 5.7). Oxygen and other elements in the same family can gain two electrons to form anions with eight outer electrons and −2 charges. The number of electrons an atom can gain equals eight minus the column number of the element. Anions with more than eight outer electrons on a single atom are not found in nature.

The formation of compounds with ionic bonding can involve cost-benefit accounting at an atomic level. For example, the reaction of sodium with chlorine to form sodium chloride can be imagined as a series of steps. Many of the steps either are costly or pay small dividends. Sodium metal can be vaporized to form individual sodium atoms (cost). Each atom of sodium can be stripped of an electron to form a

Figure 5.8
The formation of NaCl from sodium metal and chlorine gas can be imagined as a series of steps. The formation of a lattice with strong attractive forces between positive and negative ions provides the driving force for the reaction.

Step		Energy cost or benefit
Vaporization	Na(s) → Na(g)	Small cost
Cation formation	Na(g) → Na$^+(g)$ + e$^-(g)$	Small cost
Breaking Cl-Cl bond	1/2 Cl$_2(g)$ → Cl(g)	Small cost
Anion formation	Cl(g) + e$^-(g)$ → Cl$^-(g)$	Small benefit
Lattice formation	Na$^+(g)$ + Cl$^-(g)$ → NaCl(s)	Very large benefit

Figure 5.9
The forces between ions in an ionic crystal depend on distance, magnitude of charge, and the ratio of positive to negative ions.

For ions with the same charges, the attraction between opposite charges is greater for smaller ions

At a fixed distance, the force attraction between a +2 ion and a –2 ion is 4 times the attraction between a –1 ion and a +1 ion

Lattices that contain ions that differ in the magnitude of the opposite charges are more complex than are those in which the cation and the anions have the same magnitude of charge. For example, in a compound containing two +1 ions for every –2 ion, some positive ions are close together in the lattice. The repulsions between nearby ions with the same charge offsets some of the attractive forces between oppositely charged ions

sodium ion (cost). Bonds holding the diatomic chlorine molecules together can be cleaved (cost). Adding electrons to individual chlorine atoms can form chloride ions (dividend). When the energies in these steps are added, it is found that it is not favorable to form separated Na$^+$ and Cl$^-$ ions from the elements.

The big payoff in this imagined pathway for NaCl formation is the formation of the ionic lattice. Since opposite electric charges strongly attract one another, bringing positive Na$^+$ and negative Cl$^-$ ions together releases large amounts of energy. Because the benefits of lattice formation exceed the combined costs of the other steps, the formation of NaCl is favorable (see Fig. 5.8).

The charges on ions and the formulas of ionic compounds reflect the need for a favorable imbalance between the forces joining positive and negative ions, and the cost of forming those ions (see Fig. 5.9). The higher the charges on the ions, the greater the forces between them. For example, the force of attraction between +2 and –2 ions is four times as great as that between +1 and –1 ions at the same distance. Although it is more difficult to form +2 cations and −2 anions than it is to form singly charged ions, the increased forces of attraction in the lattice can pay the added cost if it is not too great. For example, the lattice of CaO contains Ca^{2+} and O^{2-} ions, rather than Ca^+ and O^-. However, the cost of forming cations increases rapidly as more electrons are removed; indeed, ions with charges higher than +3 are not stable.

■ Formulas and the Names of Compounds

The formulas of ionic compounds give the ratio of cations to anions. In NaCl, there is one Na^+ for every Cl^-. If the charges on the cation and anion differ, the ions do not combine in a 1:1 ratio. This is illustrated by the formulas Na_2O and $CaCl_2$. It requires two +1 sodium ions to balance the charge on a −2 oxygen anion. Similarly, the charge on a +2 calcium cation is balanced by two −1 chloride anions.

The charges on ions in ionic compounds may not be apparent from the formulas; consider the compound SrSe which contains strontium cations and selenium anions. To find the charge on these ions, one must look at the periodic table. Because Sr is in column II, it has a +2 charge while selenium in column VI has a −2 charge.

The naming of binary compounds (compounds formed between a pair of elements) is illustrated by the names of some common compounds. In the case of ionic elements, the name of the metallic element is given first, and the modified name of the nonmetal follows. The suffix *-ide* usually indicates a binary compound. (Two notable exceptions to this rule are sodium hydroxide, NaOH, and sodium cyanide, NaCN.) Some examples of binary compounds are NaBr (sodium bromide), K_2S (potassium sulfide), CaO (calcium oxide), and HI (hydrogen iodide).

■ Covalent Bonding

Nonmetallic elements vary greatly in their properties, and they therefore combine to form compounds spanning a very wide range of properties, including the gas carbon dioxide, the liquid water, and the mineral quartz. Atoms in diatomic elements such as hydrogen, as well as those in compounds such as water and carbon dioxide, are held together by *covalent bonds*. Covalent bonding, the sharing of electrons, can lead to the formation of discrete molecules with fixed formulas, or it can bind together a very large number of atoms as in a diamond or a grain of sand.

Atoms of nonmetallic elements can share electrons to form covalent bonds. By sharing electrons, these atoms may combine to form lower-energy, more stable molecules. Consider the approach of two isolated hydrogen atoms to one another. As the atoms come closer, the electron on each atom becomes attracted to both hydrogen nuclei. Because the electronic shell of each hydrogen atom is not filled, each electron is free to move about both nuclei. This sharing of electrons results in bonding because the hydrogen nuclei are attracted to the shared electrons, so they are bonded to one another (see Fig. 5.10).

It is more precise to say that a bond is formed when there is a net increase in attraction between atoms, for there are opposing forces when atoms combine. There are attractive forces between the opposite charges of positive nuclei and negative electrons, and there are repulsive forces between like charges of positive nuclei and negative electrons.

Covalent bonding can only occur when low energy atomic orbitals are available to hold shared electrons. In a molecule of H_2, each of the hydrogen atoms can be considered to have a filled shell containing two electrons. No molecules having the

Figure 5.10

Atoms form covalent bonds by sharing electrons.

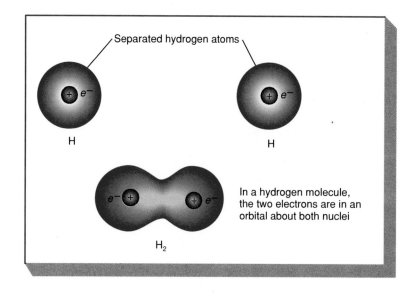

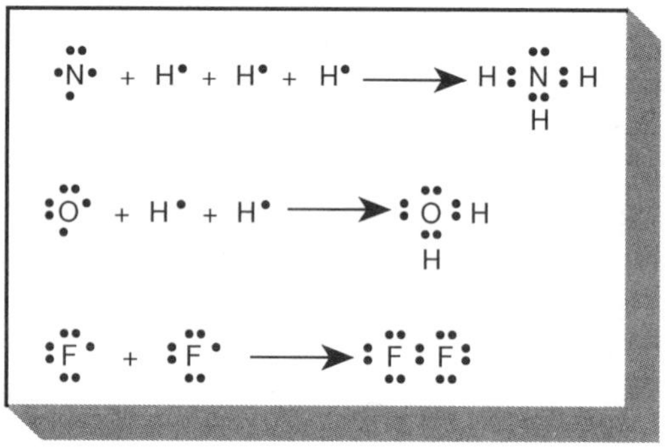

Separated hydrogen atoms

H

H

In a hydrogen molecule, the two electrons are in an orbital about both nuclei

H₂

Figure 5.11

Bonding in covalent compounds is represented by electron dot structures. A shared electron pair represents a covalent bond. Pairs on a single atom are called lone pairs.

formula H_3 are observed because the electron from the third hydrogen atom would have to go into an outer orbital of the H_2 molecule where it has no attraction for the other atoms.

Note that the noble gases in column VIII do not readily form compounds. Atoms of these elements have an octet of electrons about them (helium, the exception, has a filled shell of two electrons).

■ Lewis Structures

Electron dot structures are used to represent bonding in covalent compounds, although lines are sometimes used to represent shared pairs of electrons. These structures are often referred to as *Lewis structures* after G. N. Lewis (1875–1946), a chemist at the University of California who developed this model for covalent bonding. In a Lewis structure, all outer electrons are represented. Each hydrogen atom can share two electrons, and each second-row element can share eight electrons to form filled shells.

The bonding in ammonia, water, and fluorine is shown in figure 5.11. In ammonia, there are covalent bonds between N and each H, and there is one pair of electrons on N that does not form a bond. It is called a *lone pair*. Water has two O—H bonds and two lone pairs of electrons on oxygen. In F_2, there is a covalent bond joining fluorine atoms, and there are three lone pairs on each fluorine atom.

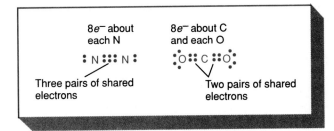

Figure 5.12
A molecule of nitrogen has a triple bond linking two nitrogen atoms. In carbon dioxide, double bonds link each oxygen atom to the carbon atom.

Note that atoms of the second-row elements N, O, and F have eight electrons about them in these molecules when shared pairs and lone pairs are counted. It is important to represent all outer electrons, both shared and unshared.

In some compounds, a pair of atoms shares more than one pair of electrons to form a *multiple bond.* A structure in which two pairs of electrons are shared between two atoms has a *double bond.* The Lewis structure of CO_2 has double bonds between the carbon atom and each of the two oxygen atoms. The bonding in nitrogen, N_2, cannot be accounted for by the sharing of a single pair of electrons. To attain an octet about each nitrogen atom, three pairs of electrons must be shared. Since three pairs of electrons hold the two nitrogen atoms together, chemists say that the nitrogen molecule has a *triple bond* (see Fig. 5.12).

The formulas of many covalent compounds can be determined by constructing Lewis structures. For example, what is the formula of the compound formed between the elements phosphorus and chlorine? Phosphorus is in column V of the periodic table, so an atom of phosphorus has five outer electrons. An atom of phosphorus can complete an octet by sharing three electrons from other atoms. Atoms of chlorine have seven outer electrons so each chlorine atom can complete an octet by sharing one additional electron with another atom to form a covalent bond. Because three chlorine atoms can bond to each phosphorus atom, the compound formed between phosphorus and chlorine has the formula PCl_3.

Incorrect Lewis structures can often be recognized, for they fail to provide maximum bonding (see Fig. 5.13). When there are fewer than eight electrons on adjacent atoms, then a more stable arrangement of electrons can be reached by forming an additional bond. Note that in Lewis structures of compounds with multiple bonds, there is an octet of electrons about each carbon, nitrogen, or oxygen atom.

■ Polyatomic Ions

Covalent bonding is not only found in neutral molecules; it also occurs in polyatomic ions. For example, the sulfate ion has the formula SO_4^{2-} and a structure in which a central sulfur atom is surrounded by four oxygen atoms as shown in figure 5.14. In the nitrate ion, NO_3^-, three oxygen atoms surround a central nitrogen atom. The negative charges on sulfate and nitrate ions are not localized on single atoms; they are shared by atoms in the anions. Although the bonding within a polyatomic ion is covalent, there are ionic bonds between a polyatomic ion and an ion of opposite charge in a compound.

The names and formulas of some compounds containing polyatomic anions are given below. Note that parentheses are used when there is more than one polyatomic ion in a formula. In these cases, the formula of the ion is written within the parentheses, and the subscript outside the parentheses indicates the number of anion units present. For example, $Ca(NO_3)_2$ has two nitrate ions per calcium ion or two nitrogen atoms and six oxygen atoms per calcium atom.

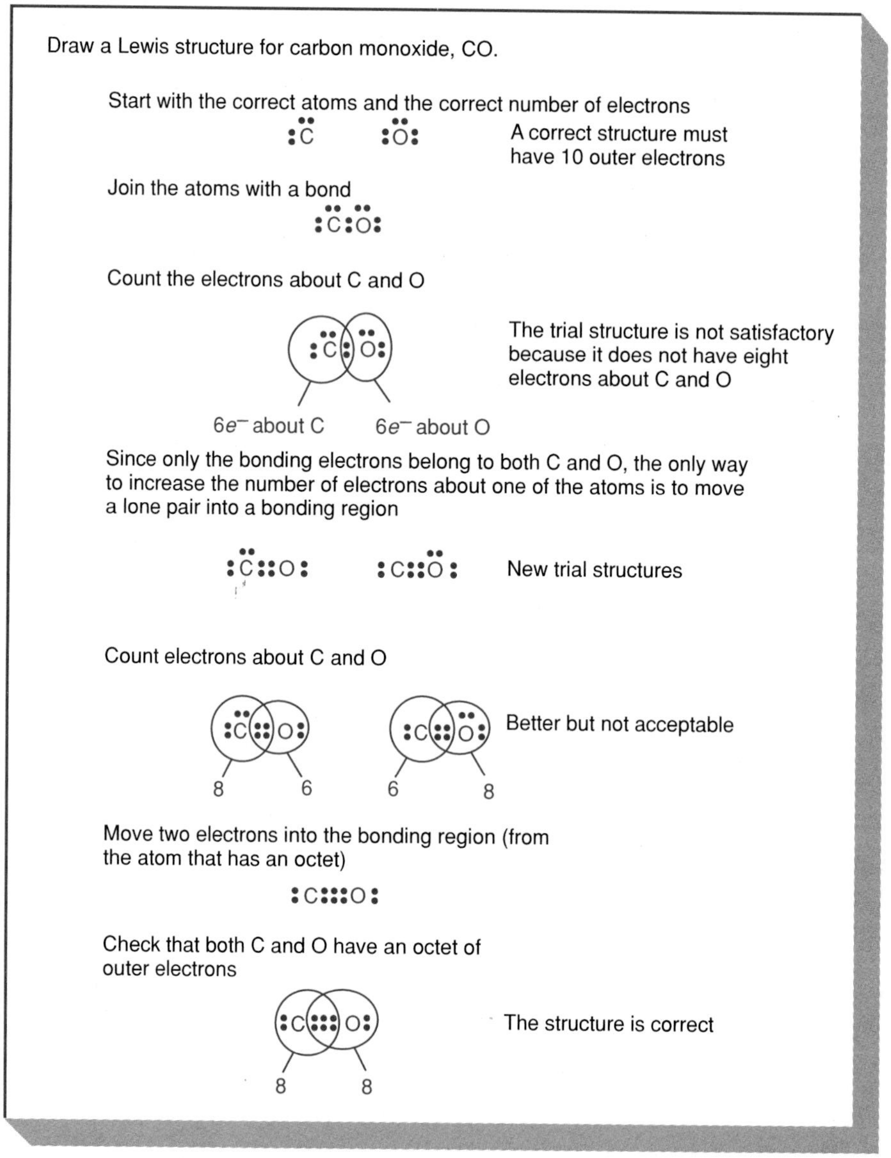

Figure 5.13
A student should be able to write correct Lewis structures for simple molecules. One must begin with the correct number of electrons, but can proceed by trial-and-error, checking proposed structures and seeking to correct their deficiencies until a satisfactory structure is achieved.

Draw a Lewis structure for carbon monoxide, CO.

Start with the correct atoms and the correct number of electrons

A correct structure must have 10 outer electrons

Join the atoms with a bond

Count the electrons about C and O

 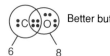

The trial structure is not satisfactory because it does not have eight electrons about C and O

6e⁻ about C   6e⁻ about O

Since only the bonding electrons belong to both C and O, the only way to increase the number of electrons about one of the atoms is to move a lone pair into a bonding region

New trial structures

Count electrons about C and O

Better but not acceptable

Move two electrons into the bonding region (from the atom that has an octet)

Check that both C and O have an octet of outer electrons

The structure is correct

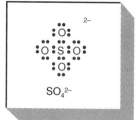

Figure 5.14
The sulfate ion has a central sulfur atom covalently bonded to four oxygen atoms. In addition to 6 outer electrons from S and 6 outer electrons from each O for a total of 30 electrons from neutral atoms, there are two additional electrons to give the ion a total of 32 outer electrons and a net charge of -2.

Name	Formula	Polyatomic Ion
Ammonium chloride	NH_4Cl	NH_4^+
Calcium nitrate	$Ca(NO_3)_2$	NO_3^-
Potassium sulfate	K_2SO_4	SO_4^{2-}
Sodium carbonate	Na_2CO_3	CO_3^{2-}

■ Shapes of Covalent Molecules

A simple theory that gives good predictions of the shapes of most molecules is called the *valence shell electron-pair repulsion (VSEPR) model*. The shapes of molecules are predicted on the basis of the number of outer-shell electron pairs about each atom. According to this theory, electron pairs repel one another so they stay as far apart as possible.

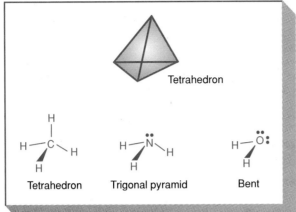

Figure 5.15
In the methane, ammonia, or water molecule, four electron pairs are located at the corners of a tetrahedron about a central C, N, or O. The shape of a molecule describes the location of nuclei.

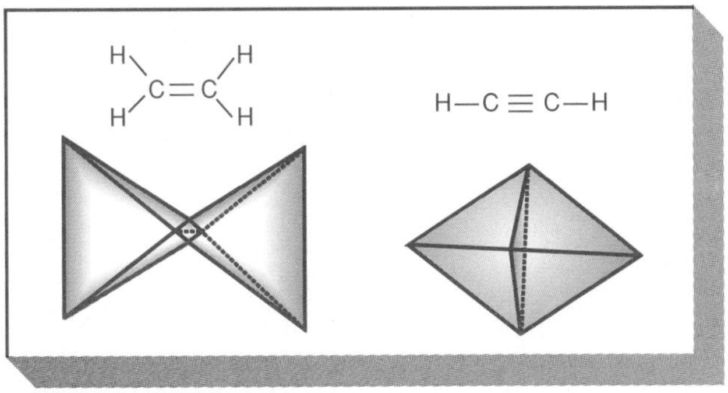

Figure 5.16
The atoms attached to the carbon atoms in a double bond lie in a plane. A double bond can be considered to be two tetrahedra sharing two corners or an edge. In the case of a triple bond, the atoms lie in a line. A triple bond can be considered as two tetrahedra sharing three corners or a face.

Methane, ammonia, and water are alike in that each molecule has four pairs of electrons about the central C, N, or O atom. The four electron pairs in molecules of these compounds are located at the corners of a tetrahedron. (A regular tetrahedron has four identical faces; each face is an equilateral triangle. The four corners of a tetrahedron are equally far apart and represent the most favorable separation of four objects.) Because measurements of the shape of a molecule give positions of the nuclei rather than of the electrons, the shape of a molecule is named to indicate the locations of the nuclei. A methane molecule is tetrahedral, an ammonia molecule is shaped like a triangular pyramid; and a water molecule is bent (see Fig. 5.15).

This simple theory also describes the molecular shapes of compounds with multiple bonds as shown in figure 5.16. In a double bond, two pairs of electrons or an edge of a tetrahedron are shared. In C_2H_4, all six nuclei lie in a single plane. In a triple bond, three pairs of electrons or a face of a tetrahedron are shared. The nuclei in C_2H_2 lie in a straight line. Single, double, and triple bonds differ in length. When additional bonding pairs of electrons are shared between carbon atoms, the atoms are closer.

Some compounds are formed that do not have octets of electrons about central atoms. The VSEPR theory also correctly predicts the shapes of these compounds. Elements in the third and higher-numbered rows of the periodic table can accommodate more than eight electrons in their outer shells. Phosphorus pentachloride, PCl_5, with ten electrons about phosphorus is a triangular bipyramid. Sulfur hexafluoride, SF_6, with twelve electrons about sulfur is octahedral (see Fig. 5.17).

Figure 5.17
Elements in the third row of the periodic table can hold more than eight electrons in their outer shells. Valence shell electron-pair repulsion theory helps to predict the shape of these compounds.

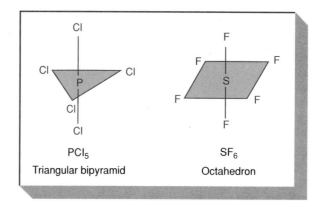

PCl_5
Triangular bipyramid

SF_6
Octahedron

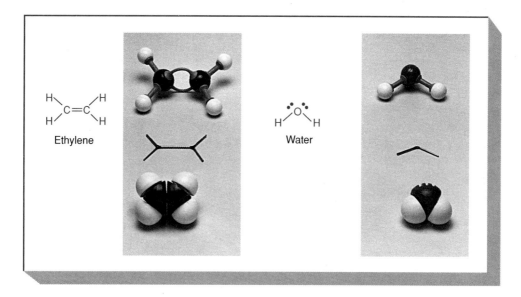

Ethylene

Water

Figure 5.18
Three different physical models for the compounds ethylene and water are shown. The top model emphasizes connectedness by showing bonds. The middle model, a Dreiding model, emphasize bond lengths and angles. The bottom model emphasizes the space occupied by electrons. (Randy Matusow.)

■ Physical Models of Covalent Compounds

A variety of molecular models are used by chemists to represent the three dimensional structures of compounds as shown in figure 5.18. Some resemble tinker toys, some are made of steel rods, and still others are made of a variety of interlocking molded plastic pieces.

It is instructive to compare different models of molecules. Models of two compounds are used to illustrate some of their variety. Ethylene (C_2H_4) has a carbon–carbon double bond. Water is a compound that has both shared and lone pairs of electrons about oxygen. In a simple model, atoms of carbon are represented by balls with four holes directed at the corners of a tetrahedron. Oxygen is represented by a ball with two holes, and hydrogen by a ball with one hole. Pegs connecting the balls represent chemical bonds. In this model, the shape of each molecule is indicated, however, no provision is made to represent the lone pairs of electrons in a water molecule.

Dreiding models emphasize internuclear distances and angles. Rigid steel pieces are scaled to molecular dimensions. The planarity of ethylene is apparent but the fact that the planarity results from the double bond is not. Dreiding models of large molecules can be assembled and used to predict important spatial relationships between atoms.

A different scale model is made of interlocking molded plastic pieces. In this model, interatomic distances and angles are represented accurately. Unlike the Dreiding models, these are space-filling models. The size of a unit is proportional to the size of the electron clouds. Electron pairs are not represented distinctly, but their contribution to the shape and size of a molecule is.

■ Aside

Reflections on Models

Which model is correct? Each model closely represents some features of a molecule and is incorrect on others. Molecular models are designed to select and emphasize the features of real compounds. By emphasizing some features and excluding others, a model may be simple or complex. Scientific theories are models of reality. Consider Lewis structures—are electrons really particles that can be represented by dots? Lewis structures ignore the wave character of electrons in atoms; however, they are very useful for an understanding of covalent bonding. Is Dalton's atomic theory an abstraction? It is formulated to account for combining weights and ratios? Properties of a compound such as bonding, taste, or smell are totally excluded. The proper question might be, "Which model is more useful for which given purpose?"

Scientists often choose to emphasize the utility of a model rather than its accuracy. A model may or may not be useful for a given purpose. For example, either Dreiding or spacefilling models would be preferred for estimating the dimensions of a complicated molecule. However, for illustrating Lewis structures, the simple ball and stick models prove superior.

It is apparent that a plastic peg is not an electron pair, or that a bent metal tube is not water. Molecular models are clearly artificial. A model is an imperfect representation of a part of reality. It simplifies. It can be used to gain insight, and it can be manipulated to predict new relationships. A good model may have high predictive power, or it may serve to relate many seemingly different phenomena in an intellectually satisfying way. The ultimate test of any model is always the degree to which it agrees with experimental observations of the properties of real substances. The greater the agreement, the more useful the model.

■ Questions—*Chapter 5*

1. What is the charge of the underlined ion in each of the following compounds?
    (a) $K_3\underline{PO}_4$
    (b) $Na\underline{HCO}_3$
    (c) $\underline{Fe}SO_4$
    (d) $\underline{Fe}_2(SO_4)_3$

2. Use the periodic table to predict which of the following compounds are ionically bonded.
   (a) NaI
   (b) CCI_4
   (c) BaO
   (d) P_4O_6
   (e) HCl
   (f) SiH_4
3. Using the periodic table, write formulas for the following ionic compounds.
   (a) ithium iodide
   (b) potassium oxide
   (c) calcium sulfide
   (d) barium chloride
   (e) aluminum fluoride.
4. Name the following compounds.
   (a) AgCl
   (b) $CaBr_2$
   (c) Al_2O_3
   (d) Na_2S
   (e) Mg_3N_2
5. Sketch a portion of the ionic lattice for the compound MgO.
6. Compare the ionic compounds NaF and MgO. In NaF, the cation is (smaller, larger), the anion is (smaller, larger), and the lattice forces are (smaller, larger).
7. Rank the following cations of Group IIA metals in order of increasing size.
   Ba^{2+}, Ca^{2+}, Mg^{2+}, Sr^{2+}
8. The following species have the same number of electrons. List them in order of increasing size.
   Ar, Ca^{2+}, Cl^{-}, K^{+}, S^{2-}
9. Which of these ionic compounds of alkali metals with halogens should have the lowest melting point? Why?
   LiF, NaCl, KBr, CsI
10. Name the following compounds.
   (a) NH_4Cl
   (b) $BaSO_4$
   (c) $CaCO_3$
   (d) $AgNO_3$
11. Write formulas for the following compounds.
   (a) potassium nitrate
   (b) lithium carbonate
   (c) magnesium sulfate
   (d) ammonium bromide.
12. Which of the following compounds are covalently bonded?
   (a) KI
   (b) PI_3
   (c) BaS
   (d) AIF_3
   (e) $GeCl_4$

13. Using the periodic table, write formulas for the following covalent compounds.
    (a) silicon fluoride
    (b) carbon sulfide
    (c) nitrogen chloride
    (d) phosphorus oxide.

14. Write Lewis structures for the following molecules or ions.
    (a) CH_4 and H_2O
    (b) CH_3CH_3 and CH_3OH
    (c) C_2H_4 and CH_2O
    (d) N_2 and CO
    (e) CO_2
    (f) H_3O^+ and OH^-

15. Draw Lewis structures of the following compounds. How many lone pairs of electrons are present in each?
    (a) SiH_4
    (b) PH_3
    (c) H_2S
    (d) HCl

16. Why are there neither ionic nor covalent compounds of the element neon (Ne)?

17. What is the shape of each of the following molecules?
    (a) PCl_3
    (b) H_2S
    (c) CF_4
    (d) HF

18. What is the shape of each of the following ions?
    (a) OH^-
    (b) H_3O^+
    (c) NH_4^+
    (d) NH_2^-

6

Chemical Structure and Physical Properties

Chemists seek to relate the physical and chemical properties of substances to their molecular structure, the approach used in the introduction to bonding presented in chapter 5. The ability of metals to conduct electricity was related to the mobility of electrons in metallic bonding, and the rigidity and strength of ionic solids was presented as a consequence of the forces between ions. In this chapter, we continue and extend that approach as we explore reasons for chemical differences between compounds with similar bonding. Why do metals differ in their physical properties and reactivities? Why do ionic compounds differ in their solubility in water? We also explore the relationship between structure and properties for covalent compounds. The gases in the air and the grains of quartz on a sandy beach both have covalent bonding, yet their properties are totally different. How do chemists account for the widely different properties of covalent compounds?

Much of our everyday chemical experience involves substances in aqueous solution. How can we explain the fact that ionic table salt dissolves in water but not in oil? And why is it that oil does not mix with water even though the bonding in both is covalent? In this chapter, we consider the very special properties of water as a pure substance and as a solvent for other chemicals.

■ Metallic Elements Differ Widely in Properties

Metals differ greatly in density. Some of these changes reflect trends in the periodic table. For example, the densities of metals increase dramatically going across a row of the periodic table from columns I to II to III. These densities reflect atomic sizes, which decrease from left to right across the periodic table. Size differences are greatest in moving from the family containing sodium to that containing aluminum. Because aluminum atoms are much smaller than sodium atoms, aluminum has nearly three times the density of sodium. In contrast, the atomic sizes of the *transition metals*, titanium through zinc, having partially or totally filled *d* subshells, show more gradual changes (see Fig. 6.1).

Element	Sodium	Magnesium	Aluminum	Potassium	Calcium
Density (g/cm^3)	0.97	1.74	2.7	0.86	1.55

Element	Titanium	Chromium	Iron	Nickel	Copper	Zinc
Density (g/cm^3)	4.5	7.2	7.9	8.9	9.0	7.1

Figure 6.1
The metals whose symbols are given are used to illustrate changes in properties occurring from family to family across the periodic table and those occurring within families going down columns.

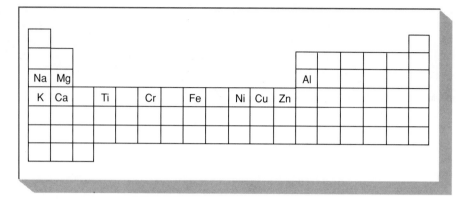

The temperatures at which metals melt also vary greatly. Indeed, mercury at the end of a row of transition elements is a liquid. The column I elements are soft and have low-melting points, while other metals have high strength even at elevated temperatures. The stronger the bonding in a metal, the higher the temperature required to melt it. The strength of bonding in a metal depends in part on the sizes of its atoms and the way they pack in crystals, as well as its kernel charge and the number of its outer electrons. In melting points, as in densities, many of the transition metals exhibit similarities.

Element m.p. (°C)	Sodium	Magnesium	Aluminum	Potassium	Calcium	
	98	651	660	64	846	

Element m.p. (°C)	Titanium	Chromium	Iron	Nickel	Copper	Zinc
	1675	1890	1535	1453	1083	419

For some people, the coinage metals, copper, silver, and gold, may seem the most "metallic" of metals. They are excellent conductors, very malleable, and quite unreactive. Atoms of copper, silver, and gold have one outer electron and a filled d subshell. They are considerably smaller and more tightly packed in the metal than their cousins, the alkali metals.

In contrast, a chemist usually considers the alkali metals, lithium, sodium, and potassium, to be the most "metallic" of elements. They, too, are excellent conductors, but they are the most reactive of metals. In the alkali metals, the outer electrons are, on the average, much further from the kernel and less tightly held. A chemist looks at these metals as potential electron donors in chemical reactions.

■ Forces Between Covalent Molecules

The boiling points of compounds vary greatly, because the strengths of intermolecular forces differ greatly. The boiling point of a compound provides a measure of the forces of attraction between molecules. When a liquid boils, molecules go from the liquid state where they are close to one another into the vapor state where they are far apart. Energy must be supplied to overcome the forces of attraction between molecules in the liquid. The greater the forces of attraction in the liquid, the more energy must be supplied, and the higher the boiling point.

Weak forces between covalent molecules depend, in part, on the total number of electrons in each molecule, which is given by the sum of the atomic numbers. For example, the total number of electrons is 18 in F_2, 34 in Cl_2, 70 in Br_2, and 106 in I_2. The boiling points in degrees Celsius are −188 for F_2, −35 for Cl_2, 59 for Br_2, and 184 for I_2. At room temperature F_2 and Cl_2 are gases, Br_2 is a liquid, and I_2 is a solid.

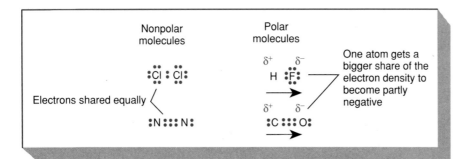

Figure 6.2
Covalent bonds may be either polar or nonpolar.

Attractive forces between molecules increase with increasing molecular size. We do not need to add up the atomic numbers to see that CH_4, $C_{10}H_{22}$, and $C_{20}H_{42}$ contain increasing numbers of electrons. At room temperature CH_4 is a gas, $C_{10}H_{22}$ is a liquid, and $C_{20}H_{42}$ is a waxy solid. Each carbon atom and each hydrogen atom has only a weak attraction for a neighboring molecule, but the attractions accumulate over all of the carbon and hydrogen atoms in a molecule.

Atoms of different elements do not share electrons equally. A more positive kernel or a smaller kernel of the same charge will exert a greater pull on a shared pair of electrons. The atom getting the greater portion of the shared electrons becomes slightly negative, and the atom losing some electron density becomes slightly positive. Chemists describe the covalent bond between such atoms as *polar*, since it has a more positive end and a more negative end (see Fig. 6.2).

The *dipole moment* of a compound is a measurement of its net polarity. Both polar bonds and lone pairs of electrons can contribute to the dipole moment. Although bonds between different elements are always polar to some extent, not all compounds formed between different elements are polar. Molecules of a compound may have a symmetric shape that causes the bond dipoles to cancel. If that happens, the molecule is nonpolar even though it has polar bonds.

Carbon monoxide has a dipole moment; carbon dioxide does not. Both compounds have polar carbon-oxygen bonds. However, in the linear carbon dioxide molecule, the carbon-oxygen bonds point in opposite directions and the two bond dipoles cancel each other out.

Stronger intermolecular forces exist within compounds made up of polar molecules. Because the positive end of one dipole is attracted to the negative end of another molecule, polar molecules tend to have higher boiling points than nonpolar molecules of similar size and composition.

■ CO_2 and SiO_2—Intramolecular Forces versus Intermolecular Forces

It is important to make a sharp distinction between the strong covalent bonds within a molecule and the weak forces between molecules. The properties of carbon dioxide and of silicon dioxide dramatically illustrate this distinction. Carbon dioxide is a gas at room temperature. Under high pressures, it liquefies. When liquid CO_2 comes out of a fire extinguisher, it forms a solid called "dry ice" that passes directly from the solid state to the gas state at $-78°C$. In contrast, SiO_2 is sand or quartz. It is a solid at temperatures up to over $1600°C$, and is both hard and strong.

These very different compounds have some similarities in their bonding. Both CO_2 and SiO_2 are covalent compounds, and their elements, carbon and silicon, are in the same family of the periodic table. Atoms of both carbon and silicon have four

Figure 6.3
Carbon dioxide and silicon dioxide have very different physical properties even though both compounds have covalent bonding. Carbon dioxide is a gas at room temperature, and silicon dioxide is a high-melting solid.

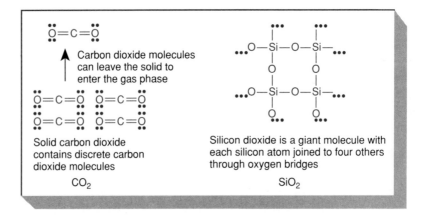

Carbon dioxide molecules can leave the solid to enter the gas phase

Solid carbon dioxide contains discrete carbon dioxide molecules

CO_2

Silicon dioxide is a giant molecule with each silicon atom joined to four others through oxygen bridges

SiO_2

outer electrons and form four covalent bonds to oxygen. Furthermore, the oxygen atoms form two covalent bonds in both CO_2 and SiO_2. The intramolecular forces, the covalent bonds, are strong in both carbon dioxide and in silicon dioxide.

However, carbon dioxide and silicon dioxide have very different structures as shown in figure 6.3. The two bonds formed by each oxygen atom in SiO_2 go to two different silicon atoms. Each oxygen atom forms a bridge between two silicon atoms so a silicon dioxide crystal is one giant covalent molecule. In contrast, carbon dioxide consists of discrete CO_2 molecules. The oxygen atoms in CO_2 do not form bridges because both bonds of each oxygen atom go to the same carbon atom. Although strong double bonds link each oxygen atom to a carbon atom, the forces between different CO_2 molecules are weak. Little work against intermolecular forces is required to separate molecules, and thermal energy is sufficient to overcome intermolecular attractions even at low temperatures.

■ Hydrogen Bonding and the Properties of Water

Water is a remarkable compound. Its boiling point is hundreds of degrees higher than those of the heavier diatomic gases O_2 and N_2, and it is higher even than those of H_2S and H_2Se, the corresponding compounds of elements below oxygen in the periodic table. Water is a good solvent for ionic compounds and for many polar molecules; and it is the solvent for life processes occurring both inside and outside of cells. More energy is required to melt ice or vaporize water than is commonly required for the corresponding phase transitions of most other simple covalent compounds.

Water molecules are bent and both the polar oxygen-hydrogen covalent bonds and the lone-pair electrons on oxygen atoms contribute to the dipole moment of water. The hydrogen atoms in each water molecule are relatively positive, and the oxygen atom is relatively negative.

Strong intermolecular forces contribute to the special properties of water. There is a strong attraction between each positive hydrogen of one molecule and a lone pair of electrons on an adjacent molecule. These intermolecular attractions are called *hydrogen bonds*. Hydrogen bonds are the strongest intermolecular force between covalent molecules. Attractive forces in water are especially large because each water molecule can be hydrogen-bonded to four neighbors (see Fig. 6.4).

Water is not the only compound to exhibit hydrogen bonding. Both ammonia, NH_3, and hydrogen fluoride, HF, form strong hydrogen bonds. To form a hydrogen bond, both a relatively positive hydrogen atom and a lone pair of electrons on a small kernel (N,O, or F) are required. Hydrogen bonding contributes to the properties of many organic compounds having OH or NH groups as a part of their structures. The chemistry of some of these compounds is discussed in later chapters.

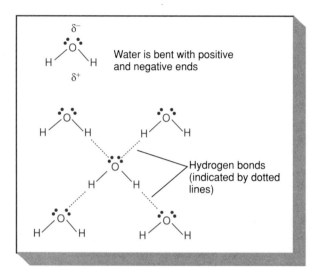

Figure 6.4
Hydrogen bonds are strong intermolecular forces which contribute to the special properties of water. In a hydrogen bond, the partially positive hydrogen atom of one water molecule is attracted to the lone pairs of electrons on an adjacent water molecule.

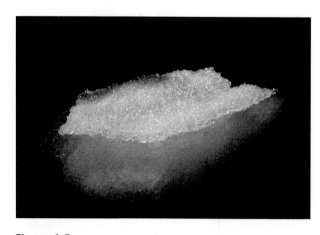

Figure 6.5
Ice floats on water because its lattice of hydrogen-bonded molecules form cage-like structures in which water molecules are farther apart than in liquid water.

■ **Aside**

Why Ice Floats

Water's expansion upon freezing is essential to life in cold climates. If ice could sink and the water above it could continue to freeze, entire lakes would solidify during the winter. Less warming could occur during the summer because only the surface waters would warm initially and the accumulated ice at lower levels would not be exposed to the sun (see Fig. 6.5).

Water is one of a small number of substances that expand upon freezing. For most substances, the solid is more dense than the liquid. In a crystal lattice, molecules or ions are usually packed closely together and there is no room for particles to move. Upon melting, most substances expand, forming holes, and the presence of these holes permits the particles to move.

Why does ice float on water? Ice has a greater number of hydrogen bonds arranged in a more regular pattern than does water. Each molecule of H_2O can participate in up to four hydrogen bonds. In the lattice of ice, molecules are lined up to form strong hydrogen bonds rather than to be as close together as possible. This gives the ice lattice a cage-like structure in which there are small holes. Because water has a less regular structure and fewer holes, its molecules are closer together, making the liquid the denser substance.

Figure 6.6
This periodic table indicates electronegativities of elements on a scale devised by Linus Pauling. On this scale, electronegativities range from a low of 0.7 for cesium to a high of 4.0 for fluorine.

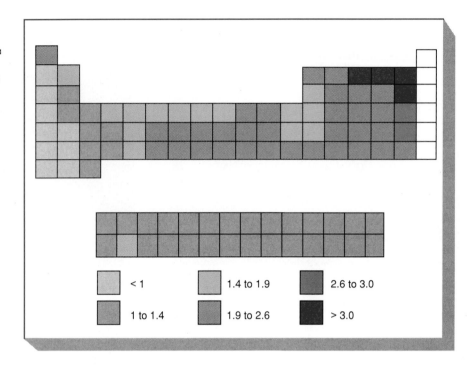

< 1	
1 to 1.4	
1.4 to 1.9	
1.9 to 2.6	
2.6 to 3.0	
> 3.0	

■ Ionic versus Covalent Bonding

Some elements can form both ionic and covalent compounds; for example, oxygen gains electrons from calcium to form oxide ions in CaO, but it shares electrons with hydrogen to form the compound H_2O. In each case, oxygen atoms are surrounded by eight outer electrons. But what determines whether the bonding between oxygen and another element is ionic or covalent?

Oxygen forms ionic compounds with metal atoms that have low kernel charges and loosely held outer electrons. Ionic bonding is favored when an element on the far right side of the periodic table combines with an element on the left side. Atoms of these elements differ greatly in kernel charge, and hence, in their attraction for electrons.

Oxygen forms covalent compounds with nonmetal atoms that have higher kernel charges and more closely held outer electrons. In general, covalent bonding is found in compounds formed between elements on the right side of the periodic table where atoms of each element have high kernel charges and strong attractions for electrons.

The term *electronegativity* has been introduced to describe the attraction that the atoms of an element have for electrons. The higher the electronegativity of an element, the greater the pull its atoms exert on electrons shared with a second element. Except for the noble gases, elements in the upper right of the periodic table have high electronegativities while those in the lower left have very low electronegativities. Fluorine with its small, highly positive kernel is the most electronegative element. Cesium, the largest naturally occurring alkali metal, has the lowest electronegativity of all elements. Within a family, the electronegativities decrease with the increasing atomic numbers (see Fig. 6.6).

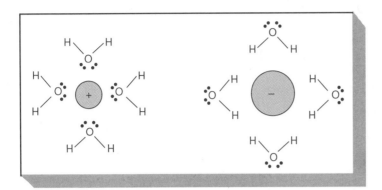

Figure 6.7
In a solution of a salt, the negative ends of water molecules are attracted to cations and the positive ends of water molecules are attracted to anions. The attractive forces between water molecules and ions help to overcome the lattice forces of an ionic compound, causing it to dissolve.

Rules for bonding in binary compounds can be stated in terms of electronegativity differences. When the difference in electronegativity between two elements is large, the bonding is ionic. When the difference in electronegativity is small and one or both elements are nonmetals, the bonding is covalent. And if the difference in electronegativity is small and both elements are metals, the bonding is metallic.

■ Water as a Solvent

Although salt is held together by strong ionic lattice forces, it readily dissolves in water. In contrast, although forces between hydrocarbon molecules are weak, oil and water don't mix. The polar character of water and its strong network of hydrogen bonds contribute to these special properties of water as a solvent.

Many ionic compounds dissolve in water because there are strong attractive forces between water molecules and both positive and negative ions, as shown in figure 6.7. Water molecules surround each cation, with the negative end of the water dipole close to the positive ion. Around each anion, the orientation of the water dipole is reversed since positive hydrogen atoms are attracted to the negative ion. Chemists say that water *solvates* ions because favorable forces between the water molecules and the ions help bring the ions into solution. In addition, the polar nature of water serves to shield the forces of attraction between separated, oppositely charged ions in solution.

Ionic compounds dissolve when the solvation of ions by water is stronger than the forces between ions in the lattice. For example, all ammonium, potassium, and sodium compounds, all nitrates, and most chlorides, bromides, and iodides are soluble. These compounds contain ions that have only +1 or −1 charges, and many of which are relatively large. When the lattice forces are much greater, solvation of ions by water cannot overcome them, and the solids tend to be insoluble. For example, $CaCO_3$ and $BaSO_4$ are both insoluble. Compounds in which +2 and +3 cations combine with −2 or −3 anions generally have low solubility.

Some molecular compounds dissolve readily in water. Compounds such as sucrose (table sugar) and ethyl alcohol (grain alcohol) which have —OH groups like those found in water readily dissolve. Strong hydrogen bonds between water molecules and the dissolved molecules provide the attractive forces that hold these compounds in solution. Many other polar compounds, particularly those containing nitrogen or oxygen atoms, readily dissolve in water.

Figure 6.8
To dissolve a large nonpolar molecule in water, many hydrogen bonds between adjacent water molecules would be broken. There is no attraction between water molecules and nonpolar molecules to overcome the cost of breaking hydrogen bonds.

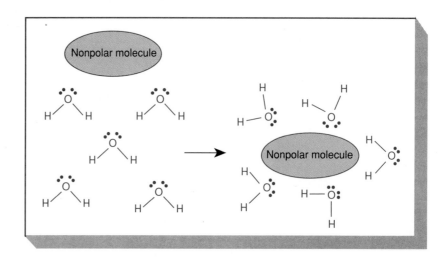

In contrast, water is a poor solvent for most nonpolar compounds. Many hydrogen bonds between adjacent water molecules would have to be broken for a large molecule of a nonpolar compound to enter water. Because there are no strong forces between nonpolar molecules and water molecules to help pay this cost, oil-like compounds do not dissolve in water (see Fig. 6.8).

■ Molecules at the Interface—Soaps and Detergents

Many interesting chemical changes take place at surfaces. Surfaces of liquids and of solids differ from their interiors, and these surface differences often contribute to the chemistry of a substance. We can begin to explore the special features of surfaces by looking at soap and detergent molecules found at interfaces. Soaps are sodium or potassium salts of long chain carboxylic acids obtained from animal fats. Soap molecules combine a *hydrophilic* (water-loving) ionic head—CO_2^-, and a *hydrophobic* (water-hating) tail—$(CH_2)_nCH_3$ (the integer n is an even number, frequently 12, 14, or 16). Molecules of synthetic detergents resemble those of soaps in that they also consist of a polar, hydrophilic head joined to a nonpolar, hydrophobic tail (see Fig. 6.9).

Does a soap or detergent dissolve in water like salt, or does it remain insoluble like oil? Both—when a tiny sliver of soap is placed on a fresh water surface, it moves about as a monolayer of soap molecules spread over the water. The hydrophilic heads of the soap particles dissolve in the water like salt and the hydrophobic tails float on the water surface like a film of oil. If the water is agitated, spherical micelles form. In the interior of a micelle, the hydrocarbon tails of the soap molecules are dissolved in one another while the ionic heads on the outer surface of the micelle cause it to be soluble in water.

Soaps clean by dissolving grease in the interior of micelles. Although oils and grease are insoluble in water, they dissolve in the hydrophobic interior of micelles and can be carried away in water. Detergents perform better than soaps in hard water since hard water contains dissolved calcium and magnesium salts. Soaps precipitate with these +2 ions, forming films or scum. Detergents do not precipitate with Mg^{+2} or Ca^{+2}, so they retain their cleaning power in hard water.

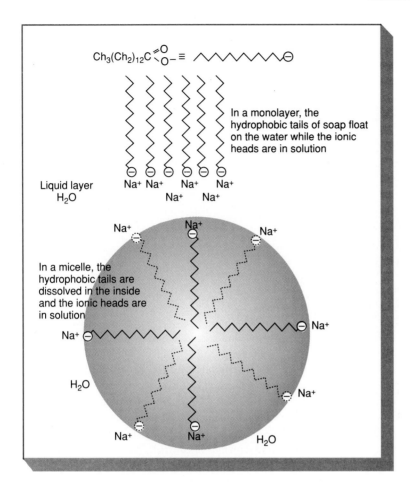

$Ch_3(Ch_2)_{12}C \overset{O}{\underset{O^-}{\lessgtr}} \equiv$

In a monolayer, the hydrophobic tails of soap float on the water while the ionic heads are in solution

Liquid layer H₂O

Na⁺ Na⁺   Na⁺   Na⁺
Na⁺   Na⁺

Na⁺   Na⁺   Na⁺

In a micelle, the hydrophobic tails are dissolved in the inside and the ionic heads are in solution

Na⁺   Na⁺

H₂O

Na⁺   Na⁺   H₂O

Figure 6.9
Soaps have a hydrophilic head and a hydrophobic tail. The hydrophilic head of a soap dissolves in water, but the hydrophobic tail does not. Soaps can form monolayers or micelles when put into water.

■ Aside

Molecular Approaches to Cleaning Up Oil Spills

Giant oil spills have occurred in pristine Alaskan waters, off the coast of Spain, in the Mediterranean, and in the Persian Gulf. The transportation of large quantities of petroleum and petroleum products has become a central necessity of modern industrial society, and spills do occur. While preventing oil spills should be a priority, it is probable that there will be future spills requiring large scale cleanups (see Fig. 6.10).

Spills of crude petroleum have definite stages of life. At first, the oil floats in a thick layer on the water. As time passes, lighter, more volatile components of petroleum vaporize, and the remaining oil becomes thicker and denser. Still later, the density of oil approaches that of water, so that the action of wind and wave create a mousse, a thick emulsion of oil and water. Finally, the mousse breaks down to form thick, sticky tar balls.

The faster an oil spill can be cleaned up, the less long-term environmental damage it causes. The volatile compounds that are first to evaporate or dissolve are often the most toxic portion of the petroleum spill, so that it is desirable to collect or destroy them. Strategies for doing so range from burning or skimming the spill to using microbes to digest it. Since burning a spill is often problematical, skimming or microbial digestion are considered to be the more likely strategies.

Figure 6.10
Booms are used to encircle oil spills
and aid in cleanup efforts.

Companies have developed compounds that aid in the cleanup of oil spills. If the strategy is to be skimming, additives comprised of tightly-wound, polymer beads are used. These beads unwind to provide a fibrous network that helps hold the oil together and slows the aging of the spill that forms a mousse. More oil can be recovered by skimming, and less remains to do environmental damage. In contrast, if the strategy calls for microbial digestion, then it is desirable to break apart the spill to speed up the biological treatment. Detergent additives are used to speed the breakup of the slick into tiny globules that can more readily be ingested by oil-eating microbes. These examples illustrate the strategy of employing chemical approaches to the remediation of environmental problems.

■ Questions—*Chapter 6*

1. Which of the following compounds are covalently bonded? Which are ionically bonded? KI, PI_3, BaS, AIF_3, $GeCl_4$
2. Use the periodic table to determine which element in each of the following pairs is more electronegative.
    (a) Br and Cl
    (b) F and N
    (c) Al and Na
    (d) O and P
3. Silver fluoride, alone of the silver halides, is soluble in water. Why is the bonding in AgF ionic, while that in AgCl, AgBr, and AgI is covalent?
4. The compound ICl is polar. Draw a Lewis structure of this compound. Would I or Cl attract elements more? Which end of the ICl molecule is more negative?
5. The compound BF_3 has a zero dipole moment even though each of the three B—F bonds is polar. What is the shape of BF_3?

6. Which of the following compounds has no net dipole moment? (In each compound, a carbon atom is at the center with bonds directed toward the corners of a tetrahedron.)

CH_2Cl_2 $CHCl_3$ CCl_4

7. Although the bonding in elemental carbon (diamond) and in elemental nitrogen (N_2) is covalent, these elements differ greatly in their physical properties. Why?

8. Why is the boiling point of NH_3 much higher than that of CH_4?

9. Why is ammonia, NH_3, much more soluble in water than is methane, CH_4?

10. Illustrate the solvation of Ca^{2+} and of Cl^- ions that enables the ionic compound $CaCl_2$ to dissolve in water.

11. If the water molecule were linear rather than bent, would ocean water be as salty? Why or why not?

12. Methane (CH_4), the major component of natural gas, is much more soluble in water than C_8H_{18}, a component of gasoline. Why should the smaller hydrocarbon molecule be more soluble in water?

13. Sodium chloride dissolves in liquid HF but not in C_8H_{18}. Why should this ionic solid be more soluble in HF than in C_8H_{18}?

14. Seawater contains salts of sodium, potassium, and calcium. Why does a clam use less abundant calcium ions to help build its shell rather than the more abundant sodium or potassium ions?

15. Why would the sodium salt of a short chain (four to eight carbon atoms) carboxylic acid fail to be a satisfactory soap?

7

Radiation and Molecules

On a clear day, we might see a distant mountain, city skyline, or ship along the horizon. In contrast, on a cloudy night we cannot see the stars and may see only a diffuse glow of moonlight. Because we see visible light, we have a direct experience of the interaction of light with the atmosphere. If we could only see at other wavelengths, the view would appear to be quite different. The atmosphere is almost opaque to radiation in the high-energy ultraviolet portion of the solar spectrum, so the sun would cast only a dim light even at midday and surface objects would merely be faintly lit. In contrast, the atmosphere is translucent to radiation in the lower-energy infrared portion of the solar spectrum, so both the earth and the sky would glow by day and by night.

Invisible compounds injected into the atmosphere in this industrial age are changing the way that the atmosphere interacts with both ultraviolet (UV) and infrared (IR) radiation. These changes may have dire consequences on the environment and on the future of life on the planet. In this chapter, we explore the interactions of radiation and matter and consider the environmental threats posed by the depletion of the ozone layer and by climatic warming brought about by the "greenhouse" effect.

■ Visible Light and Colored Compounds

Colored compounds absorb certain wavelengths of light and reflect or transmit other wavelengths. The plant pigment chlorophyll appears green, because chlorophyll molecules absorb red light as shown in figure 7.1. What we see in chlorophyll is the complementary color of red, which is the green light that has been reflected by the sample. The color of a compound depends both on which wavelengths of light are absorbed and the amount of light absorbed at each wavelength. For example, the red compound lycopene, found in tomatoes, is more intensely colored than its chemical cousin carotene, the yellow-orange compound found in carrots.

Just as the emission of light by excited atoms gives information about the arrangement of electrons in atoms, the absorption of light by a compound gives information about the arrangement of electrons in molecules. A quantum of light can lift an electron from an occupied, lower-energy orbital to a vacant, higher-energy orbital. Only a relatively small number of compounds have a filled and a vacant orbital close enough in energy so that a quantum of visible light can excite an electron.

Figure 7.1
Absorption spectrum of chlorophyll a, one of two similar chlorophylls found in higher plants.

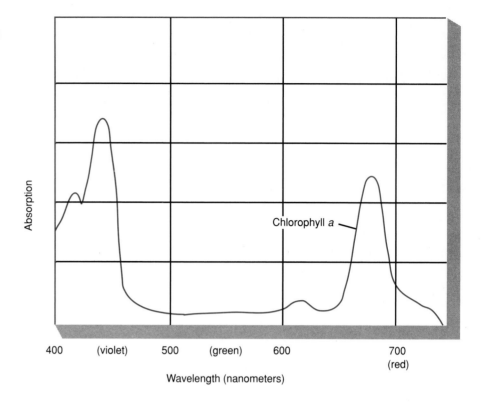

Therefore most chemicals appear white as solids and colorless in solution. In contrast, dye molecules absorb light so strongly that a small amount of dye can impart its color to a large quantity of material.

■ Ultraviolet Radiation and the Ozone Layer

High energy ultraviolet radiation from the sun does not readily penetrate the atmosphere. Instead, two forms of oxygen gas, O_2 and O_3, absorb incoming photons high above the earth in the stratosphere. Photons of ultraviolet radiation, with more energy than those of visible light, cause chemical changes to occur. When absorbed, they supply sufficient energy to break some chemical bonds and initiate subsequent chemical reactions.

Incoming solar radiation initiates the process that produces ozone in the stratosphere when a molecule of oxygen O_2 absorbs short-wavelength, ultraviolet radiation. The added energy of the photon breaks the oxygen-oxygen bond and produces two highly reactive oxygen atoms.

$$O_{2(g)} + h\upsilon \rightarrow 2\ O_{(g)}$$

When an atom of oxygen collides with a molecule of O_2, it can then react to form ozone, O_3.

$$O_{(g)} + O_{2(g)} \rightarrow O_{3(g)}$$

This overall process accounts for the formation of ozone in the upper atmosphere.

Ozone in the stratosphere absorbs a large portion of incoming ultraviolet radiation. The absorbed radiation initiates a reaction that destroys ozone molecules but provides the reactants to generate new ozone molecules. When a molecule of ozone absorbs a photon of ultraviolet radiation, it breaks apart, forming a molecule of O_2 and an oxygen atom.

$$O_{3(g)} + h\upsilon \rightarrow O_{2(g)} + O_{(g)}$$

A second reaction in which ozone is re-formed completes the cycle. The oxygen atoms reacts with a molecule of oxygen to yield a molecule of ozone.

$$O_{(g)} + O_{2(g)} \rightarrow O_{3(g)}$$

Just as it required the addition of energy to break a chemical bond, the formation of a new chemical bond releases energy. The "hot" molecule initially formed must lose some of its excess energy in collisions. As the "hot" ozone molecules collide with other molecules, the energy of bond formation is converted into the energy of motion. What is the net effect of this cyclic process? Dangerous incoming ultraviolet radiation is converted into harmless heat.

Far more ultraviolet radiation would reach the surface of the earth if only the O_2 absorbed it. Why should O_3 capture ultraviolet radiation so much more efficiently than O_2? Molecules of O_2 absorb only the shorter-wavelength portion of the incoming ultraviolet radiation. The bonding electrons in O_3 are more loosely held than those in O_2, and they require less added energy to break. Consequently, molecules of O_3 also absorb longer-wavelength, lower-energy ultraviolet radiation, a part of the solar spectrum where molecules of O_2 are transparent. The outer electrons in N_2 (the major component of the atmosphere) are still more tightly held than those in O_2, so that N_2 is transparent to the wavelengths of ultraviolet radiation that are absorbed by O_2.

■ Chlorofluorocarbon Emissions Threaten the Ozone Layer

Chlorofluorocarbon gases are readily vaporized, non-toxic, non-corrosive, non-flammable compounds used as heat transfer materials in refrigerators and air conditioners and as solvents by manufacturers of computer chips. These compounds are part of the wide variety and large number of halogenated hydrocarbons that have been placed in the environment by man. (Few compounds containing only carbon, hydrogen, and halogen atoms are found in nature. Those that do occur are synthesized by plants living in the sea where halide ion concentrations are high.)

The use of chlorofluorocarbons (CFCs) and of many other halogenated hydrocarbons has come under criticism as we become aware of the environmental costs associated with their continued use. While many compounds react readily with water or with oxygen, CFCs are very resistant to reaction. Most halogenated hydrocarbons are not susceptible to oxidation, and have little solubility in water because they do not form hydrogen bonds with it. Therefore, they cannot be readily disposed of by simple dilution. Microorganisms which degrade most organic compounds lack the enzymes to react with these man-made compounds. The net result is that CFCs are long lived in the environment.

Chlorofluorocarbons pose a particular threat to the ozone layer. When discarded or released in spills and leaks, these remarkably stable compounds remain airborne for many years. They slowly diffuse into the upper atmosphere where chlorofluorocarbon molecules absorb high-energy ultraviolet radiation, breaking carbon—chlorine bonds to produce reactive species with odd numbers of electrons.

$$CF_3Cl_{(g)} + h\upsilon \rightarrow CF_{3(g)} + Cl_{(g)}$$

The chlorine atoms initiate a pair of reactions that convert ozone, O_3, back to the more normal form of oxygen, O_2. In the first reaction, a chlorine atom takes an oxygen atom from ozone to form a molecule of oxygen and a molecule of ClO as products.

$$Cl_{(g)} + O_{3(g)} \rightarrow O_{2(g)} + ClO_{(g)}$$

In the second reaction, ClO reacts with atoms of oxygen (from the cleavage of ozone initiated by ultraviolet radiation) to form a molecule of oxygen and to regenerate a chlorine atom.

$$ClO_{(g)} + O_{(g)} \rightarrow Cl_{(g)} + O_{2(g)}$$

These two reactions take place again and again so that a small number of chlorine atoms can bring about the destruction of a very large number of ozone molecules. By adding together the last two reactions and subtracting the species that occur on both sides of the arrow, one can see that the net result is the removal of ozone by reaction with the oxygen atoms.

$$Cl_{(g)} + O_{3(g)} + ClO_{(g)} + O_{(g)} \rightarrow O_{2(g)} + ClO_{(g)} + Cl_{(g)} + O_{2(g)}$$
$$O_{3(g)} + O_{(g)} \rightarrow O_{2(g)} + O_{2(g)}$$

Chemists describe the overall process by saying that Cl and ClO catalyze the destruction of ozone.

The atmospheric chemistry involved in ozone depletion is complex. Some other pollutant gases contribute to the depletion of ozone in the stratosphere, for example, nitrogen oxides from high flying supersonic airplanes also catalyze the destruction of ozone. Some compounds react to generate reactive fragments and still others remove reactive fragments from the atmosphere. Increasing the concentrations of ozone-depleting species shifts the balance of these fragments in one direction, and decreasing their concentrations shifts this balance in the opposite direction.

The thinning of the ozone layer is a global threat. Scientists estimate that for each one percent depletion of ozone in the stratosphere, there is a two percent increase in the amount of ultraviolet radiation reaching the surface of the earth. Because it takes several years for CFCs to diffuse from the surface of the earth to the stratosphere, we have yet to experience the full damage to the ozone layer resulting from past emissions. Even if CFC production were to stop today, the thinning of the ozone layer will continue for several years but at a slowing rate.

Newer measurements suggest that the threat to the ozone layer is greater than was initially realized in the mid-1980s. In 1985, scientists first discovered that holes in the ozone layer develop over polar regions during the coldest seasons and then dissipate when the weather warms (see Fig. 7.2). In the winter months, there is little exchange of cold polar air and warmer air from the temperate regions. As temperatures drop, thin clouds consisting of ice crystals (and particles of crystallized pollutants) form at very high altitudes. Scientists believe that the reactions that remove chlorine radicals from the stratosphere are reversed by the reactions occurring on the surfaces of these crystals in the clouds. During the coldest months, the concentrations of ozone-destroying chlorine atoms and of ClO increase and the concentration of ozone drops. In the spring when the weather begins to warm up, polar air and warmer air mix to a greater extent, the ice clouds dissipate, and the concentrations of ozone-destroying species drop.

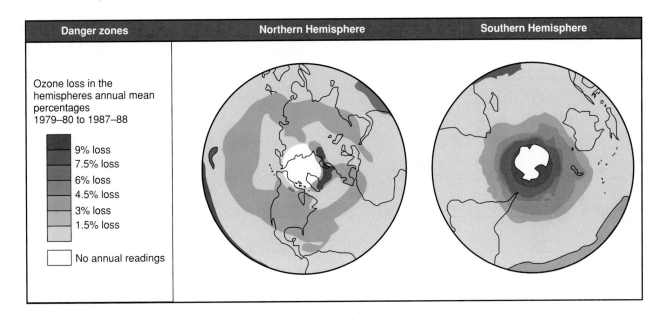

| Danger zones | Northern Hemisphere | Southern Hemisphere |

Ozone loss in the hemispheres annual mean percentages 1979–80 to 1987–88

9% loss
7.5% loss
6% loss
4.5% loss
3% loss
1.5% loss

No annual readings

Figure 7.2

Ozone depletion in polar regions.

Sources: U.S. Environmental Protection Agency (1986); NASA Ozone Trends Panel; and John Gille, U.S. National Center for Atmospheric Research. Data compiled by Nigel Dudley.

■ Ultraviolet Radiation Threatens Plants and Animals

The thinning of the ozone layer brought about by CFCs is dangerous to living organisms since it permits more ultraviolet radiation to reach the surface of the earth. When ultraviolet radiation is absorbed by molecules in surface tissues, it can cleave chemical bonds to form reactive fragments. These reactive fragments can cause cancer tumors as well as cell deaths in the form of severe sunburns. Dermatologists recognize these destructive effects of overexposure to sunlight and they recommend that people exposed to the sun protect themselves by using sun-screen lotions containing compounds that efficiently capture ultraviolet light (see Fig. 7.3).

Plants, too, can suffer the effects of too much exposure to ultraviolet radiation, so increased exposure threatens to reduce agricultural productivity. Some scientists advocate an accelerated program to develop plant strains that have enhanced resistance to ultraviolet radiation. Because water does not absorb ultraviolet radiation, some marine plants and animals may also be at risk.

■ Actions to Protect the Ozone Layer

The countries of the world are negotiating agreements that provide for a coordinated international effort to reduce this threat to the ozone layer. The signing in 1987 of an international agreement known as the Montreal Protocol on Substances That Deplete the Ozone Layer was an important first step. The Montreal Protocol set the goal of a fifty percent reduction of world-wide chlorofluorocarbon emissions by the year 1998.

After this agreement was signed, countries began to take significant steps to slow the growth of the manufacturing and use of CFCs. Substitute compounds have replaced CFCs as the propellants in aerosol dispensers and as the foaming

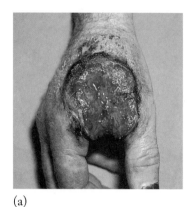

(a)

(b)

Figure 7.3
There is a growing awareness of the health risks associated with over-exposure to UV radiation from the sun. (a) Skin cancer; (b) Sunscreen lotions contain compounds that efficiently absorb UV light before it strikes the skin.
(a) © Ken Greer/Visuals Unlimited, (b) © Jim Shaffer

agents in the manufacturing of insulating materials. Manufacturers are developing new substitute compounds to replace CFCs as solvents in the manufacture of computer chips and as refrigerant gases.

In response to the discoveries of polar ozone holes and the realization that the thinning of the ozone layer is occurring faster than had been previously thought, nations have agreed to take stronger actions to limit CFC emissions. In 1989, eighty-one nations signed a Helsinki accord pledging to phase out five CFCs by the year 2000, and, in 1990, ninety-three countries agreed to halt CFC production by the end of the 1990s. Nations further agreed, in 1992, to accelerate the phasing out of many ozone-depleting chemicals.

■ Aside

Economic Justice and Pollution Rights

How are the world's developed and undeveloped nations to share responsibilities and to respond with justice? Industrial nations are responsible for a great deal of past and present environmental damage. Do undeveloped countries have a right to pollute as they seek to industrialize? If they industrialize following current technologies, they can only duplicate and exacerbate environmental problems facing the developed nations. What should be the roles of wealth and poverty on global environmental policy?

What technologies should be available to countries that trail in their development? Do industrialized countries have a responsibility to share newer, cleaner, and possibly more profitable technologies? As wealthier developed nations replace chlorofluorocarbons with more expensive but environmentally safer compounds, should they subsidize the use of these substitute compounds? Poorer undeveloped countries are not responsible for the destruction that has already occurred so should they enjoy the right to use chlorofluorocarbons and to contribute to the further depletion of the ozone layer as they pursue industrialization? An answer was suggested when, as part of the 1990 agreement, a $240 million fund was established to provide developing countries technical assistance in finding substitutes for CFCs.

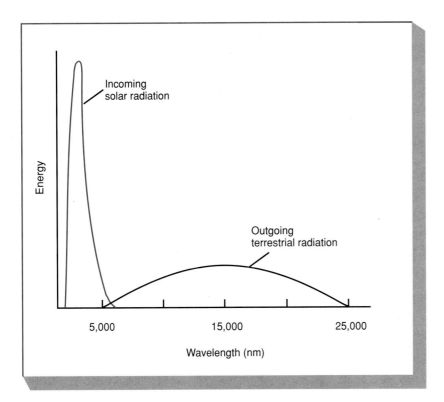

Figure 7.4
Spectra of the incoming radiation from the sun and the outgoing radiation from Earth's surface.

■ Radiant Energy and Molecular Motion

When low-energy infrared radiation (radiant heat) is absorbed, it excites molecules to higher vibrational and rotational energies. One might think of a molecule as a mobile with various sized balls joined by springs of different lengths and thicknesses. Each molecule jiggles and turns in a unique way when struck by a photon. Patterns are present because the masses of atoms of a given element are alike and because many chemical bonds are similar in strength. For example, the O—H bond found in water and in alcohols strongly absorbs infrared light at a characteristic wavelength.

Polar molecules and molecules with polar bonds capture photons of infrared radiation much more efficiently than their nonpolar cousins. It is as if the presence of a dipole moment provides a foothold and a grip for a photon to push these atoms further apart and cause them to vibrate still more. All molecules with three or more atoms, including those with no polar bonds, absorb infrared radiation since they all have some vibrations that change the polarity of the molecule.

■ Greenhouse Gases and Global Warming

Daily temperatures rise as solar radiation heats the surface of the earth. At night the earth cools by radiating energy into space. Both the energy gain from incoming solar radiation and the energy loss by radiative cooling contribute to the earth's daily, annual, and long-term weather. The temperature of the earth's surface warms during the day, because the gain of solar energy is greater than the loss by radiant cooling. Most of the radiant energy reaching the surface of the earth is in the visible range of the spectrum. In contrast, the earth radiates lower-energy, longer-wavelength infrared radiation (see Fig. 7.4).

Figure 7.5

Increases of atmospheric carbon dioxide can lead to the warming of the earth's atmosphere by a greenhouse effect. The earth cools by sending energy into space in the form of infrared radiation. Molecules such as carbon dioxide and water that absorb infrared radiation cause a portion of that radiation to be retained in the atmosphere and converted into heat.

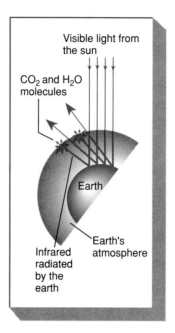

The earth's atmosphere acts to moderate the daily temperature swings through a greenhouse effect. Carbon dioxide, water vapor, and other so-called greenhouse gases absorb infrared radiation. These gases capture part of the energy radiated from the surface of the earth and they, in turn, radiate it in all directions, returning a portion of it to the earth's surface as shown in figure 7.5. The important role of water in the natural greenhouse effect is familiar to most people because they know that temperatures drop more on dry, clear nights than on cloudy nights.

The quantity of carbon dioxide entering the atmosphere is increasing as shown in figure 7.6. The use of fossil fuels has increased dramatically since the start of the Industrial Revolution. Carbon dioxide produced by the burning of these fuels is entering the atmosphere faster than plants can utilize it in photosynthesis. In addition, the destruction of tropical rain forests contributes significantly to the rising levels of CO_2 in the atmosphere.

There are also increases in the quantities of other important greenhouse gases, including chlorofluorocarbons, hydrocarbons such as methane, and nitrogen oxides from auto emissions. Although the quantities of CFCs, methane, and nitrogen oxides in the atmosphere are small compared to the quantity of CO_2, these compounds absorb infrared radiation more efficiently than CO_2 (and at different wavelengths) and so contribute to the overall problem as shown in table 7.1. (Note also that the warming effects of CFCs occur throughout the atmosphere and not just in the stratosphere.)

The shift in the balance of radiation gains and losses is expected to have profound effects on the climate of the earth. As the atmosphere warms, some ice in the polar regions will melt. In the Northern Hemisphere, a large polar ice sheet is floating on the Arctic Ocean. When this ice melts, the exposed water will absorb more solar radiation than did the ice sheet. This increase in the absorption of solar energy will amplify the warming of the earth initiated by increases in atmospheric CO_2.

Many scientists believe that if the consumption of fossil fuels at current high levels is continued into the next century, the earth would warm, the sea level would

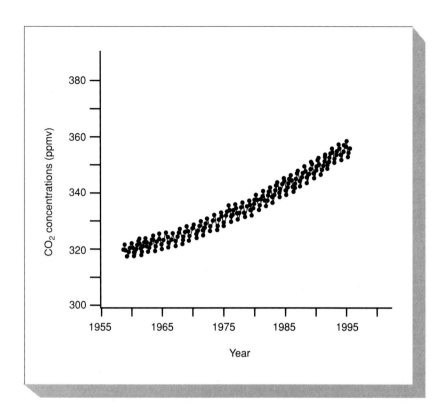

Figure 7.6
Mauna Loa atmospheric CO_2 measurements constitute the longest continuous record of atmospheric CO_2 concentrations available in the world. The methods and equipment used to obtain these measurements have been essentially unchanged over the 32-year record.

Source: Thomas A. Boden, Robert J. Sepanski, and Frederick W. Stoss, (Eds.), *Trends '91 Highlights: A Compendium of Data on Global Change.* Publication No. ORNL/CDIAC-49. ESD Publication No. 3797.

• **Table 7.1** •

Major Greenhouse Gases and Their Characteristics

Gas	Atmospheric Concentration (ppm)	Annual Increase (percent)	Life Span (years)	Relative Greenhouse Efficiency (CO_2 = 1)	Current Greenhouse Contribution (percent)	Principal Sources of Gas
Carbon Dioxide (Fossil Fuels)	351.3	.4	x[1]	1	57 (44)	Coal, Oil, Natural Gas, Deforestation
(Biological)					(13)	
Chlorofluoro-carbons	.000225	5	75–111	15,000	25	Foams, Aerosols, Refrigerants, Solvents
Methane	1.675	1	11	25	12	Wetlands, Rice, Fossil Fuels, Livestock
Nitrous Oxide	.31	.2	150	230	6	Fossil Fuels, Fertilizers, Deforestation

[1] Carbon dioxide is a stable molecule with a 2–4 year average residence time in the atmosphere.

Figure 7.7

These variations in global annual temperatures show little trend during the nineteenth century, marked warming to 1940, relatively steady conditions to the mid-1970s, followed by a rapid warming during the 1980s. The 1980s was a decade of unprecedented warmth, with 1990 the warmest year since comparable records began in the middle of the 19th century.

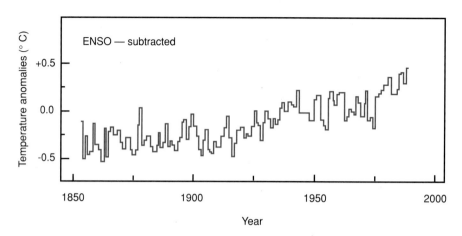

rise several feet, and climates would change in unpredictable ways. These climatic changes would produce profound social consequences. Flooding will endanger coastal cities on all continents, and floods in some densely populated, low-lying agricultural lands would lead to large migrations of displaced persons. Reduced rainfall would change some productive agricultural land to desert. If the consumption of fossil fuel is slowed, these changes would occur more slowly. These scientists believe that it is imperative to increase the use of alternate forms of energy.

■ Uncertainties in Modeling Climate Change

How serious is the threat of global warming? Global temperatures are increasing as shown in figure 7.7. However, the task of estimating the impacts of changing concentrations of greenhouse gases on the global climate is formidable. Scientists use powerful computers to estimate the outcomes of future scenarios. All of the calculations involve assumptions—people's best estimates of the interactions in a complex, dynamic system. Reputable scientists disagree, a fact that many lay people find confusing.

The difficulties encountered in modeling studies can be seen in the opposing effects of changes in cloud cover. Because clouds reflect incoming solar radiation, increased cloudiness should lead to lower surface temperatures. But what portion of the sun's energy is reflected? And what portion of the earth's surface will be covered by clouds? Do clouds at equatorial regions reflect more energy than clouds in polar regions? Water in the clouds absorbs infrared radiation escaping the surface of the earth. This capture of radiant heat warms the atmosphere, an effect opposite to the cloud's reflecting of solar energy. What is the magnitude of this effect? The answer is complex, for low-lying clouds at moderate temperatures have a smaller overall impact on global warming than do cold, high-level clouds.

While projections of global warming differ in magnitude, there is a growing scientific consensus that greenhouse gases in the atmosphere are contributing to global warming. This warming will have impacts on the climates of many global regions in the next several decades.

■ Combating Global Warming

International negotiations to slow the introduction of greenhouse gases into the atmosphere are underway. However, it is proving far more difficult to reach agreements in this area than it has been to agree on steps to protect the ozone layer. (It should be noted however that the elimination of CFC emissions is an important and necessary step toward slowing the rate of increase of global warming.)

Industrialized countries in Europe agreed in 1991 to set targets for first, the stabilization of CO_2 emissions, and then, their reduction. At that time the United States had proposed targets for its reductions in the production of greenhouse gases (chiefly CFCs), but it had made no commitment to reducing CO_2 emissions. At an United Nations sponsored "Earth Summit" meeting held in Brazil in June 1992, nations signed a treaty setting goals for limiting CO_2 emissions. Some people hailed this as an important step toward combatting global warming. Others criticized it for its failure to include enforcement provisions.

■ *Reflection*

The Investigation of Color and the Growth of Science

Scientists solve puzzles; the practice of science often resembles the play of a game. Solved puzzles and answered questions become a part of scientific knowledge, and scientists move on to new questions. The practice of science occurs at the horizon of our knowledge; it acts to expand that horizon. What lies over the horizon? What new directions might science take next? The answer to such questions depends, in part, on the practice of science. Scientists choose new problems that are difficult enough to present challenges yet which offer prospects of finding a successful solution. The ability to solve a problem depends greatly on the experimental techniques that are available to the investigators.

Why does science grow so very rapidly? We can explore an aspect of that growth by tracing the evolution of two laboratory tools that are widely used in chemistry. Chromatography and spectroscopy were first developed in investigations directed toward gaining an understanding of the chemistry of colors. These techniques have evolved to become the powerful methods used by present day chemists to isolate individual components from a mixture and to unravel the structures of compounds.

Chemists set out to separate and study the pigments responsible for colors in nature—chlorophyll, responsible for the color of green leaves; yellow, orange, and red pigments seen in the autumn leaves. These early chemical studies featured a fascination with color, a curiosity about the workings with nature, and a quest for practical knowledge. Would a knowledge of the structures of leaf pigments lead to a better understanding of the processes of photosynthesis? Would it enable chemists to manufacture new dyes for the textile industry?

Chromatography (color writing) is the name given to a separation method first developed to use on leaf pigments. In 1906, Mikhail Tswett (1872–1919) described the separation of an extract of plant leaves into bands of individual pigments. He first dissolved the pigments from crushed leaves in a hydrocarbon solvent. This extract was passed through a column consisting of a tube of glass packed with finely divided calcium carbonate (limestone). The pigments remained bound to the limestone at the top of the column. When alcohol was introduced at the top of the column and allowed to flow slowly downward, the green, yellow, and red pigments separated and migrated down the column in colored bands. By collecting the individual bands and then evaporating the alcohol present in each of the separate fractions, Tswett successfully isolated the colored compounds (see Fig. 7.8).

Figure 7.8
Chromatography is a method used to separate individual components from mixtures. It was first applied to the separation of mixtures of colored compounds.

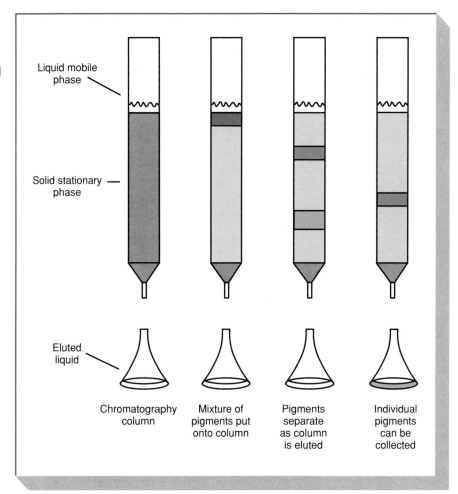

Liquid mobile phase

Solid stationary phase

Eluted liquid

Chromatography column

Mixture of pigments put onto column

Pigments separate as column is eluted

Individual pigments can be collected

In principle, chromatography can be used in the separation of any chemical mixture. Some chromatographic separations are exceedingly simple. One can use a pen to make a mark on a paper tissue. If the edge of the paper (but not the part with the ink mark) is then placed in water or rubbing alcohol, the liquid ascends the paper and, depending on its complexity, may separate the ink into its component compounds. In chromatography, the components of a mixture are separated by passing a liquid (or gas) slowly over and through a stationary phase. Separations are more favorable if the solvent moves slowly and the stationary phase has a high surface area. We seek conditions such that the individual component compounds spend part of the time dissolved in the mobile phase and part of the time bound to the stationary phase. One can separate the components of a mixture if they differ slightly in their solubility in a solvent or in the tightness of their binding to a stationary phase.

Chemists have developed a repertoire of powerful chromatography methods for the separation of mixtures. They modified the materials used as stationary phases and they varied the solvents used to elute mixtures. (A stream of helium gas is used to carry volatile compounds through packed columns in a modification known as gas chromatography.) To extend further the range of compounds that could be separated by chromatography, chemists developed newer, more sensitive methods for their detection. Dyes were used to stain some compounds; fluorescent markers were attached to others. With each improvement to the chromatographic process, chemists opened new avenues for investigation and they could study smaller samples and separate more complex mixtures.

The study of the interaction of matter and light is known as spectroscopy. Colored compounds absorb certain wavelengths of light and reflect or transmit other wavelengths. If visible light is first separated into a rainbow of colors and the various wavelengths of light are then passed through a solution of a compound to be studied, one can measure the amount of light absorbed at each wavelength. The absorption spectrum of a compound is a graph of the relative amount of light absorbed by a compound versus the wavelength of light.

Spectroscopy has also been extended to cover the interaction of compounds with electromagnetic radiation of wavelengths both shorter (hence higher in energy) and longer (hence lower in energy) than those of visible light. Chemists use visible and ultraviolet spectroscopy to gain information about the arrangement of electrons in molecules. They use infrared spectroscopy to gain information about the strengths of bonds and the shapes of molecules. They also use it as a criterion for the identity of samples, since each compound has a unique infrared spectrum.

Other types of spectroscopy are best known by their applications. In microwave cooking, radiation in the microwave region of the spectrum causes water molecules to spin faster. When the excited water molecules collide with other molecules, the organized energy of spinning is converted to the random kinetic energy of heat. A different type of spectrum is obtained in magnetic fields, because electrons and some nuclei have magnetic properties. This magnetic resonance spectroscopy (or imaging) has been applied to the field of medicine.

These various types of spectra, patterns of light and shadows, are used to detect the presence of compounds, to identify known compounds and to probe the structures of unknown compounds. The improvements in electronics that have helped to make it possible to detect faint radio signals from satellites exploring distant planets have permitted the construction of more sensitive spectrometers. The time between the building of a prototype instrument to probe a new region of the spectrum and the development of more powerful, commercially available instruments continues to shrink. State of the art instruments may offer automated sample handling and computer-enhanced data acquisition. The tools of science continue to evolve, becoming more powerful (and more expensive) each decade.

■ Questions—*Chapter 7*

1. Halons, compounds containing only carbon, fluorine, and bromine, are used for extinguishing fires in computer facilities, telephone exchanges, and grain elevators. These dense gases can form a blanket that smothers the fire by excluding oxygen. Why would water be unsuitable for fighting such fires?
2. The C—Br bond of a halon molecule is cleaved by ultraviolet radiation. What is the reactive species that poses a threat to the ozone layer?

3. Proposed replacement compounds for CFCs contain hydrogen as well as carbon, chlorine, and fluorine. The destruction of a molecule of one of these compounds is initiated by attack of an OH radical, present in the lower atmosphere, on a susceptible C—H bond. To be an effective replacement for CFCs, how much shorter an atmospheric lifetime should the substitute compound have?

4. One proposal for lowering CFC emissions is to place a large tax on CFCs. Would such a tax spur recycling of these compounds? Why or why not?

5. The space shuttle uses small amounts of CFCs as a propellant when maneuvering in orbit. Why should this use of CFCs be more destructive of the ozone layer than the use of an equal quantity as a refrigerant? Should this use of CFCs be discontinued?

6. Use the library to learn how the size and duration of the ozone holes have changed during the past three years.

7. Although replacement compounds for CFCs will not remain aloft long enough to reach the stratosphere and contribute to ozone depletion, they will be potent greenhouse gases. What steps should be taken to minimize releases of these gases?

8. Should developed countries be as concerned with the proliferation of use of CFCs as they are with nuclear proliferation? Justify your position.

9. Inhalers using CFCs as propellant are widely used to deliver medicines for sufferers of asthma and bronchitis. Should this use of CFCs be exempt from the ban on the production and use of CFCs?

10. Does uncertainty in estimating global temperature change justify postponing limitations on carbon dioxide emissions at the present time? Justify your position.

11. Some major volcanic eruptions have caused global temperatures to drop in the year following the eruption. How might volcanic eruptions contribute to global cooling?

12. Does the United States have a legitimate basis to criticize the buildup of atmospheric carbon dioxide from the burning of forests in the Amazon basin? Why or why not?

13. Does India have a legitimate basis to criticize the buildup of atmospheric carbon dioxide accompanying the generation of large amounts of electrical power in the United States? Why or why not?

14. Nuclear power supporters argue that the increased use of nuclear power would help reduce carbon dioxide emissions and slow global warming. Should greater use of nuclear power be considered in the United States? Why or why not?

15. Some scientists argue that global warming is inevitable and that society should immediately undertake a major scientific effort to develop heat resistant varieties of plants important to agriculture. Do you agree with this premise and conclusion? Why or why not?

16. For each of the following countries, indicate whether the thinning of the ozone layer or global warming would be perceived as the greater environmental threat
    (a) Australia
    (b) Bangladesh
    (c) Brazil
    (d) Chile
    (e) Netherlands
    (f) Norway

Acids and Bases

The chemistry of acids and bases was first explored by early alchemists. One branch of alchemy that related to medicinal chemistry studied organic acids and bases. These compounds could be prepared by fermentation and putrification. Early descriptions suggest a tame and familiar nature for acids and bases. Acetic acid in vinegar and citric acid in lemon juice typify acid behavior. They taste sour, cause certain vegetable dyes to change color[1], and fizz with the base sodium bicarbonate, better known as baking soda. Baking soda and ammonia typify bases. Bases taste bitter, reverse the color change that vegetable dyes undergo with acids, and react with acids. Organic acids and bases play essential roles in the chemistry of living organisms and in the technology of the food industry.

Another branch of alchemy that related to metallurgy explored mineral acids and bases. These acids and bases had more powerful actions and could be obtained by the heating or burning of certain mineral compounds. Mineral acids could be used to dissolve metal ores and they reacted with some metals to bring them into solution. These mineral acids and bases with their more powerful chemical actions played an essential role in the development and expansion of technology.

The exploration of acid-base chemistry continues to play a central role in the development of the modern science of chemistry. Chemists classify reactions to describe chemical changes, and they assign roles to chemical participants. They then use the classification schemes to show patterns in chemical change just as they use the periodic table to show patterns in chemical structure. The reactions of acids with bases comprise a major class of chemical reactions.

In chemistry, as in other disciplines, more than one classification scheme is used. A structural approach to chemistry was presented in chapters 5 and 6. Compounds were grouped into two major classes, those with ionic bonding and those with covalent bonding, and the physical properties of representative compounds were discussed in relation to their bonding type. Biologists and political scientists also use both structural and behavioral classifications. When discussing a plant or animal, a biologist may emphasize either its taxonomy or its ecological niche. Similarly, a political scientist might discuss either constitutional checks and balances or voting behavior when talking about democratic government. Chemists, too, use both structural and behavioral criteria when discussing acids and bases.

[1]The juice of a red cabbage can serve as a readily available, dramatic indicator for investigating acid-base properties of familiar substances.

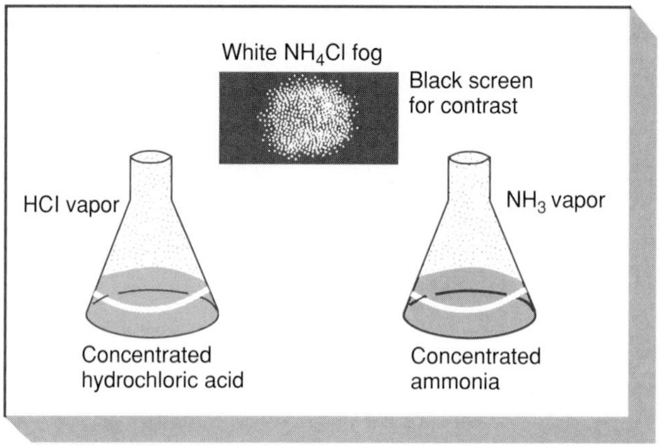

Figure 8.1

HCl vapor and NH_3 vapor react in an acid base reaction to form solid NH_4Cl.

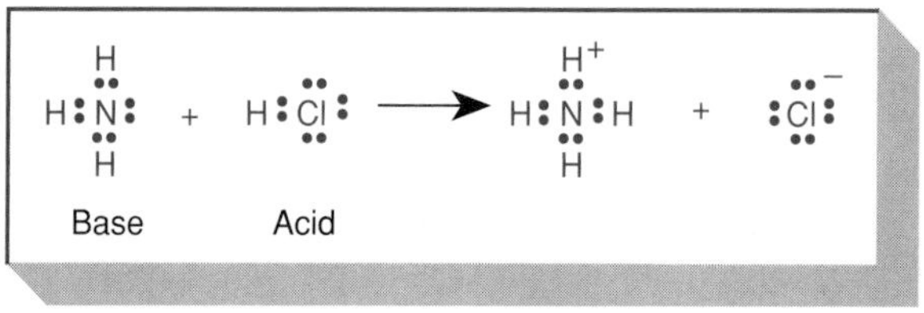

White NH_4Cl fog

Black screen for contrast

HCl vapor

NH_3 vapor

Concentrated hydrochloric acid

Concentrated ammonia

Figure 8.2

In the acid-base reaction of hydrogen chloride and ammonia, a bond between H^+ and Cl^- is broken, and a new bond is formed between H^+ and NH_3. The NH_4^+ and the Cl^- ions are combined in the ionic solid NH_4Cl.

Base   Acid

■ Reactions of Acids and Bases

What structural changes occur when an acid reacts with a base? Consider the following reaction. When fumes from a bottle of hydrochloric acid (a solution of HCl in water) mix with fumes from a bottle containing a solution of ammonia in water, a white fog forms and a powder settles out as shown in figure 8.1. In the reaction, the covalently bonded gases react to form an ionic solid. The reactants are hydrogen chloride and ammonia and the product is ammonium chloride.

$$HCl_{(g)} + NH_{3(g)} \rightarrow NH_4Cl_{(s)}$$

Hydrogen chloride   Ammonia   Ammonium chloride

In this acid-base reaction, the acid HCl donates a proton (the nucleus of a hydrogen atom) to the base NH_3, leaving chloride ion Cl^- as the anion in ammonium chloride. The base NH_3 accepts a proton to form ammonium ion, NH_4^+, which is the cation in ammonium chloride. In this acid-base reaction, a covalent bond between H and Cl is broken, and a new covalent bond is formed between H and N (see Fig. 8.2).

According to a definition formulated by J. N. Brønsted (1879–1947) and Thomas Lowry (1874–1936), *an acid is a proton (hydrogen ion) donor,* and *a base is a proton acceptor.* According to this definition, all acids contain a hydrogen atom that is donated as H^+, and all bases have an unshared pair of electrons to bond to the incoming proton. In every acid-base reaction, a covalent bond is broken and a new covalent bond is formed.

Brønsted and Lowry extended their definitions to the products formed in an acid-base reaction. The species formed from an acid by the loss of a proton is called the *conjugate base* of the acid. Similarly, the species formed when a base gains a proton is called the *conjugate acid* of the base. Consider the reaction of HCl with NH_3 to form NH_4Cl. In this example, Cl^- is the conjugate base of HCl, and NH_4^+ is the

• • • • • • • • • • • • • • • Table 8.1 • • • • • • • • • • • • • • •

Common Strong Acids

Name of Acid	Formula	Conjugate Base	Name of Base
Hydrochloric acid	HCl	Cl$^-$	Chloride ion
Nitric acid	HNO$_3$	NO$_3$$^-$	Nitrate ion
Sulfuric acid	H$_2$SO$_4$	SO$_4$$^{2-}$*	Sulfate ion

*Sulfate ion is formed by the loss in step sequence of two acidic protons from sulfuric acid. Strictly speaking, HSO$_4$$^-$ is the conjugate base of sulfuric acid. However, in dilute solutions of sulfuric acid, HSO$_4$$^-$ reacts extensively with water to form hydronium ions and sulfate ions.

• • • • • • • • • • • • • • • Table 8.2 • • • • • • • • • • • • • • •

Common Strong Bases

Name	Formula
Sodium hydroxide	NaOH
Potassium hydroxide	KOH
Barium hydroxide	Ba(OH)$_2$

conjugate acid of NH_3. Note that the products of acid-base reactions have the potential to act as acids and bases. The conjugate base of every acid has an unshared pair of electrons with which it can accept a proton. The conjugate acid of every base contains a hydrogen atom which it can donate as a proton.

■ Strong Acids and Bases in Water

A solution of the gas hydrogen chloride (HCl) in water is called hydrochloric acid. Hydrochloric acid is a good conductor of electricity since it contains ions formed by the reaction of HCl with water.

$$\underset{\text{Acid}}{HCl_{(g)}} \quad + \quad \underset{\text{Base}}{H_2O_{(l)}} \quad \rightarrow \quad \underset{\text{Hydronium ion}}{H_3O^+_{(aq)} + Cl^-_{(aq)}}$$

Hydrogen chloride reacts as an acid to donate a proton to water. Water is the base in this reaction because it accepts a proton to form a hydronium ion, H_3O^+. Only a few molecules of HCl remain in a dilute solution of hydrogen chloride. Acids that react almost completely with water to form hydronium ions are called *strong acids*. Common strong acids are hydrochloric acid, nitric acid, and sulfuric acid (see Tab. 8.1). Hydronium ions are the most powerful proton donors that can exist in the presence of water. Acid-base reactions of aqueous strong acids are thus reactions of H_3O^+.

The best proton acceptor found in dilute aqueous solutions is the hydroxide ion, OH^-. Bases that dissolve in water or react almost completely with water to form hydroxide ions are called *strong bases*. Sodium hydroxide (NaOH) is the most common strong base (see Tab. 8.2). The bonding in NaOH is ionic, and a solution of sodium hydroxide contains sodium ions and hydroxide ions. Other strong bases are potassium hydroxide and barium hydroxide. If any base stronger than OH^- is added to water, it reacts with the water to remove a proton and form hydroxide ion.

Strong mineral acids and bases occur only in a few places in nature, and their quantities are small and their solutions often dilute. Stomach acid is an example of a strong acid found in living organisms. A dilute solution of hydrochloric acid is excreted into the stomach to assist in the digestion of foods. There it reacts to help unwind proteins and nucleic acids to facilitate enzymatic attack. It also helps in the cleavage of the bonds in carbohydrates and fats. This acid action is not entirely benign, for the formation of ulcers is associated with excess stomach acid and it has been recently found that the hydrochloric acid needed for digestion can cause the loss of tooth enamel in long-term bulimics who frequently regurgitate. Nitric acid, another strong acid, can be formed by a series of reactions that begin when the oxygen and nitrogen in air react (to a small extent) when heated in lightning strikes and fires. This formation of nitric acid is an important source of "fixed nitrogen" for the enriching of soils. Sulfuric acid, formed by the reaction of oxygen in air with certain compounds of sulfur, is found in waters draining from some abandoned coal mines. (The impacts associated with widespread introduction of large quantities of strong acids in the form of acid rain are an important environmental concern which is discussed in chapter 12.)

■ Reactions of Strong Acids with Strong Bases

The reaction of an acid with a base is called a *neutralization* reaction. When solutions of hydrochloric acid and sodium hydroxide are mixed, an acid-base reaction occurs—hydronium ions and hydroxide ions react to form water. In this example, one might say that the strong acid has neutralized the strong base or that the base has neutralized the acid.

$$H_3O^+ + Cl^- + Na^+ + OH^- \rightarrow Na^+ + Cl^- + 2\,H_2O$$

Chemists write *net ionic equations* to emphasize the species that are actually undergoing reaction. If equal molar quantities of hydrochloric acid and sodium hydroxide are mixed, the resulting solution is identical in composition to a solution prepared by dissolving sodium chloride in water. In this reaction, the sour taste of the acid and the bitter taste of the base are replaced by the salty taste of the sodium chloride solution. Sodium ions and chloride ions are present in solution both before and after the acid-base reaction. Since they do not react, they may be omitted from the equation for the reaction. The resulting equation, a net ionic equation, focuses on the chemistry of the reacting species only.

$$H_3O^+ + OH^- \rightarrow 2\,H_2O \text{ (net ionic equation)}$$

■ Weak Acids and Bases

Strong acids and bases play important roles in industry, but for many people, acetic acid in vinegar, citric acid in fruit juice, and the base ammonia in window washing fluid are far more familiar. Solutions of these compounds are poor conductors of electricity because relatively few ions are present. Acetic acid and citric acid are *weak acids,* acids that only react to a small degree with water in order to form hydronium ions and their conjugate bases. Ammonia only reacts to a slight extent with water to form hydroxide ions and ammonium ions; it is considered to be a *weak base.* Most acidic and basic substances react only to a small degree with water. The majority of acids are weak acids, and most bases are weak bases (see Tab. 8.3).

•••••••••••••• **Table 8.3** ••••••••••••••••

Common Weak Acids And Weak Bases*

Name	Formula	Conjugate Base	Name
Phosphoric acid	H_3PO_4	$H_2PO_4^-$	Dihydrogen phosphate ion
Dihydrogen phosphate ion	$H_2PO_4^-$	HPO_4^{2-}	Hydrogen phosphate ion
Hydrogen phosphate ion	HPO_4^{2-}	PO_4^{3-}	Phosphate ion
Acetic acid	$HC_2H_3O_2$	$C_2H_3O_2^-$	Acetate ion
Carbonic acid	H_2CO_3	HCO_3^-	Bicarbonate ion
Bicarbonate ion	HCO_3^-	CO_3^{2-}	Carbonate ion
Ammonium ion	$NH4^+$	NH_3	Ammonia

*The conjugate of a weak acid is a base. All of the species in the right column except dihydrogen phosphate ion form basic solutions in water.

To illustrate the slight reactions of weak acids and bases, consider acetic acid and ammonia as representative examples:

$$CH_3CO_2H + H_2O \rightleftarrows CH_3CO_2^- + H_3O^+$$

Acetic acid   Acetate ion

$$NH_3 + H_2O \rightleftarrows NH_4^+ + OH^-$$

Ammonia   Ammonium ion

Arrows pointing in both directions are used to indicate that the reaction does not go to completion in either direction. Solutions of ammonia in water are sometimes sold under the label "ammonium hydroxide" although they contain far fewer ammonium ions and hydroxide ions than ammonia molecules.

However, acetic acid will neutralize sodium hydroxide, and ammonia will react completely with hydrochloric or sulfuric acid. The reaction of a weak acid with a strong base and the reaction of a weak base with a strong acid are relatively complete:

$$CH_3CO_2H + OH^- \rightarrow CH_3CO_2^- + H_2O \text{ reaction with a strong base}$$

$$NH_3 + H_3O^+ \rightarrow NH_4^+ + H_2O \text{ reaction with a strong acid}$$

The conjugate bases of weak acids are much stronger bases than are the conjugate bases of strong acids. The more weakly an acid donates a proton, the stronger the conjugate base of the acid.

Carbonic acid (H_2CO_3) is in equilibrium with carbon dioxide and water.

$$CO_{2(g)} + H_2O \rightleftarrows H_2CO_3$$

Carbonic acid can donate one proton to water and leave a bicarbonate ion (HCO_3^-) a weak base.

$$H_2CO_3 + H_2O \rightleftarrows HCO_3^- + H_3O^+$$

A second proton can be removed to leave carbonate ion (CO_3^{2-}), a stronger base. The reactions of carbonic acid and bicarbonate ion with hydroxide are:

$$H_2CO_3 + OH^- \rightarrow HCO_3^- + H_2O$$

$$HCO_3^- + OH^- \rightarrow CO_3^{2-} + H_2O$$

Figure 8.3
Acetic acid in vinegar reacts with the base $NaHCO_3$, commonly known as baking soda, to form carbon dioxide.
© Leonard Lessin/Peter Arnold, Inc.

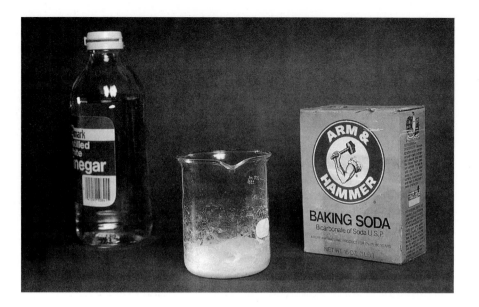

One test used to identify an acid shows that they fizz with baking soda ($NaHCO_3$), and adding acid to a bicarbonate or carbonate compound reverses the reactions above, which results in the evolution of carbon dioxide gas (see Fig. 8.3). The reactions of carbonate ion and bicarbonate ion with acid are:

$$CO_3^{2-} + H_3O^+ \rightarrow HCO_3^- + H_2O$$

$$HCO_3^- + H_3O^+ \rightarrow H_2CO_3 + H_2O$$

$$H_2CO_3 \rightarrow CO_{2(g)} + H_2O$$

Low cost and noncaloric carbonated water has a slightly acidic or sour taste. A small amount of sweetener can give a carbonated beverage an attractive sweet, yet tart, taste. (To hold the tartness even when the bubbles are gone, soft drink producers add phosphoric acid to their syrups.) Cooks use the same principal of combining sweet and sour tastes in sauces. When they use vinegar, a stronger acid, to make their sweet and sour sauces, more sweetener is required.

Why do some indicators, such as those from plant sources, change color in acids and bases? Dyes used as indicators are themselves weak acids or weak bases. The colors of an indicator acid and its conjugate base are different, so when a base reacts with the acidic form of an indicator, it removes a proton to convert the indicator to its conjugate base and produces a color change. Similarly, when an acid donates a proton to the basic form of an indicator, it brings about a different color change.

■ Acids and Bases—A Structural Definition

Pure water or a neutral water solution, one that is neither acidic nor basic, contains equal concentrations of hydronium ions and hydroxide ions. The concentration of each of these ions in pure water is very small, 1×10^{-7} mole/L at 25° C. These ions are present because the acid-base reaction of water with itself occurs to such a small extent.

$$H_2O + H_2O \leftrightarrows H_3O^+ + OH^-$$

In a commonly used definition of acids and bases, a comparison to pure water is made when labeling a solution acidic or basic. An *acidic solution* has an excess of hydronium ions over hydroxide ions; a *basic solution* has an excess of hydroxide ions over hydronium ions (see Fig. 8.4). There is a subtle shift of definition in this

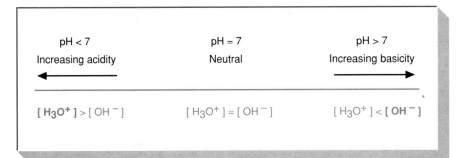

Figure 8.4
In a neutral solution, the concentration of hydronium ions equals the concentration of hydroxide ions. An acidic solution has an excess of hydronium ions, and a basic solution has an excess of hydroxide ions.

• • • • • • • • • • • • • • **Table 8.4** • • • • • • • • • • • • • • • •

The pHs of Some Representative Solutions

pH	$[H_3O^+]$	Solution
0	1.0	1.0 mole/L hydrochloric acid
1	1×10^{-1}	0.1 mole/L hydrochloric acid
2	1×10^{-2}	Lemon juice
3	1×10^{-3}	Vinegar
7	1×10^{-7}	Neutral solution (pure water)
7.4	4×10^{-8}	Blood
9	1×10^{-9}	Baking soda
11	1×10^{-11}	Household ammonia
13	1×10^{-13}	0.1 mole/L sodium hydroxide
14	1×10^{-14}	1.0 mole/L sodium hydroxide

description, for the terms "acid" and "base" were introduced earlier to classify roles played in chemical reactions. The definition of acids and bases that involves comparisons to pure water makes no reference to these reactions; instead this definition is based solely on the concentrations of chemical species present.

It is important to differentiate between the total acid present in a sample and the concentration of hydronium ions present. A strong acid reacts almost completely with water to form hydronium ions and the conjugate base of the acid, so the total concentration of a strong acid is the same as the concentration of hydronium ions. In a solution of a weak acid, both hydronium ions and unreacted acid molecules are present, and the total concentration of a weak acid is greater than the concentration of hydronium ions.

The pH scale is designed to express the concentration of hydronium ions present in a solution, which can vary over a very wide range. Concentrations of H_3O^+ from 1.0 mole/L to 1.0×10^{-14} mole/L are converted to a convenient scale with a range of 0–14 by reporting the negative logarithm of the hydronium ion concentration as pH.

$$pH = -\log[H_3O^+]$$

Each unit of pH corresponds to a tenfold change in concentration of H_3O^+.

A neutral solution has a pH of 7, with the pH of acidic solutions being less than 7, and that of basic solutions greater than 7. The lower the pH of a solution, the higher its acidity. The pHs of some representative solutions are given in table 8.4.

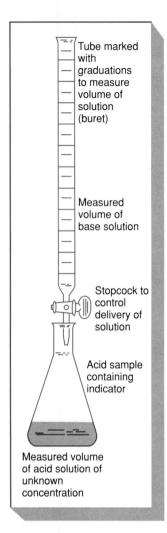

Tube marked with graduations to measure volume of solution (buret)

Measured volume of base solution

Stopcock to control delivery of solution

Acid sample containing indicator

Measured volume of acid solution of unknown concentration

Figure 8.5
In a titration of an acid, an indicator is added to a carefully measured volume of a solution of an acid. To determine the volume of base required to neutralize the acid, a solution of a base is added until the indicator just changes color. If the concentration of the acid is known, the concentration of the base can be calculated. If the concentration of the acid is not known, it can be determined by using a base of known concentration.

■ Titration of Acids and Bases

The total acid in a solution can be measured by *titration* with a strong base as shown in figure 8.5. A measured sample of an acid is titrated by adding a base of known concentration to the sample until the base has reacted with all of the acid present. A small amount of an indicator dye can be used to determine the endpoint because the indicator changes color when just enough base has been added to neutralize the acid. For every mole of protons donated by an acid, a mole of proton acceptors is required.

$$\text{no. of moles}_{acid} = \text{no. of moles}_{base}$$

For example, while one mole of either HCl or $HC_2H_3O_2$ requires one mole of $NaOH$, one mole of H_2SO_4 provides two moles of H_3O^+ that require two moles $NaOH$ for neutralization.

For a solution, the concentration (in moles per liter) of a species multiplied by the volume (in liters) of solution is equal to the number of moles present. The number of moles of the acid and of the base can be calculated by multiplying the concentration of each by the volume of each.

$$V_a C_a = V_b C_b$$

Dividing both sides of the equation by the volume of the acid gives an expression that can be solved for the concentration of the acid:

$$C_a = \frac{V_b C_b}{V_a}$$

Likewise, the total base in a solution can be measured by titration with a strong acid.

Sample Calculation 1

A 25.0 mL sample of a basic solution of unknown concentration is titrated with 0.100 mole/L hydrochloric acid. A total of 20.0 mL of acid is required to neutralize the base. What is the concentration of the base?

$$\text{no. of moles}_{base} = \text{no. of moles}_{acid}$$
$$V_b C_b = V_a C_a$$
$$C_b = \frac{V_a C_a}{V_b}$$

$$C_b = \frac{20.0 \times 10^{-3} \text{ L} \times 0.100 \text{ mole/L}}{25.0 \times 10^{-3} \text{ L}}$$
$$C_b = 0.080 \text{ mole/L}$$

■ Buffers

A titration curve is a graph which shows the pH of a solution of an acid as a function of the volume of added base. The reaction of hydrochloric acid with sodium hydroxide illustrates the reaction of a strong acid and a strong base. When hydrochloric acid is titrated with sodium hydroxide, the pH changes only slightly until almost all of the strong acid has reacted. Near the end point, the pH rises

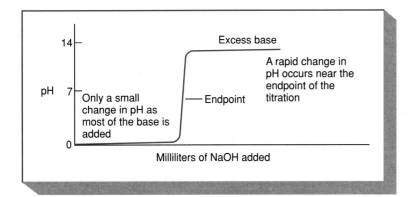

Figure 8.6
The curve obtained when 1.00 *M* hydrochloric acid is titrated with 1.00 *M* sodium hydroxide solution is an example of a titration curve for the reaction of a strong acid with a strong base.

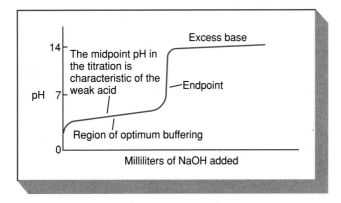

Figure 8.7
The curve for the titration of 1.00 *M* acetic acid with 1.00 *M* NaOH solution is an example of a titration curve for a weak acid with a strong base.

steeply. At the end point, the indicator changes color to signal the end of the titration. If more base is added after all the acid has reacted, the curve begins to level and it approaches the pH of the solution of strong base being added (see Fig. 8.6).

The graph for the titration of a weak acid with a strong base has a significantly different shape. The slowest rate of pH increase is at the halfway point of the titration. When half of the base required to neutralize the acid has been added, the pH of the solution has a value that is characteristic of the weak acid (see Fig. 8.7). At this midpoint pH, the concentration of the weak acid is equal to the concentration of its conjugate base. When the weak acid is acetic acid, this pH is near 4.7, and if the acid is ammonium ion, the pH is about 9.3. The weaker the acid being titrated, the higher the value of the midpoint pH. Note that the pH changes slowly on either side of the midpoint pH in the titration curve for acetic acid. In this middle region of the titration curve, the solution is said to be buffered. *Buffer solutions* change pH only slightly when a small amount of either strong acid or strong base is added.

How does a buffer work? A buffer solution contains about equal concentrations of a weak acid and its conjugate base. When a strong base is added to a buffer solution, the hydroxide ion reacts with the weak acid:

$$HA + OH^- \rightarrow A^- + H_2O \qquad \text{Addition of a strong base}$$

The concentration of the weak acid in the buffer solution is decreased, and the concentration of its conjugate base is increased, but the change in the pH is small. When a strong acid is added to a buffer solution, the hydronium ions react with the weak base:

$$H_3O^+ + A^- \rightarrow H_2O + HA \qquad \text{Addition of a strong acid}$$

The concentration of weak acid increases, but the concentration of hydronium ions increases far less. As long as both the weak acid and its conjugate base are present, the pH of a buffer changes very little.

Both intercellular fluids and blood contain buffers that maintain a pH near neutrality (7.0). These buffer systems play an essential role in living organisms because the enzymes that catalyze metabolic reactions function over narrow regions of pH. In the absence of buffering both strong acids and strong bases rapidly denature these proteins, causing them to lose their biological function. A phosphate buffer maintains a narrow pH range in living cells. It is the weak acid $H_2PO_4^-$ and the weak base HPO_4^{2-} that comprise this buffer that helps maintain a pH of nearly 7 in cells.

$$H_2PO_4^- + OH^- \rightarrow HPO_4^{2-} + H_2O$$
$$HPO_4^{2-} + H_3O^+ \rightarrow H_2PO_4^- + H_2O$$

The buffer system in blood is more complex. In addition to a phosphate buffer, there is a second buffer system involving carbonic acid (H_2CO_3) and bicarbonate (HCO_3^-) as the weak acid and the weak base.

$$H_2CO_3 + OH^- \rightarrow HCO_3^- + H_2O$$
$$HCO_3^- + H_3O^+ \rightarrow H_2CO_3 + H_2O$$

Carbonic acid in the blood is converted to carbon dioxide in water:

$$H_2CO_3 \rightarrow CO_2 + H_2O$$

Blood coming from the lungs has a higher pH than blood returning to the lungs. When we exhale CO_2, the blood becomes less acidic and its pH increases. As carbon dioxide from the oxidation of food enters the blood stream, the blood becomes more acidic and its pH decreases.

■ Periodic Trends in Acid-Base Chemistry

A long-standing generalization in chemistry concerns the oxides of metals and of nonmetals. Oxides of metals, particularly those in columns I and II, are bases in water and the oxides of nonmetals are acids in water.

The purely ionic oxides of metals such as sodium and barium react with water to produce hydroxide ions. Oxide ion O^{2-} is a stronger base than OH^-. Each O^{2-} strips a proton from H_2O to form two OH^- ions. This reaction is so favorable that these oxides react with water vapor if exposed to the atmosphere.

$$Na_2O_{(s)} + H_2O \rightarrow 2\,Na^+ + 2\,OH^-$$
$$BaO_{(s)} + H_2O \rightarrow Ba^{2+} + 2\,OH^-$$

The covalent oxides of many non-metals react with water to form acids. For example, carbon dioxide and sulfur trioxide react with water to form carbonic and sulfuric acids, respectively:

$$CO_{2(g)} + H_2O \rightarrow H_2CO_3$$
$$SO_{3(g)} + H_2O \rightarrow H_2SO_4$$

These reactions are examined in detail later in this chapter.

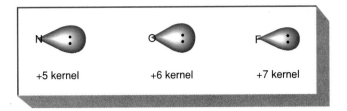

Figure 8.9
The higher the kernel charge, the closer the unshared pair of electrons is held. Ammonia (NH_3) is the strongest base and HF is the weakest because the unshared pair on nitrogen is the most available for sharing and that on fluorine is the least available.

The acid-base behavior of the covalent hydrides of the second row elements, CH_4, NH_3, H_2O, and HF shown in figure 8.8 illustrates a trend occurring across the periodic table. Methane (CH_4) has no acid-base chemistry except under the most exotic conditions. Ammonia (NH_3) reacts as a weak base with water. Hydrofluoric acid (HF) reacts as a weak acid with water. (Note a convention: in an acidic compound containing hydrogen, the hydrogen is usually written first in the formula, but if the compound is not as acidic as water, the hydrogen is not written first.)

In the Lewis structure for each of these compounds, there are eight electrons about the central atom. Methane cannot act as a base because it has no unshared pair of electrons to accept a proton. The other compounds have unshared pairs of electrons and could act as bases. The presence of lone pairs of electrons does not explain the differences between NH_3 and HF; however, the kernel charge does. Going across the periodic table from nitrogen to fluorine, the charge on the kernel of the central atom increases. Because a more positive kernel pulls electrons closer, lone-pair electrons become less available for sharing as shown in figure 8.9. Ammonia is more basic than water because the lone pair on nitrogen extends further and is more available for sharing than a lone pair on oxygen or fluorine. (Note that acid-base chemistry is not simply statistical in nature. It is wrong to think that water is a better base than ammonia because it has two unshared pairs and ammonia has only one.)

To explain the difference in acidity across the periodic table, a similar argument can be made. The higher the kernel charge, the stronger the acid. The +7 kernel charge on a fluorine atom pulls electrons from the hydrogen in HF more than the +6 kernel charge on an oxygen atom does in water. The hydrogen atom in HF is more positive, more exposed, and more available to react with a base than is a hydrogen atom in H_2O.

■ Lewis Acids—Broadening the Scope of Acid-Base Reactions

The reaction of copper ions with ammonia shown in figure 8.10 is an example of the reaction of transition metal ions with bases. When ammonia is added to a pale-blue solution of copper sulfate, the color changes to deep blue. If alcohol is added to the solution, an intensely colored, deep-blue compound precipitates. This compound has the formula $[Cu(NH_3)_4]SO_4$. Ammonia has reacted with copper ions to form a new species responsible for the deep color:

$$Cu^{2+} + 4\,NH_3 \rightarrow Cu(NH_3)_4^{2+}$$

If a strong acid is added to the copper-ammonia solution, the color fades to a pale blue. The hydronium ions of the acid react with the ammonia to form ammonium ions so that ammonia molecules are no longer bonded to copper.

$$Cu(NH_3)_4^{2+} + 4\,H_3O^+ \rightarrow Cu^{2+} + 4\,NH_4^+ + 4\,H_2O$$

Figure 8.10

Both ammonia and water molecules can act as bases and bond to 2+ copper ions. Lone pairs of electrons on the bases are used to form covalent bonds to the Lewis acid.

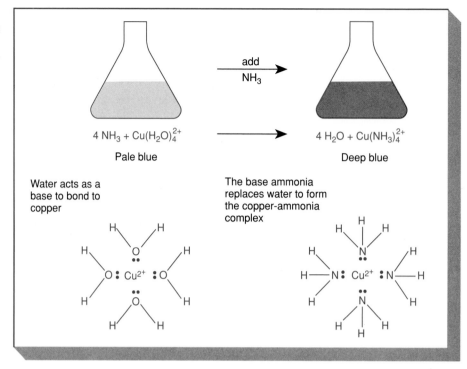

The reaction of Cu^{2+} with NH_3 is called a *Lewis acid-base reaction* in honor of G. N. Lewis, cited in chapter 5, for his contributions to the understanding of bonding. Ammonia has a lone pair of electrons while a copper ion has empty orbitals. In the reaction, the base ammonia donates a share in its unshared pair to the copper ion to form a new covalent bond. This reaction is similar to the acid-base reactions discussed earlier in the chapter, for the base donates an electron pair to form a new covalent bond. Only the acid is different—rather than being a proton donor, the acid accepts a share in an electron pair donated by the base. (Brønsted-Lowry acids then become a subset of Lewis acids, namely, those that donate protons to react with the unshared electron pairs of bases.)

The molecules CO_2 and SO_3 react as Lewis acids with water. With CO_2, a water molecule furnishes an electron pair to form a new covalent bond to the carbon atom of carbon dioxide. Then one of the protons from H_2O shifts to one of the oxygen atoms on the carbon, giving the structure of H_2CO_3. The reaction of water with sulfur trioxide is similar. First H_2O donates an electron pair to the S atom, and then a proton shifts to give the structure of H_2SO_4 (see Fig. 8.11).

Chemists enlarged the class of acid-base reactions to include the reactions of Cu^{2+} with NH_3 and of CO_2 with H_2O rather than consider these reactions to be a completely separate class. This choice to expand the concept of acid-base reactions is characteristic of science. By broadening definitions, scientists seek to emphasize similarities in phenomena and to develop broad unifying models.

■ Strong Acids, Strong Bases, and the Growth of Technology

The number and kinds of acids and bases found in both homes and automobiles provide a window for viewing the chemical changes accompanying industrialization. The properties of sulfuric acid (a strong acid) in a car battery are far different

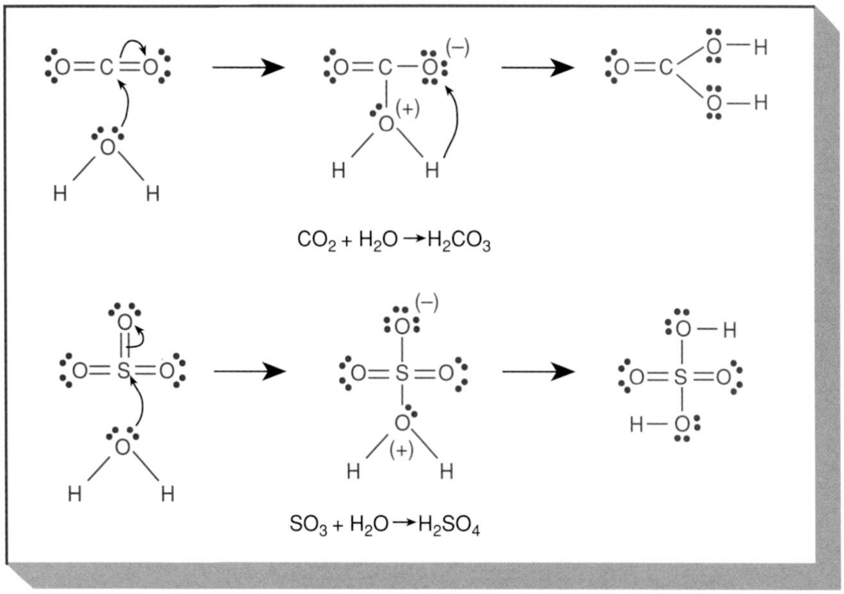

Figure 8.11
Molecules of CO_2 and SO_3 act as Lewis acids in their reactions with water. The base H_2O furnishes an electron pair to form a new covalent bond. After shifts of protons, the acids H_2CO_3 and H_2SO_4 are formed.

than those of the weak acids in fruits and vinegar. Battery acid must be treated with care, for an accidental spill can quickly corrode metal as well as cause chemical burns. In addition to the ammonia in window cleaning solutions and baking soda, there are far stronger bases found in products for cleaning drains or ovens. These caustic products, labeled as poisons and best used wearing gloves and goggles, are not to be tasted. Rather than tingling the taste buds for bitterness, they attack the membrane linings of the mouth and esophagus.

The production of large quantities of strong acids and strong bases was essential to the development and expansion of technology. The quantities of sulfuric acid and sodium hydroxide used in industry provide a measure of the extent of chemical change accompanying industrialization. More sulfuric acid is produced than any other industrial chemical. Among bases, sodium hydroxide is the most abundantly used, and it also ranks in the top ten chemicals in industrial production.

Consider a sampling of the uses of these chemicals in technology. Sulfuric acid is the workhorse of the chemical industry (see Fig. 8.12). It is used in the production of fertilizers for agriculture, in the processing of metals, and in the refining of petroleum. Because it is less volatile than the other common mineral acids, it can be used in their manufacture. For example, the reaction of sulfuric acid with brine is used to make hydrochloric acid. Similarly, sulfuric acid reacts with very insoluble phosphate-containing minerals to give the more soluble hydrogen phosphates used as fertilizers. Sulfuric acid is used to clean metals at many stages of their processing. It also serves as a catalyst for the conversion of low boiling hydrocarbon gases from petroleum refining into high octane components of gasoline. Finally, sulfuric acid is used, together with various bases, in the manufacture of a variety of detergents.

Strong bases are used both in the production of paper from wood pulp and in the manufacture of glass, although the role of the base is quite different in each process. A solution of sodium hydroxide is used in the conversion of wood pulp to paper. Long, strong cellulose fibers in wood are imbedded in a matrix of a complex organic material called lignin. The strong base reacts with the lignin to break it into smaller organic molecules, freeing the cellulose fibers used to produce paper. Sodium hydroxide is not incorporated into the end product, and so the disposal of this basic solution rich in organic compounds derived from lignin poses environmental hazards. The dilution and neutralization of the remaining base is only a part

Figure 8.12
The transportation of sulfuric acid is done by tanker trucks or by (a) railroad tank cars. These vehicles frequently carry (b) a warning sign denoting a corrosive load.
(a) © Jim Shaffer; (b) © Yoav Levy/PhotoTake, Inc.

(a)

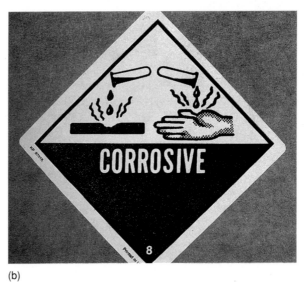

(b)

of the problem. If the basic waste solution is disposed of in lakes or rivers, the reaction of large quantities of organic material derived from lignin with oxygen can deplete downstream waters of this needed oxygen, and cause the death of large numbers of fish.

Reflection

The Disposal of Acids and Bases and Its Environmental Impact

Both the use and disposal of strong acids and bases can generate demands on the environment. Acids and bases are highly concentrated when first manufactured and become progressively diluted or neutralized as used. Many industrial waste streams carry dilute solutions of acids or bases that pose few problems. For these industries, the best and the least expensive approach is that of dilution. Each ten-fold dilution of a strong acid solution decreases the concentration of hydronium ion ten-fold and raises the pH by one unit. Correspondingly, each ten-fold dilution of a base lowers the pH of the basic solution by one unit.

Neutralization, although it sounds desirable, may not be. Consider, for example, the neutralization of strong acids. Sodium hydroxide is too expensive to use, both because of its relatively high monetary cost and because of the adverse environmental impact produced by burning fossil fuels to generate the electrical energy used in its production. Where it is readily available, the less expensive and more abundant base calcium carbonate (limestone) can be used to neutralize acids, but the by-products of its use for neutralization reactions can also pose problems. At best, each neutralization introduces mineral ions needed for plant growth. The accumulation of mineral ions from many sources can convert pristine water into a rich broth for the growth of algae if the water flow is low or the dilution incomplete. The disposal of strong bases presents a more complex chemistry, but a simpler problem. Over long periods of time, carbon dioxide from the air reacts with exposed strongly basic hydroxides to first produce less basic carbonates and then still less basic bicarbonates, partially neutralizing the strong base and lowering the pH of the solution.

Chemical spills of acids or bases in transport and the disposal of mixed waste streams pose particular dangers to the environment. In spills, the risks depend upon the particular compounds involved, the actual quantities, and one's ability to contain the spilled material until treatment is complete. The dilution of large quantities of concentrated acids to the pHs near 6 found in natural waters may require very large amounts of water, and, when chemicals are spilled, dilution is often incomplete. At worst, waste acid and base streams carry with them other, more destructive chemicals that have been introduced in processes such as cleaning metal or digesting wood pulp. These compounds can pose far greater risks to the environment than the acids or bases accompanying them.

Has society entered into a Faustian bargain with chemical technology? The wide use of chemicals such as sulfuric acid and sodium hydroxide is essential to the technologies that characterize developed societies. Yet, with the gains brought by technology, there have come new dangers. Many people blame chemistry for serious environmental problems while failing to recognize or acknowledge that they both want and depend on the products of technology. Did such phrases as "better living through chemistry" promise too much? Was society all too ready to believe that progress carried no price tag? Do the real risks and dangers match those perceived? Is our response to be one of fear and distribution of blame or is it to be one of care and readiness? The answers to these questions are to be determined by our actions now and in the future.

Questions—*Chapter 8*

1. Draw Lewis structures for the reactants and the products of the following acid-base reaction:

$$HCl + H_2O \rightarrow H_3O^+ + Cl^-$$

2. A solution of hydrogen bromide (HBr) is a good conductor of electricity. Hydrogen bromide is a (weak, strong) acid in water. The ions conducting electricity in solution are_____ and_____.

3. Hydrogen cyanide (HCN) is a weak acid in water. A solution of sodium cyanide (NaCN) is (acidic, neutral, basic) because the cyanide ion (CN^-) is (a weak acid, neither an acid nor a base, a weak base.)

4. What is the pH of each of the following solutions?
   (a) 1.0 mole/L hydrochloric acid
   (b) 1.0×10^{-5} mole/L hydrochloric acid
   (c) 1.0 mole/L sodium hydroxide
   (d) 1.0×10^{-4} mole/L sodium hydroxide

5. What is the concentration of hydronium ions (in mole/L) in a solution with the following pH?
   (a) 1.0
   (b) 5.0
   (c) 7.0
   (d) 12.0

6. In the titration of hydrochloric acid solution (initial pH = 1.0) with sodium hydroxide, what fraction of the original concentration of hydronium ions remains when the pH is 2.0? 3.0? 4.0?

7. A 50.0 mL sample of a base is neutralized by 30.0 mL of 0.100 mole/L hydrochloric acid. What is the concentration of the base?

8. What volume of 0.200 mole/L acid would neutralize 40.0 mL of 0.550 mole/L base?

9. If a container of sodium hydroxide solution is left open to the air for a long time, the pH decreases. What minor component of air reacts with hydroxide ion to decrease the pH?

10. When lemon juice is added to tea, the color of the tea lightens. Does tea contain an indicator? Is the indicator molecule an acid, or is it a base? Explain your answer.

11. Write the reaction, if any, that occurs when hydrochloric acid is added to each of the following solutions. When a solution of sodium hydroxide is added? Which of these solutions acts as a buffer?
    (a) NH_3 (an aqueous solution of ammonia)
    (b) NH_4^+ (an aqueous solution of ammonium chloride)
    (c) NH_3 and NH_4^+ (an aqueous solution of ammonia and ammonium chloride)

12. Which of the following oxides react with water to form acidic solutions? Which react with water to form basic solutions?
    (a) K_2O
    (b) MgO
    (c) SO_2
    (d) P_4O_{10}

13. Use arguments based on Lewis structures and the periodic table to identify the stronger base in each of the following pairs:
    (a) PH_3 and H_2S
    (b) SiH_4 and PH_3
    (c) OH^- and F^-

14. Use arguments based on the position of elements in the periodic table to identify the stronger acid in each of the following pairs:
    (a) H_2S and HCl
    (b) NH_3 and H_2O
    (c) NH_4^+ and H_3O^+

15. When young, the author's children delighted in making "volcanos" by adding vinegar to sodium bicarbonate. What chemical reaction did they observe?

16. In the following acid-base reactions, underline the acid. Which of the acids are Lewis acids?
    (a) $H_2CO_3 + H_2O \rightarrow HCO_3^- + H_3O^+$
    (b) $CO_2 + H_2O \rightarrow H_2CO_3$
    (c) $SO_2 + H_2O \rightarrow H_2SO_3$
    (d) $H_2SO_3 + NH_3 \rightarrow HSO_3^- + NH_4^+$
    (e) $Zn^{2+} + 4\,NH_3 \rightarrow Zn(NH_3)_4^{2+}$
    (f) $Zn(NH_3)_4^{2+} + 4\,H_3O^+ \rightarrow Zn^{2+} + 4\,NH_4^+ + 4\,H_2O$

17. Compare hydrogen bonding between water molecules to the acid-base reaction of water molecules to produce H_3O^+ and OH^-. What is the role of an unshared pair of electrons in each case? Of a partially positive hydrogen atom?

18. Cyanide ion (CN^-) binds tightly to Fe^{3+} in a Lewis acid-base reaction. (This reaction is responsible for the toxicity of cyanide-containing compounds.) Predict whether cyanide binds to iron using its unshared pair of electrons on carbon or its unshared pair on nitrogen. Justify your prediction.

9

Oxidation–Reduction Reactions

Technology based on chemical change is central to the history of civilization. From prehistory people have used fire to transform the raw materials of the world. By cooking food, they gained access to concentrated sources of energy—the seeds of plants with their high concentrations of nutrients and the flesh of animals. Fire provided heat to warm caves and shelters and enable people to move to new and harsher climates. With fire, people forged metal tools for planting and harvesting, for hunting, and for aggression and defense. Chemical changes occur in the burning of fuels, in the cooking of foods, and in the forging of metals.

Chemistry plays large roles in both ancient and modern technologies since people have always used chemical reactions as sources of energy. They use fires or electricity to cook their food and have battery-powered flashlights to light their way. The reactions occurring in fires and those used to produce electricity in flashlight batteries are called oxidation-reduction reactions. This important class of chemical reactions includes such diverse phenomena as the rusting of an abandoned car and the metabolic conversion of sugar to carbon dioxide and water. In this chapter, we shall explore various ways to describe and classify these chemical reactions, and we shall consider a variety of their applications.

Lavoisier was the first to recognize that the reactions occurring in combustion involved chemical combination with oxygen. He introduced the term *oxidation* to describe reactions in which molecular oxygen combined with other elements.

$$CH_{4(g)} + 2\ O_{2(g)} \rightarrow CO_{2(g)} + 2\ H_2O_{(l)}$$

$$4\ Fe_{(s)} + 3\ O_{2(g)} \rightarrow 2\ Fe_2O_{3(s)}$$

The term *reduction* was introduced to describe reactions with hydrogen or carbon that converted metal oxides to the free metals.

$$H_{2(g)} + CuO_{(s)} \rightarrow H_2O_{(l)} + Cu_{(s)}$$

$$3\ C_{(s)} + 2\ Fe_2O_{3(s)} \rightarrow 3\ CO_{2(g)} + 4\ Fe_{(s)}$$

Figure 9.1

Zinc metal reacts with copper ions to form zinc ions and copper metal in a redox reaction.

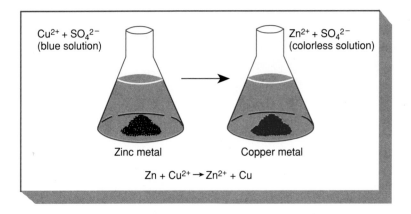

$Cu^{2+} + SO_4^{2-}$ (blue solution)

$Zn^{2+} + SO_4^{2-}$ (colorless solution)

Zinc metal

Copper metal

$$Zn + Cu^{2+} \rightarrow Zn^{2+} + Cu$$

■ Some Chemical Reactions Involve Electron Transfer

When finely divided zinc is added to a solution of copper sulfate, the blue color characteristic of copper ions in solution fades, and a red deposit of copper metal forms as shown in figure 9.1. When the reaction is complete, zinc sulfate can be isolated from the solution by evaporating the water. The reaction of zinc with copper sulfate is described by the following equations:

$$Zn_{(s)} + CuSO_{4(aq)} \rightarrow ZnSO_{4(aq)} + Cu_{(s)}$$

$$Zn_{(s)} + Cu^{2+}_{(aq)} \rightarrow Zn^{2+}_{(aq)} + Cu_{(s)} \text{ net ionic equation}$$

The reaction of zinc metal with a solution of copper sulfate is an electron transfer reaction. Zinc metal donates electrons to the copper ions, and copper ions accept electrons to form copper. A reaction in which electrons are transferred is called an *oxidation-reduction* reaction or *redox* reaction. No student who can remember the mnemonic "LEO the lion says GER" should confuse the definitions of oxidation and reduction—*L*oss of *e*lectrons is *o*xidation, *G*ain of *e*lectrons is *r*eduction. Zinc, the electron donor, is oxidized to form zinc ions, and the copper ions are reduced. Because the zinc brought about the reduction of copper ions, zinc is said to be the *reducing agent*. Note that the reducing agent was oxidized in the reaction. Copper ions acted as the *oxidizing agent* for the zinc and were themselves reduced.

The reaction of zinc with a strong acid to produce hydrogen is also a redox reaction.

$$Zn_{(s)} + 2 H_3O^+ \rightarrow Zn^{2+} + H_{2(g)} + 2 H_2O$$

Since zinc loses electrons, it is oxidized. Since every electron donor must give its electrons to an electron acceptor, hydronium ions must be reduced.

■ Oxidation Numbers

The use of oxidation numbers, a formal chemical convention, provides a more general system for describing redox reactions. Oxidation numbers are assigned to elements themselves and to elements in compounds. Reactions that involve changes in oxidation numbers are then classified as oxidation-reduction (redox) reactions.

A simplified list of rules for assigning oxidation numbers follows:

1. The oxidation number of an uncombined element is zero.
2. The oxidation number of a monatomic ion is the charge on the ion.
3. Oxygen has an oxidation number of −2 in its compounds.
4. Hydrogen has an oxidation number of +1 in its compounds with non-metals.
5. The sum of the oxidation numbers of the atoms in a neutral compound is zero.
6. The sum of the oxidation numbers of the elements in a polyatomic ion is equal to the net charge on the ion.

Sample Calculation 1

Find the oxidation number of N in HNO_3.

First note that according to the simplified list of rules, the oxidation numbers of both H (rule 4) and O (rule 3) are fixed. Since HNO_3 is neutral, the sum of the oxidation numbers of the atoms is zero according to rule 5. We can write an equation for the sum of the oxidation numbers in HNO_3 and solve for the oxidation number of nitrogen.

$$\text{Sum of the oxidation numbers} = 0$$
$$H + N + 3 \times O = 0$$
$$+1 + N + 3 \times (-2) = 0$$
$$N - 5 = 0$$
$$N = +5$$

Some examples of oxidation numbers are:

Ion or Compound	Oxidation Numbers	Rules Applied
Cu	Cu = 0	1
Cu^{2+}	Cu = +2	2
H_2O	H = +1, O = −2	3, 4
H_3O^+	H = +1, O = −2	3, 4, 6
NO_3^-	N = +5, O = −2	3, 6
HNO_3	H = +1, N = +5, O = −2	3, 4, 5
KCl (K^+/Cl^-)	K = +1, Cl = −1	2, 5
$K_2SO_4(2K^+/SO_4^{2-})$	K = +1, S = +6, O = −2	2, 3, 5
Fe_3O_4	Fe = +8/3, O = −2	3, 5

■ Redox Reactions

By definition, a reaction involving a change in oxidation numbers is a redox reaction. Oxidation is an *increase* in oxidation number, and reduction is a *decrease* in oxidation number. The use of oxidation numbers serves to broaden the definition of oxidation and reduction, for reactions in which there is no apparent electron transfer may also be classified as involving oxidation and reduction.

Consider the reactions first used to introduce the terms oxidation and reduction. In the burning of methane, carbon is oxidized from the −4 oxidation state to the +4 state, and oxygen is reduced from the 0 oxidation state to the −2 state:

$$CH_{4(g)} + 2\,O_{2(g)} \rightarrow CO_{2(g)} + 2\,H_2O_{(l)}$$

In the rusting of iron, the metal goes from the 0 oxidation state to the +3 state, and oxygen goes from the 0 oxidation state to a −2 state:

$$4\,Fe_{(s)} + 3\,O_{2(g)} \rightarrow 2\,Fe_2O_{3(s)}$$

The burning of methane in natural gas and the rusting of iron are both redox reactions. The conversion of metal oxides (in which metallic elements have positive oxidation numbers) to the free metals (with oxidation numbers of zero) are also redox reactions:

$$H_{2(g)} + CuO_{(s)} \rightarrow H_2O_{(l)} + Cu_{(s)}$$

$$3\,C_{(s)} + 2\,Fe_2O_{3(s)} \rightarrow 3\,CO_{2(g)} + 4\,Fe_{(s)}$$

Note that the name of a reactant may not indicate whether a given reaction is redox or acid-base. For example, dilute sulfuric acid may react either as an acid or as an oxidizing agent. The following equations illustrate the two types of reactions:

$$H_3O^+ + OH^- \rightarrow 2\,H_2O_{(l)} \quad \text{acid-base}$$

In the acid-base reaction of H_3O^+ with OH^-, there is no change in oxidation number.

$$2\,H_3O^+ + Zn_{(s)} \rightarrow H_{2(g)} + Zn^{2+} + 2\,H_2O_{(l)} \quad \text{redox}$$

In the redox reaction of H_3O^+ with Zn, hydrogen changes from the +1 oxidation state in hydronium ion to the 0 oxidation state in H_2, and zinc goes from the 0 oxidation state to the +2 state.

■ Redox Reactions and the Periodic Table

The elements at the far left side of the periodic table readily donate electrons in redox reactions. The outer electrons of these metals are weakly held because they are relatively far from kernels with charges of +1 or +2. Sodium, potassium, barium, and calcium are all excellent, but expensive and dangerously reactive, reducing agents.

Among the elements, the strongest oxidizing agents are fluorine, oxygen, and chlorine, all found in the upper right portion of the periodic table. Atoms of these elements are small and have high kernel charges. These elements readily accept additional electrons from reducing agents.

The chemistry of the halogen family can be used to illustrate redox trends within a family of the periodic table. Chlorine water, a solution of chlorine in water, can be prepared by mixing dilute hydrochloric acid with common household bleach, a solution of sodium hypochlorite. Note that the chlorine is both oxidized and reduced in this reaction. Chlorine has a −1 oxidation number in Cl^-, +1 in OCl^-, and 0 in Cl_2.

$$2\,H_3O^+ + Cl^- + OCl^- \rightarrow 3\,H_2O + Cl_2$$

When chlorine water is added to a solution of sodium bromide, the chlorine oxidizes bromide ions to bromine and is itself reduced to chloride ions:

$$Cl_2 + Br^- \rightarrow 2\,Cl^- + Br_2$$

When either chlorine water or bromine water is added to a solution of sodium iodide, the iodide ions are oxidized to elemental iodine:

$$Cl_2 + 2\,I^- \rightarrow 2\,Cl^- + I_2$$
$$Br_2 + 2\,I^- \rightarrow 2\,Br^- + I_2$$

Color changes accompanying these reactions can be enhanced by extracting the halogen into chloroform or carbon tetrachloride. In these solvents, chlorine is a faint yellow, bromine is red, and iodine is purple.

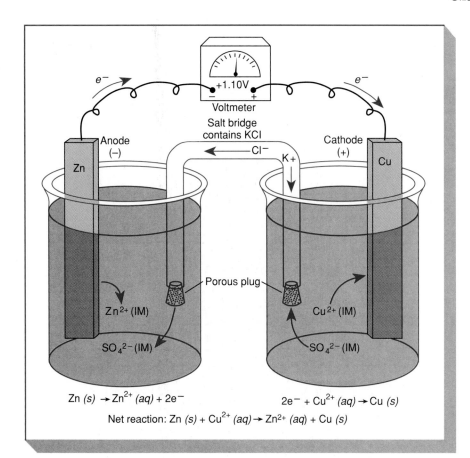

Figure 9.2
In a voltaic cell, the oxidation and reduction steps of a redox reaction are separated. Electrons from the oxidation of zinc flow through the external circuit to reduce cupric ions and plate out copper. A connection through a salt bridge is necessary to complete the electric circuit.

The smaller a halogen molecule, the more readily it accepts added electrons. Fluorine (F_2) is the strongest oxidizing agent of the halogens. The larger the halide ion, the less tightly the outer electrons are held. Thus I^- is more easily oxidized than Br^- or Cl^-.

■ Voltaic Cells

In a voltaic cell, the oxidation and reduction steps of a redox reaction are separated. In the cell shown in figure 9.2, the solution containing Cu^{2+} is not in contact with the zinc, so the reaction of Cu^{2+} with Zn cannot occur directly. Reduction occurs at the cathode and oxidation occurs at the anode.

$$Cu^{2+} + 2\ e^- \rightarrow Cu_{(s)} \quad \text{cathode reaction}$$

$$Zn_{(s)} \rightarrow Zn^{2+} + 2\ e^- \quad \text{anode reaction}$$

Electrons produced at the anode by the oxidation of zinc go through an external circuit to the cathode where they reduce copper ions and plate out copper.

A connection through a salt bridge is necessary to complete the electric circuit. A solution of a salt such as KCl permits the motion of ions between the two solutions to balance the flow of electrons through the external circuit. Cations always flow toward the cathode, anions toward the anode. The connection of an electrode in contact with a readily reduced metal ion in solution to an electrode in contact with an easily oxidized metal produces a voltage. That voltage is a measurement of the combined strengths of the oxidizing and reducing agents. Because the voltage

Figure 9.3
The Leclanche dry cell is widely used in flashlight batteries. The voltage developed by this cell is +1.5 V.

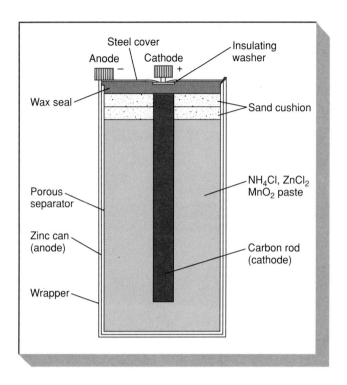

of a cell depends on the concentrations of ions in solution, standard conditions are chosen for comparisons. When the idealized concentrations of copper sulfate and zinc sulfate are both 1.00 M, the voltage of the illustrated cell is +1.10 V.

■ Dry Cells

Familiar flashlight batteries and the small disc batteries used to power calculators and watches are examples of dry cells, in which redox reactions produce an electric current at a nearly constant voltage. A variety of dry cells are in use; two common cells are discussed here.

The best known dry cell is the common flashlight battery, a Leclanche cell shown in figure 9.3. A zinc jacket serves as the anode and is oxidized. The cell is filled with a paste containing manganese dioxide and ammonium chloride. (Dry cells are not really dry, for they contain water in a paste to allow the movement of ions to complete the circuit.) A carbon rod serves as an electrode, and manganese dioxide is reduced on discharge. Until the reactants are consumed, the Leclanche cell has a constant potential of 1.5 V.

$$Zn_{(s)} \rightarrow Zn^{2+} + 2\ e^-$$

$$2\ e^- + 2\ MnO_{2(s)} + 2\ NH_4^+ \rightarrow Mn_2O_{3(s)} + 2\ NH_3 + H_2O$$

These equations are a simplification, for one would expect a buildup of ammonia and Zn^{2+} ions. Ammonia and zinc ions combine in a Lewis acid-base reaction, and insoluble zinc compounds are produced in reactions not involving redox.

The small button-shaped dry cell batteries have mercuric oxide and zinc as electrodes and potassium hydroxide as the electrolyte. The compact cells deliver current at 1.35 V.

$$Zn_{(s)} + 2\ OH^- \rightarrow ZnO_{(s)} + H_2O + 2\ e^-$$

$$2\ e^- + HgO_{(s)} + H_2O \rightarrow Hg_{(s)} + 2\ OH^-$$

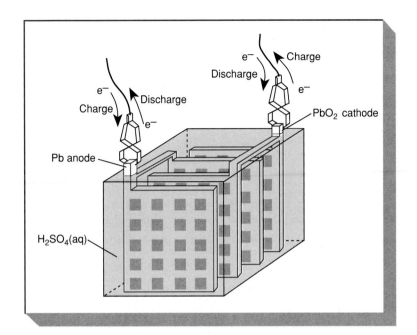

Figure 9.4
A lead storage battery consists of a series of cells. Each cell produces a 2.0-V potential. In each cell, there are parallel plates of large surface area in which a honeycomb grid is packed with either lead or lead dioxide. A solution of sulfuric acid provides ions to carry a current through the solution. As the battery discharges, lead sulfate builds up on the surface of the plates.

■ Rechargeable Electrochemical Cells—Lead Storage Batteries

Lead storage batteries are widely used as an energy source in automobiles. Current from the battery is used to start the engine, and part of the energy from the running engine is then used to recharge the battery. Lead storage batteries provide a high current at a constant voltage, and because they are rugged and can be recharged, they have a long, useful lifetime.

A lead storage battery consists of a series of cells. Each cell has a 2.0V potential, so that a 12V battery consists of six cells. In a cell, there are parallel plates of large surface area in which a honeycomb grid is packed with either lead or lead dioxide. The anode plates are filled with lead (Pb), and the cathode plates are filled with lead dioxide (PbO_2). A dilute aqueous solution of sulfuric acid (SO_4^{2-}) fills the cells (see Fig. 9.4).

As a cell discharges, producing current, the following half reactions occur:

$$Pb_{(s)} + SO_4^{2-} \rightarrow PbSO_{4(s)} + 2\ e^-\ \text{anode reaction}$$

$$2\ e^- + PbO_{2(s)} + 4\ H_3O^+ + SO_4^{2-} \rightarrow PbSO_{4(s)} + 6\ H_2O\ \text{cathode reaction}$$

$$Pb_{(s)} + PbO_{2(s)} + 2\ SO_4^{2-} + 4\ H_3O^+ \rightarrow 2\ PbSO_{4(s)} + 6\ H_2O\ \text{net reaction}$$

Lead sulfate ($PbSO_4$) forms on both sets of plates and the concentration of sulfuric acid decreases. The density of the sulfuric acid solution in a cell can be used as a measure of the charge on a battery, for as the battery discharges, the acid solution becomes more dilute and less dense.

By applying an external voltage greater than 2.0 volts per cell, the flow of electrons and the chemical reactions occurring at each electrode can be reversed. As long as the lead sulfate clings tightly to each electrode, the battery can be recharged. With age and use, some $PbSO_4$ flakes and falls to the bottom. When this occurs in a battery, not as much Pb and PbO_2 can form during the recharging, and the battery cannot supply as much energy between recharges.

Because heat is produced during the charging and discharging of a battery, some water may evaporate from the cells. Distilled water should be used to restore the fluid level, for some tap waters contain impurities that can contribute to the loss of charge on a battery. The use of water containing iron impurities can lead to the loss of charge in a battery since iron, a transition metal, can have either the +2 or +3 oxidation state in solution. In a lead storage battery, the ferric ion (Fe^{3+}) gains an electron to form ferrous ion (Fe^{2+}):

$$2 \, Fe^{3+} + Pb_{(s)} + SO_4^{2-} \rightarrow 2 \, Fe^{2+} + PbSO_{4(s)}$$

Fe^{2+} can carry that electron through solution to the lead dioxide electrode where it is deposited:

$$2 \, Fe^{2+} + PbO_{2(s)} + SO_4^{2-} + 4 \, H_3O^+ \rightarrow 2 \, Fe^{3+} + PbSO_{4(s)} + 6 \, H_2O$$

The regenerated ferric ion can then return to pick up another electron. Because the redox reactions of Fe^{2+} and Fe^{3+} can occur over and over again, small amounts of iron salts will discharge a battery. This is known as the "ferriboat" process for battery discharge.

■ The Photographic Process

The light-induced redox reduction of silver halides is the basis of the photographic process. Photographic film contains an emulsion of insoluble, fine-grained silver bromide that is activated for reduction upon exposure to light. Activated silver bromide crystals, symbolized by $AgBr°$, are more easily reduced to silver metal than are unactivated crystals. The greater the number of photons absorbed by a grain of $AgBr$, the more readily it is reduced:

$$AgBr_{(s)} + light \rightarrow AgBr°_{(s)} \text{ activation}$$

In the development process, the latent image is converted to a negative. By using a very mild reducing agent, selective reduction of only the activated silver bromide produces a dark image consisting of finely divided silver. Silver atoms present in activated crystals catalyze the further reduction of $AgBr$:

$$AgBr°_{(s)} + developer \rightarrow Ag_{(s)} + Br^- + oxidized \ developer$$

To preserve the image, unreacted $AgBr$ must be removed, for it would darken during continued exposure to light. Photographers use a solution of sodium thiosulfate ($Na_2S_2O_3$) to stop the development reactions and to fix the image on the negative. (An old name for sodium thiosulfate is sodium hyposulfite and the fixing solution is sometimes called "hypo.") A Lewis acid-base reaction with $S_2O_3^{2-}$ ions dissolves unreacted $AgBr$ so it can be washed away.

$$AgBr_{(s)} + 2 \, S_2O_3^{2-} \rightarrow Ag(S_2O_3)_2^{3-} + Br^-$$

The photographic process is repeated to make a print that reverses the image in the negative. When light is passed through the negative onto film, the dark regions of the negative block light and give rise to light regions on the print. The development and fixation steps are repeated to complete the process (see Fig. 9.5).

■ Redox Reactions For Making Sulfuric Acid and Sodium Hydroxide

Sulfuric acid is the most widely used strong acid; indeed, it is the largest volume product of the chemical industry. The manufacture of sulfuric acid starts with the direct oxidation of sulfur by oxygen to give sulfur dioxide:

$$S_{(s)} + O_{2(g)} \rightarrow SO_{2(g)}$$

(a)

(b)

Sulfur dioxide is then oxidized using a catalyst, vanadium pentoxide (V_2O_5) to form SO_3:

$$2 SO_{2(g)} + O_{2(g)} \rightarrow 2 SO_{3(g)}$$

The reaction of sulfur trioxide (SO_3) with water gives sulfuric acid:

$$SO_{3(g)} + H_2O \rightarrow H_2SO_{4(l)}$$

In practice, sulfur trioxide is first dissolved in concentrated sulfuric acid. The resulting solution is cooled and then reacted with water.

The most widely used strong base is sodium hydroxide. Its manufacture requires large amounts of electrical energy, for it is prepared by the electrolysis of *brine,* a concentrated salt solution. Chlorine, an important oxidizing agent, is produced at the anode, and sodium hydroxide and hydrogen gas are produced at the cathode:

$$2 Cl^- \rightarrow Cl_{2(g)} + 2e^- \quad \text{anode reaction}$$

$$2 e^- + 2 Na^+ + 2 H_2O \rightarrow H_{2(g)} + 2 Na^+ + 2OH^- \quad \text{cathode reaction}$$

It is desirable to keep the products of electrolysis separated in order to obtain pure hydrogen and chlorine and to avoid dangerous mixtures of these gases. A second reason for this separation is that sodium hydroxide and chlorine react with one another to form sodium hypochlorite (NaOCl) or bleach:

$$2 Na^+ + Cl_{2(g)} + 2 OH^- \rightarrow Na^+ + OCl^- + Na^+ + Cl^- + H_2O \quad \text{bleach formation}$$

One way the separation is accomplished involves the use of liquid mercury as a barrier as shown in figure 9.6. In these electrolytic cells, sodium ions are reduced at the mercury's surface to produce sodium amalgam, a solution of sodium metal in liquid mercury. Sodium amalgam then reacts with fresh water to form NaOH and hydrogen gas:

$$2 Na^+ + 2 e^- \rightarrow 2 Na_{(Hg)} \quad \text{sodium amalgam}$$

$$Na_{(Hg)} + 2 H_2O \rightarrow 2 Na^+ + 2 OH^- + H_{2(g)}$$

Figure 9.5
The light and dark areas of (a) a photographic negative and (b) its positive print are reversed.
© Jim Shaffer

Figure 9.6
The mercury cell for the electrolysis of brine separated the production of chlorine from the production of sodium hydroxide by using liquid mercury as a barrier. The loss of mercury from these cells was followed by the production of toxic methyl mercury by microorganisms living in sediments where waste water was discharged.

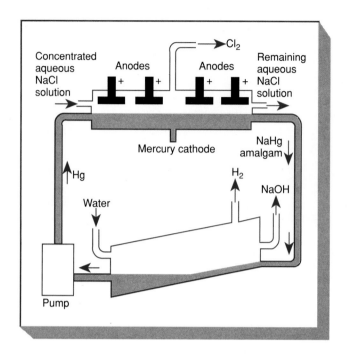

■ Methyl Mercury—A Toxic By-Product of Technology

The chemical industry discovered that over the years mercury was lost from the electrochemical cells used to produce chlorine and sodium hydroxide. Periodically, more mercury was added. While this was of some economic concern, people were not alarmed even though mercury compounds were known to be poisonous. It was common knowledge that mercury is an unusually unreactive element and that mercurous chloride, an expected product of the reaction of chlorine and mercury, is exceedingly insoluble.

It was, therefore, a most unpleasant surprise when high mercury levels were found in fish caught downstream from chlor-alkali plants. Scientists discovered that microorganisms at the bottom of the rivers and lakes had converted the mercury to a soluble compound, methyl mercury, that entered the food chain and accumulated in the fat tissue of fish. Because of this mercury pollution, fish caught from these streams and lakes are unfit for human consumption.

The story of methyl mercury pollution is a sobering one. A seemingly benign technology was suddenly discovered to have a substantially adverse environmental impact. It was not enough to assume a process was safe because there were no known risks. Nor could technology itself be rejected, for it was only through scientific inquiry that the injection of mercury compounds into the food chain was discovered. Neither ignorance nor the rejection of scientific inquiry is an acceptable approach to controlling technology.

■ *Reflection*

A Chemical Creation Story

The earth is a rich and varied chemical laboratory. The sun supplies
energy to heat the world and to make photosynthesis possible. The earth's
gravity retains an atmosphere rich in oxygen. Water, required to sustain
life, falls to the earth as rain and flows in rivers to the seas. With life come
both simple and complex organic compounds and a rich variety of
biochemical transformations. The world's temperatures and atmosphere,
the existence and composition of seas, and the forms of life have changed
over geological time. With the changes wrought by geological forces and
living organisms, the chemistry possible on earth has changed as well. To
consider the role of technology in the course of human prehistory and
history, it is useful to begin with a description of an even earlier creation.

Astronomers tell us that the solar system was formed by the
condensation of gases. At an early stage of coalescence, much of the
elemental composition of the earth was fixed—those elements we call
metals and those called nonmetals were already present. However, the
relative quantities of the various elements were different then than now,
for in its early history the earth lost large portions of the gaseous elements
present in its atmosphere, particularly the lightest element, hydrogen.
There were other, smaller changes. Quantities of radioactive elements
decreased as they decayed over the earth's history of more than 4.5 billion
years. Impacting meteors and meteorites rained down other elemental
particles.

As the earth cooled, the distribution of elements became segregated.
Large quantities of the metals nickel and iron formed the core of the
molten planet. Rocks containing complex combinations of various metallic
and nonmetallic elements floated on the liquid core. The earth was
surrounded by an atmosphere of nonmetal elements and their compounds.
Only later, as the surface of the earth continued to cool, did water appear.

The processes of geology produced our mineral resources from the
structural elements of chemistry. The forces of geology shuffled atoms of
elements of metals and of nonmetals into various compounds and
concentrated them. Throughout geological time, the combinations of
elements and the distribution of compounds continued to evolve. In the
interior of the earth, rocks crystallized in succession from molten magmas,
separating minerals one from another, and concentrating them in veins of
varying composition. The processes of crystallization first selected the
combination of elements that formed a crystalline compound having the
highest melting temperature and then another combination having a lower
melting temperature, and still later, another (see Fig 9.7). At high
temperatures and pressures, trapped superheated water shuffled elements
by dissolving some compounds and then depositing others when the
trapped water escaped. On the surface of the earth, the processes of
weathering turned rocks into soils of various compositions and erosion

Figure 9.7
This close up of granite shows the different colored grains that result from crystallization processes.
© Doug Sherman/Geofile

deposited those soils in floodplains, river deltas, and inland seas. Flowing water dissolved salts and carried them to the seas, and then, over time, these salts were concentrated and deposited when shallow seas evaporated (see Fig. 9.8).

During recent decades, the theory of plate tectonics has enabled geologists to unify the historic record of geological change with ongoing processes of this change. The processes of mountain-building, volcanism, and the formation of new seafloor continue to compete with the forces of erosion in shaping and reshaping the land. As one plate sinks below another, the rocks and soils in the subsiding plate can again be turned to magma to enter into a new cycle of crystallization or volcanic mountain building. And where plates spread apart, magma moves up to create a new ocean floor.

The actions of living organisms did much to create and store fossil fuels and oxygen. Both the charcoal or coal and the oxygen consumed by fire are ultimately the products of photosynthesis. Neither fossil fuels nor atmospheric oxygen were present when life first appeared on the planet more than three billion years ago, for the atmosphere had a far different composition than at present. The element oxygen was found in combined states in minerals in the earth, in water on its surface, and in carbon dioxide gas in the atmosphere. It was only with the advent of algae and green plants, and with them, the processes of photosynthesis, that oxygen gas appeared in the atmosphere. Animals, requiring oxygen in the air they breathe, could begin to evolve only after plants had produced an atmosphere rich in oxygen.

The growth and decay of plants are part of a cycle. Green plants capture energy from the sun to drive a series of chemical reactions that convert the plant nutrients carbon dioxide and water to sugars and oxygen:

$$6 \, CO_{2(g)} + 6 \, H_2O \rightarrow C_6H_{12}O_{6(s)} + 6 \, O_{2(g)}$$

These sugars, in turn, are converted to the cellulose of wood and leaves, and to the starches of seeds and tubers. Of the many complex reactions that are to be found in living organisms, those of photosynthesis are the

Figure 9.8
Collecting salts by evaporation at a Great Salt Lake site.
© Doug Sherman/Geofile

most widespread, and the enzymes that catalyze photosynthetic reactions are among the most abundant. When a forest burns, grains are eaten, or plants die and decay, the overall result of photosynthesis is reversed. Starch and cellulose react with oxygen to reform carbon dioxide and water. The cycle described in the sprouting of an acorn, the growth of an oak, the death of the tree, and its final decay contributes no net change to the amounts of oxygen or carbon dioxide in the atmosphere.

In billions of years of biological history, however, the pattern is not simply a cycle of death and renewal. The processes of growth and decay have not been in balance—far more energy from the sun has been captured and stored in the growth of plants than has been lost by decay. That excess energy from the sun remains stored as chemical energy, preserved in the form of reactive chemicals—atmospheric oxygen and fossil fuels.

Coal and natural gas are the chemical skeletons of the trees and forests of long ago. Oil, believed to have been formed from decaying marine microorganisms, has molecules that echo intricate patterns found in molecules present in organisms living today. In the formation of coal, the accumulation of oxygen from photosynthesis was accompanied by the accumulation of partially decayed plant material. In bogs and swamps, water covered fallen plant material and slowed the reactions of decay. Organic material accumulated, and slow-moving waters were depleted of oxygen to further disrupt the decay of organic material. Complex reactions slowly converted the organic material, first to peat, and then to coal.

How might we relate this chemical view of creation and the recognition of the inherent human participation in technology with a religious view of creation? If the use of technology has always played an integral role in human experience, it may be appropriate to reconsider and reinterpret the biblical creation stories. Consider two creation stories found in Genesis. In the first and more familiar story, God creates the world in six days and rests on the seventh. Man is created in the image of God, and he is given dominion over the earth. In the second creation story, God forms man from the dust of the earth. In this account, man is materially one with both the inanimate world and the animate, and he is charged to be their steward.

In the book of Genesis, Adam and Eve are cast out of Eden after having eaten of the fruit of the knowledge of good and evil. Science has changed our description of the world and has extended our abilities to manipulate materials using technology, yet it has not changed the nature of the human predicament. We live in an imperfect world, and we are called to be responsible for our actions. Again and again, as if in a dialectic process, the solution to one problem has brought with it a new, sometimes smaller, and sometimes larger, problem. Science deepens our knowledge. Science informs and extends the technology that is a part of man's participation in the world. It may be that science necessarily is the knowledge that can be used for both good and evil and that technology necessarily is the power that can be used for good and evil. In the future, will we only strive to have dominion over creation, or will we also seek to be its good stewards?

■ Questions—*Chapter 9*

1. Calculate the oxidation number of the indicated element in each of the listed compounds or ions:
   (a) P in PF_5, PO_4^{3-}, PCl_3
   (b) S in SO_4^{2-}, H_2SO_3, $Na_2S_2O_3$, H_2S
   (c) Mn in Mn, Mn^{2+}, MnO_2, MnO_4^-
   (d) Cl in HCl, Cl^-, OCl^-, ClO_4^-
2. When Na and Cl_2 react to form NaCl, Na is (oxidized, reduced) and Cl_2 is (oxidized, reduced).

3. In the following equations for redox reactions, underline the species being oxidized. Circle the species being reduced. (The equations for the reactions are not balanced.)

   (a) $CH_{4(g)} + O_{2(g)} \rightarrow CO_{2(g)} + H_2O$

   (b) $Cl_{2(g)} + Fe^{+2} \rightarrow Cl^- + Fe^{3+}$

   (c) $H_3O^+ + Fe^{2+} + MnO_4^- \rightarrow Fe^{3+} + Mn^{2+} + H_2O$

   (d) $Br_2 + I^- \rightarrow Br^- + I_2$

   (e) $Cl_{2(g)} + OH^- \rightarrow Cl^- + OCl^- + H_2O$

4. The compound sodium hydride is ionic; it is a solid consisting of both Na^+ and H^- ions. Sodium hydride reacts with water to form hydrogen gas:

$$NaH + H_2O \rightarrow NaOH + H_{2(g)}$$

   The reaction of sodium hydride with water can be classified as _____.

   (a) An acid-base reaction

   (b) A redox reaction

   (c) Both an acid-base and a redox reaction

5. Which of the coinage metals is most easily oxidized? Least easily oxidized?

6. Which of the following reactions involves oxidation-reduction?

   (a) $Al_{(s)} + H_3O^+ \rightarrow Al^{3+} + H_{2(g)} + H_2O$

   (b) $Zn^{2+} + NH_3 \rightarrow Zn(NH_3)_4^{2+}$

   (c) $H_2CO_3 + NH_3 \rightarrow HCO_3^- + NH_4^+$

   (d) $Fe_2O_{3(s)} + CO_{(g)} \rightarrow FeO_{(s)} + CO_{2(g)}$

   (e) $SO_{2(g)} + H_2O \rightarrow H_2SO_{3(l)}$

   (f) $H_2SO_3 + O_2 \rightarrow H_2SO_4$

7. Alchemists found that *aqua regia*, a mixture of hydrochloric acid and nitric acid would oxidize and dissolve gold, even though neither acid acting alone would do so:

$$Au_{(s)} + 4\,H_3O^+ + 4\,Cl^- + NO_3^- \rightarrow AuCl_4^- + NO_{(g)} + 6\,H_2O$$

   Answer the following questions concerning this reaction.

   (a) What is the oxidation state of gold in $AuCl_4^-$?

   (b) Does the nitric acid or the hydrochloric acid act as the oxidizing agent?

   (c) What ion acts as a base toward gold in its positive oxidation state?

8. Why is oxygen a stronger oxidizing agent than sulfur?

9. Why is an iodide ion more easily oxidized than a chloride ion?

10. Why would current stop flowing in the voltaic cell shown in Figure 8.2 if the salt bridge was removed?

11. In a Leclanche dry cell, current is carried by the movement of NH_4^+ and Cl^- ions. Immediately after discharge, the concentration of _____ ions is higher near the zinc electrode and the concentration of _____ ions is higher near the MnO_2 electrode.

12. Batteries differ in the quantity of lead and lead dioxide in each cell. Cells containing greater amounts of these compounds should provide (a greater voltage, more current) at constant voltage.

13. As a lead storage battery discharges, the quantity of lead in the anode (decreases, increases), the lead oxide in the cathode (decreases, increases), the lead sulfate at each electrode (decreases, increases), and the concentration of sulfuric acid (decreases, increases).

14. How would a photographic image be changed if too weak a reducing agent is used in an attempt to develop it? If too strong a reducing agent is used?

15. Sulfur compounds occur in some petroleums and coals. What chemistry might occur after sulfur dioxide is introduced into the atmosphere when these fuels are burned?

16. When an electric current is passed through molten NaCl, _____ is oxidized and _____ is reduced.

17. Brine can be used for the production of hydrochloric acid or sodium hydroxide.
    (a) What is the reaction for making hydrochloric acid from brine?
    (b) What is the reaction for making sodium hydroxide from brine?
    (c) Which of these reactions requires the larger energy input?

10

Reaction Rates and Pathways

Wheat is oxidized, but the speed with which the reaction occurs varies greatly. Ground into flour or stored in elevators, grain is insurance against lean years and famine. When eaten, its oxidation occurs by metabolic pathways, fueling our bodies under benign physiological conditions. Yet as a finely divided dust in the air, it has sometimes exploded, leveling multistory grain elevators. What factors govern the speeds at which reactions occur? The design of explosives and the preservation of food present practical problems in the control of rates of reactions. The solutions to these and similar problems often depend on an understanding of factors that affect the rates of reactions.

How do reactions occur? The study of reactants and products tells no more about the path of a reaction than a study of the building materials and the finished house tells about how the house is built. Scientists study changes in the speeds of reactions as they vary the quantities of reacting substances to gain insight into reaction pathways.

This chapter opens with a consideration of the rates of nuclear reactions. The rates of nuclear reactions illustrate a simple form of *reaction kinetics*. In addition, the rates of nuclear decay provide us with important knowledge concerning the age of our planet and of human history on earth.

The rates of chemical reactions are often more complex than those of nuclear reactions. In this chapter, factors that influence rates of chemical reactions are presented. The connection between reaction rates and reaction pathways is illustrated. The chapter closes with insights pertaining to the relationship between the actions of pharmaceutical drugs and the study of rates of chemical reactions.

■ Radioactive Isotopes Decay with a Characteristic Half-life

Scientists can measure the production of alpha particles, beta particles, or gamma rays produced by nuclear decay. For example, a beta particle can trigger an electric impulse in a detector known as a Geiger counter. By counting the electrical impulses in a time interval and dividing by the length of the interval, a rate can be calculated. For example, if 10,800 counts were recorded in four minutes, the rate of decay would be proportional to 2,700 counts per minute (cpm). (A counter does not detect all decay events, but scientists seek to arrange samples and detectors so that decays are detected with the same efficiency for any comparison.)

Figure 10.1
For nuclear decay reactions, the graph of rate of decay versus time follows the same characteristic curve. The half-life of an isotope is the time required for one-half the nuclei present in the sample to undergo decay.

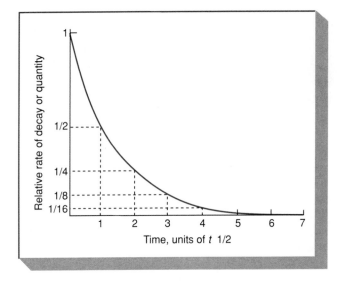

Though the rate of decay of a radioactive isotope is sometimes fast and sometimes slow, there does exist a unifying feature in all cases. *The rate of decay is proportional to the quantity of isotope present.* It is not affected by changes in temperature, pressure, or chemical state. A graph for the rate of decay of a radioactive isotope against time follows a curve characteristic of *exponential decay* (see Fig. 10.1).

The *half-life* of an isotope is the time required for one-half the nuclei present in the sample to undergo decay. The half-life is independent of the original quantity of the isotope, so that after two half-lives, one-fourth of the original quantity remains, and after three half-lives, one-eighth remains. Half-lives of radioisotopes differ greatly; some are as short as 10^{-9} seconds and, therefore, require fast methods to measure them. In contrast, the half-life of one isotope of uranium is 4.5 billion years, a number close to the age of the earth. With improvements in the ability to detect and count radioactive decay reactions, both shorter-lived and longer-lived isotopes have been detected.

We can speak of the half-life of a sample of a radioisotope, but we cannot predict the lifetime of an individual nucleus. The half-life describes average behavior, but any particular nucleus may decay after 0.1, 1, or 100 half-lives. With a very small sample or with a radioisotope with a long half-life, fewer decays are observed and the uncertainty of measurements is greater.

■ Estimating the Age of the Earth

The most abundant isotope of uranium, uranium 238, has a half-life of 4.5×10^9 years. The element formed by the alpha decay of $^{239}_{92}U$ is radioactive as are the products of several subsequent nuclear decay reactions. (The radioactivity which Becquerel discovered in a compound of uranium was due not only to the slow decay of uranium but also to the decay of such radioactive "daughter" isotopes as the polonium and radium, later isolated by the Curies.) The decay series that begins with $^{238}_{92}U$ ends with a stable isotope of lead, $^{206}_{82}Pb$. Because the daughter isotopes have much shorter half-lives than uranium, most, but not all, of the $^{238}_{92}U$ that has decayed over time is now $^{206}_{82}Pb$.

To estimate the age of rocks, scientists measure the ratio of $^{238}_{92}U$ to the sum of $^{206}_{82}Pb$ and $^{238}_{92}U$ present in samples. The sum of the $^{206}_{82}Pb$ and $^{238}_{92}U$ present in a sample is equal to the amount of $^{238}_{92}U$ present when the mineral solidified and

Figure 10.2
Apollo 11 Lunar is a volcanic rock that solidified (from lava) 3.65 billion years ago, as inferred from its content of radiogenic argon. Additional work, based on measurement of lead isotopes, indicates that the material melted to form the lava was at least 4.2 billion years old.
Photo by NASA.

cooled. From the fraction of the original $^{238}_{92}U$ remaining, scientists calculate the fraction of a half-life that has elapsed, or the age of the sample. The oldest rocks found on earth solidified about 3×10^9 years ago. Some rocks found in meteorites and some of those brought back from the moon are even older. These rocks, formed about 4.5×10^9 years ago, enable scientists to estimate the age of our solar system (see Fig. 10.2).

Prior to cooling, lead and uranium present in molten rock may have been separated by differences in migration and by chemical reactions. Since the isotope $^{206}_{82}Pb$ constitutes less than twenty-five percent of the natural abundance of lead, the absence of other isotopes of lead is evidence that a rock contains lead only from radioactive decay.

■ Carbon-14 Dating in Archeology

Archaeologists use the decay of $^{14}_6C$, an isotope of carbon, to gain new insight into our early history and prehistory. The half-life of this isotope is 5,760 years, a time comparable to that of the existence of written history. The ages of a wooden beam from an early castle, a papyrus scroll from the time of the pharaohs, and the charcoal from a prehistoric camp fire can be determined because they contain a small, but measurable, fraction of the amount of $^{14}_6C$ that is found in plants alive today (see Fig. 10.3).

Carbon 14 is produced by nuclear reactions occurring in the upper atmosphere. Cosmic rays bombarding the earth's atmosphere constantly produce neutrons, and the neutrons react with nitrogen to produce $^{14}_6C$:

$$^{14}_7N + ^1_0n \rightarrow ^{14}_6C + ^1_1H$$

In the atmosphere, newly formed $^{14}_6C$ reacts with oxygen to form CO_2.

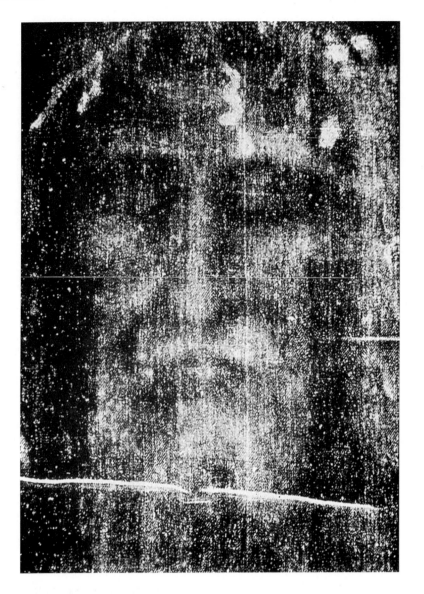

Plants incorporate CO_2 into compounds in the reactions occurring in photo-synthesis. Leaves and newly formed wood contain the same fraction of $^{14}_{6}C$ as the atmosphere. In living matter, as in the atmosphere, the amount of $^{14}_{6}C$ present reflects the balance between the rate of its radioactive decay and the rate at which it is replenished. When a plant dies or a tree is cut down, the $^{14}_{6}C$ present continues to decay while no new $^{14}_{6}C$ is incorporated:

$$^{14}_{6}C \rightarrow {}^{14}_{7}N + {}^{0}_{-1}e$$

The age of an artifact can be computed by burning a small sample of its carbon to CO_2, measuring the fraction of the total carbon present that is $^{14}_{6}C$, and compar-ing that fraction to the fraction of $^{14}_{6}C$ found in the atmosphere today (radiocarbon dating assumes that the amount of $^{14}_{6}C$ in the atmosphere has been constant over time).

The dating of artifacts using carbon 14 raised new questions. The belief civilization spread by cultural diffusion was challenged by the results of such dating. Dates calculated on the basis of the fraction of $^{14}_{6}C$ remaining supported an alternate hypothesis that certain technologies arose independently in widely separated places. Was it necessary to reconsider ideas about how cultures arose and spread, or were the assumptions made in carbon-14 dating wrong?

In order to use the decay of carbon 14 for dating, scientists assumed that the atmospheric concentration of carbon 14 is constant over time. The discovery of long-lived, well-preserved bristlecone pines provided a way to test this assumption. The growing seasons recorded in the tree's rings provided an independent way to document age, for the inner rings of these long-lived trees were formed more than 4,000 years before the outer rings. Scientists measured the ages of the inner wood samples from these trees both by counting rings and by using carbon-14 dating.

The ages determined by the two methods were very close, but not identical. Scientists concluded that there have been small variations of carbon-14 levels in the atmosphere that reflect variations in the sun's output of energy. However, the corrections to the ages obtained by radiochemical dating were in the wrong direction to give support to those arguing for cultural diffusion as the single mechanism for the appearance of civilization in distant regions.

■ **Aside**

Have Collisions with Meteors Caused Mass Extinctions?

Both geology and radioactive dating provide evidence that the earth is very old. A part of the geological evidence is a fossil record preserved in sedimentary rocks. In particular, limestone is formed in shallow seas by the slow deposition of the skeletal remains of microscopic marine organisms. Even a small piece of limestone contains vast numbers of microscopic fossils. Layers of limestone hundreds of feet thick were deposited over millions of years. By sampling the deeper layers of limestone, scientists can sample more ancient seas.

When a core is drilled from a thick bed of limestone and the fossils are examined, we find that the numbers and kinds of fossils present change very slowly throughout most of the core. However, in some cores there are abrupt changes in the relative numbers and kinds of marine fossils occurring in narrow zones. In these regions of discontinuity, many species present in lower parts of the core disappear completely to be replaced by species found only in the higher segments. Paleontologists refer to these abrupt changes in fossil composition as mass extinctions.

What events could trigger the mass extinctions found in the fossil record? Scientists now believe that a large meteor striking the earth may have been responsible. (The absence of a large crater on the surface of the earth suggests that the collision must have occurred in the ocean.) If a large meteor had struck the earth, dust from the meteor could have darkened the skies for long periods, causing temperatures to drop and photosynthesis to slow. This large and sudden change in climate is believed to have triggered the mass extinction of vulnerable species.

Evidence for this hypothesis first came from measurements of trace amounts of osmium and iridium, elements found beside platinum in the

Figure 10.4
Drill core indicating clays where anomolies in the amounts of Os and Ir indicate a possible explanation for catastrophic extinction.

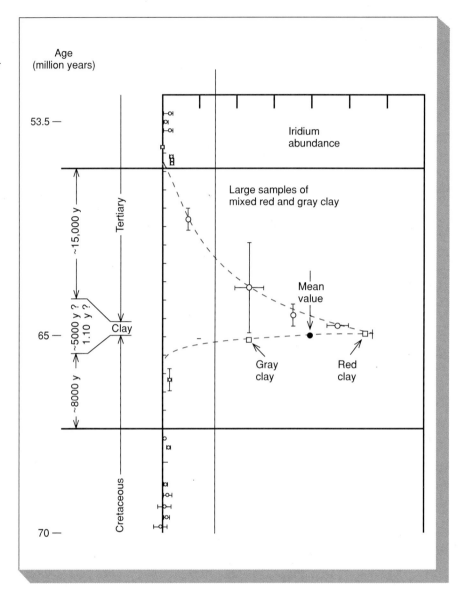

periodic table (see Fig. 10.4). A technique known as neutron activation is used to detect and measure traces of these elements. First, samples are bombarded with neutrons, and then the gamma-ray emission from the samples is studied. When a neutron is absorbed by a nucleus of an atom, a new unstable isotope of the same element is formed. These unstable isotopes lose some of their energy by emitting gamma rays. The energies of emitted gamma rays and the half-lives of the gamma-emitting isotopes are measured and compared to the emissions from known samples.

Osmium and iridium are very scarce in terrestrial rocks, but they are more abundant in known samples of meteors. Tiny meteors are bombarding the earth constantly, and are burned up as they pass through the atmosphere, leaving a trail of dust. Because dust from these meteors is deposited, traces of osmium and iridium occur throughout limestone. In limestone cores taken from widely separated regions of the globe, it has been found that the abundance of osmium and iridium in samples at the boundaries of mass extinctions is more than twenty times as high as it is in other parts of the cores. The greater abundance of these elements at the boundaries is evidence of a large meteor striking the earth.

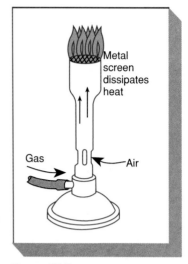

Figure 10.5
A Fisher burner used in chemistry laboratories provides a dramatic example of the effects of temperature on reaction rates. Even though the gas and air are mixed at the base of the burner, combustion occurs only at high temperatures over the burner.

■ Concentration and Temperature Affect Rates of Chemical Reactions

In contrast to the rates of nuclear reactions, the rates of chemical reactions depend on the physical states of the reactants, so the rates vary greatly as concentrations and temperatures change. The rates of nuclear decay reactions depend only on the relative instabilities of nuclei comprising the inner cores of atoms, but the rates of chemical reactions depend on the contact between reacting particles.

Chemists classify reactions that take place at surfaces and those that take place in solution differently. *Heterogeneous reactions* occur at surfaces, and *homogeneous reactions* occur in solution or in the gas phase. The effects of different surface areas on the rates of heterogeneous chemical reactions can be readily observed. For example, wood shavings used as kindling are easier to light than a Yule log. The more surface area of wood in contact with air, the speedier the combustion. The effects of different concentrations on the rates are also apparent. For example, carbon dioxide fire extinguishers slow combustion by providing a smothering blanket of CO_2 gas that lowers the concentration of oxygen at the burning surface.

Increases in concentration also speed homogeneous chemical reactions. For example, the reaction of oxygen and nitrogen to form nitrogen oxides occurs much more rapidly in the cylinder of an automobile engine where the hot gases are compressed than in an open fire where both nitrogen and oxygen are more diluted.

Temperature increases also increase the speed of most reactions. A Fisher burner shown in figure 10.5 provides a dramatic example of temperature effects. The mixture of air and natural gas coming out of the barrel of a burner burns rapidly with a very hot flame. Why doesn't the flame jump back into the barrel? A metal screen carries away heat from the flame above so that the mixture of gases below does not reach ignition temperature. Safety lamps operating on the same principle permitted the mining of coal in the nineteenth century. Wire screens that dissipated heat kept the illuminating flames of the miner's lamps from igniting flammable gases in the mine.

■ A Collision Model for Reaction Rates

A simple model to explain reaction rates is based on collisions. This model provides an explanation of the effects of changing concentrations on rates. The greater the concentration of the reacting species, the more frequent the collisions between reacting molecules, and the faster the reaction. Consider a reaction in which a molecule of A reacts with a molecule of B to form a molecule of C:

$$A + B \rightarrow C$$

According to the collision model, doubling the concentration of both A and B, the two reacting species, would quadruple the rate of reaction because the collisions of A and B would occur with four times the original frequency (see Fig. 10.6).

Figure 10.6
Collision theory provides a simple model to describe reaction rates. The rate of a chemical reaction depends both on the frequency of collisions between reaction species and the energy of the collisions.

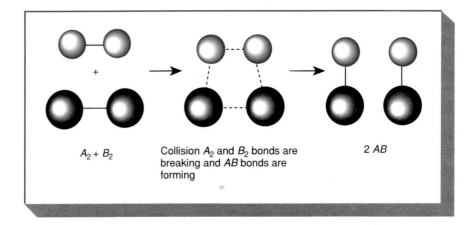

$A_2 + B_2$ Collision A_2 and B_2 bonds are breaking and AB bonds are forming $2\ AB$

Figure 10.7
At a low temperature, only a small fraction of the molecules in a sample have high kinetic energies. The fraction of high-energy molecules increases as the temperature increases.

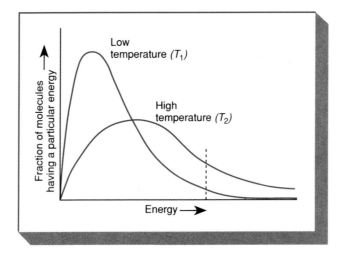

A collision model also helps explain the effect of temperature on reaction rates. Molecules can react only when they collide, but not all collisions lead to reactions. Reactions involve both the breaking of old bonds and the formation of new bonds. Productive collisions are high-energy collisions because energy from collisions is needed to weaken or break the bonds in the reacting molecules. Since the fraction of high-speed molecules in a sample increases rapidly as the temperature increases, the number of high-energy collisions and the reaction rate increase rapidly with increasing temperature also (see Fig. 10.7).

Most collisions do not involve enough energy to lead to reaction; therefore, chemists speak of a barrier to reaction. For a slow reaction, there is a high barrier to reaction; for a fast reaction, the barrier is lower. This barrier to reaction is called the *activation energy*. The magnitude of the activation energy may determine whether or not a reaction occurs. For example, at 20° C the combustion of gasoline occurs at a negligible rate because the activation energy for this reaction is too high to overcome at this temperature.

If two or more reactions are possible, the reaction with the lower activation energy occurs faster. By analogy, although Denver is higher in altitude than both San Francisco and New Orleans, the mountain barrier is too high for water to flow west toward San Francisco and the streams instead flow east from Denver, crossing lower man-made dams to empty into the Gulf of Mexico near New Orleans.

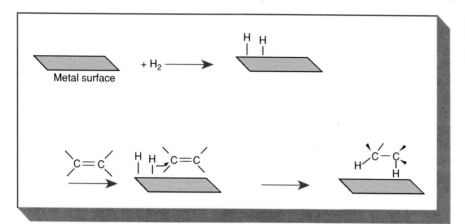

Figure 10.8
Finely divided platinum metal is used to catalyze the addition of hydrogen to carbon-carbon double bonds.

■ Catalysts

Catalysts are substances that accelerate the rates of chemical reactions without being consumed by the reaction. A catalyst does not occur as a reactant or as a product in the net equation for a chemical reaction. It may react in one step of a reaction and be regenerated in another, but it is present both at the beginning and at the end of the reaction. A catalyst provides an alternate lower energy pathway for a reaction to occur. At a fixed temperature, a greater fraction of the collisions between molecules have sufficient energy to surmount this lower barrier, and the reaction is faster.

Some catalysts speed reactions by providing surfaces, at which concentrations of reacting species are high and/or at which the bonds in one or more reactants may be weakened or broken. In the production of gasoline from petroleum, finely divided platinum and palladium catalyze the addition of hydrogen to carbon-carbon double bonds as shown in figure 10.8. These metals absorb a large amount of hydrogen, and there is a high concentration of hydrogen at their surfaces. At the metal's surface, the bond between hydrogen atoms in H_2 is weakened, lowering the activation energy for the addition reaction. Since a catalyst is not used up in a reaction, metals as expensive as platinum and palladium can be used as catalysts even though the gasoline produced sells for a relatively low price.

■ The Rate Law Describes a Chemical Reaction

The rate of a reaction is the rate of change in concentration. To measure a reaction rate, chemists measure either the rate of disappearance of a reactant or the rate of appearance of a product. One way to measure the rate is to withdraw and analyze samples at regular intervals during a reaction. The rate of a reaction is the slope of the graph of concentration versus time (see Fig 10.9).

To investigate the relationship between concentration and rate, the concentration of one reactant at a time is varied, and the rate of reaction is measured. It is found that the results of these measurements can be expressed in an equation called the *rate of law:*

$$A + B \rightarrow C + D \qquad \text{hypothetical reaction}$$

$$\text{Rate} = k[A]^a[B]^b \qquad \text{form of rate equation}$$

In the rate equation, [A] and [B] are concentrations of reactants. (The concentration of a catalyst can appear in a rate equation also.) The exponents a and b are

Figure 10.9
One way to determine the rate of a chemical reaction is to measure the concentration of a product at different times. The rate of the reaction at a given time is the slope of the graph of concentration versus time.

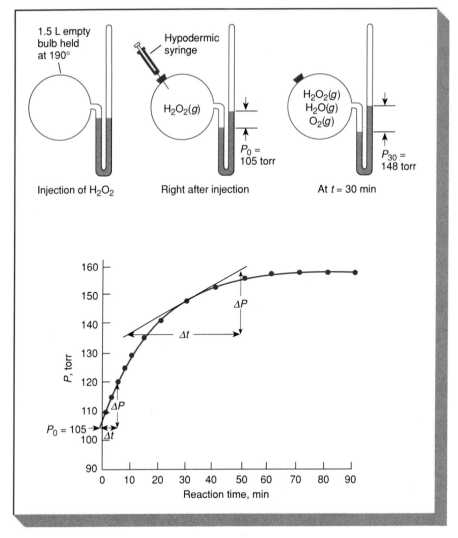

determined by experiment. The exponent of a concentration in the measured rate law is the order of the reaction. For example, if $a = 1$ and $b = 1$ for a reaction, the reaction rate is first-order in A, first-order in B, and second order overall ($1 + 1 = 2$). A rate constant k is another quantity that is determined for a reaction. The value of k increases as the temperature of the reaction increases.

To illustrate the application of a rate law, consider a hypothetical reaction in which A goes to B. To determine the reaction order in A, a chemist doubles the concentration of A and then measures the effect on the rate. Three possible outcomes of this experiment are given in the accompanying table:

$$A \rightarrow B$$

Effect of doubling [A]	Rate Law	Order in A
No increase in rate	Rate = $k[A]^0$	Zero
Rate of reaction doubles	Rate = $k[A]^1$	First
Rate of reaction quadruples	Rate = $k[A]^2$	Second

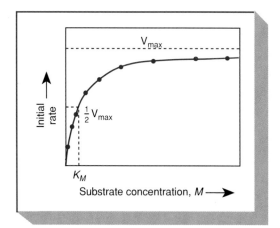

Figure 10.10
For an enzyme-catalyzed reaction, the rate of reaction first increases, then levels off as the substrate concentration increases.

Many reactions occur as a series of steps. For such a reaction, the slowest step of the overall reaction acts as a kinetic bottleneck. Just as a narrow neck on a large bottle limits the speed for emptying it, the rate of the slowest step limits the overall rate of a reaction.

■ Enzyme Kinetics

Enzymes are proteins that catalyze metabolic reactions for compounds called *substrates*. The rate equations for enzymatic reactions differ from those found for simple chemical reactions. At low substrate concentrations, the initial rate doubles when the substrate concentration is doubled, but at high substrate concentrations, adding more substrate has no effect on the rate. The order of reaction changes from first-order to zero-order with increasing substrate concentration. A plot of the initial rate of the enzyme-catalyzed reaction versus the substrate concentration at constant enzyme concentration is shown in figure 10.10. With increasing substrate concentration, the reaction rate first increases and then becomes constant.

Early in the twentieth century, Leonor Michaelis and Maude Menten, working at the University of Chicago, investigated the kinetics of reaction for the digestive enzyme invertase. Invertase cleaves sucrose or cane sugar into simpler sugars. A small amount of the enzyme catalyzes the reaction of a large quantity of its substrate, sucrose. To fit the experimental curve, Michaelis and Menten proposed a mechanism in which enzyme-catalyzed reactions occur in steps. In the first, reversible, step, an enzyme combines with a substrate to form an enzyme-substrate complex. The enzyme-substrate complex can dissociate to reform the enzyme and substrate, or it can react in a second step to give free enzymes and products.

$$E + S \rightleftarrows ES \text{ step 1}$$

$$ES \rightarrow E + P \text{ step 2}$$

Either the first or second step of their mechanism can be the slow step that is rate-limiting. The rate of the formation of the enzyme-substrate complex depends on the concentrations of both enzyme and substrate, while the rate of the second step depends on the concentration of the complex ES. When the concentration of the substrate is low, reactions of substrate and enzyme are less frequent, and step 1 is slow and rate-limiting. When the substrate concentration is high, the rate of formation of the enzyme-substrate complex is rapid, and step 2 becomes the rate-limiting step.

■ **Aside**

**Matching a Mechanism to an Experimental Curve
is a Test for Correctness**

Chemists study kinetics (the rate behavior of reactions) to understand how chemical reactions take place. Sometimes they propose mechanisms by which complex reactions occur in simpler steps. To test a possible mechanism, they examine its consequences. A correct mechanism should predict a curve identical to that which is observed.

Michaelis and Menten derived the following equation from their assumed two-step mechanism:

$$Rate = \frac{V_{max}[S]}{K_m + [S]}$$

The initial rate of reaction and the substrate concentration [S] are measured quantities. V_{max} and K_m are constants that are evaluated graphically. Their equation fits the experimental curve for a wide range of substrate concentrations, confirming their proposed mechanism for catalysis by enzymes. Had their mechanism not fit the experimental curve, the mechanism would have been rejected.

To see how the equation derived by Michaelis and Menten fits the graph of the observed rate, consider the low and high extremes of substrate concentration. At very low substrate concentration, the denominator of the rate equation is approximately equal to K_m, and the initial rate is proportional to the substrate concentration. Here the reaction is first-order with respect to [S].

$$Rate = \frac{V_{max}[S]}{K_m} \quad \text{if } [S] << K_m$$

At very high concentrations of substrate, the denominator of the rate equation is approximately equal to [S], and the initial rate remains constant. Here the rate is zero order with respect to [S].

$$Rate = \frac{V_{max}[S]}{[S]} = V_{max} \quad \text{if } [S] >> K_m$$

■ **Enzymes' Active Sites**

The formation of the enzyme-substrate complex plays a very important role in interpreting enzymatic catalysis. The site on the enzyme that binds the substrate is called the *active site*. The substrate fits into the active site as a key fits into a lock (see Fig. 10.11). In the active site, the substrate is aligned with reactive groups on the enzyme or on a second substrate. The binding of the substrate to the enzyme in the enzyme-substrate complex generates a high concentration of reactive groups in an orientation favorable for reaction. The selective and oriented binding of the substrate to an active site is a major way that enzymes speed up reactions.

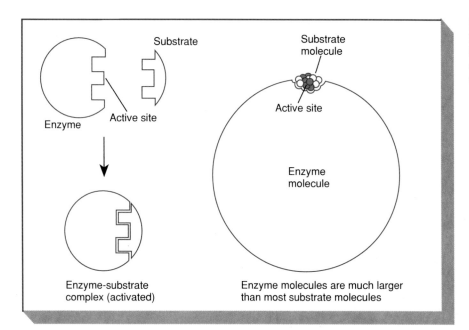

Figure 10.11
In an enzyme-catalyzed reaction, the substrate first binds to an active site on the enzyme to form an enzyme-substrate complex.

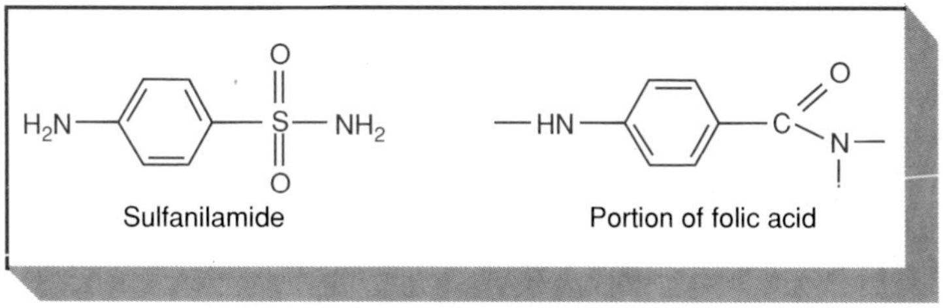

Figure 10.12
Sulfanilamide resembles a portion of the folic acid vitamin and it binds to the active site of an enzyme, catalyzing the production of folic acid. By occupying the active site, sulfanilamide blocks the production of folic acid required for bacterial growth and reproduction.

The active-site hypothesis has been extended to some proteins that are not enzymes. These proteins selectively bind small molecules as part of a chemical communication system between cells. Some of these proteins provide binding sites for hormones; others for molecules that transmit nerve impulses between cells.

■ Drugs at the Active Sites of Enzymes

Most drugs act by affecting one or more enzymes and some of them bind at the active sites of these enzymes. To do so, their molecular structures must resemble those of normal substrates. Drugs that interfere with the metabolism of invading microorganisms while not affecting the host are medically effective because there is a biochemical pathway that is unique to the invading organism. The enzymes on that pathway are potential sites for the drug's action.

Sulfa drugs, introduced in the 1930s, were the first drugs effective against a wide spectrum of bacteria. Folic acid (a B vitamin) is required for the manufacture of new DNA and is, therefore, required for cell division. While people obtain folic acid through their diets, bacteria must make their own folic acid. Sulfanilamide resembles one of the molecules needed to assemble folic acid as shown in figure 10.12. Sulfanilamide binds to the active site of the bacterial enzyme that synthesizes folic acid, blocking the binding of the substrate. A patient treated with sulfa is not affected, while the growth of infecting bacteria is stunted by a vitamin deficiency.

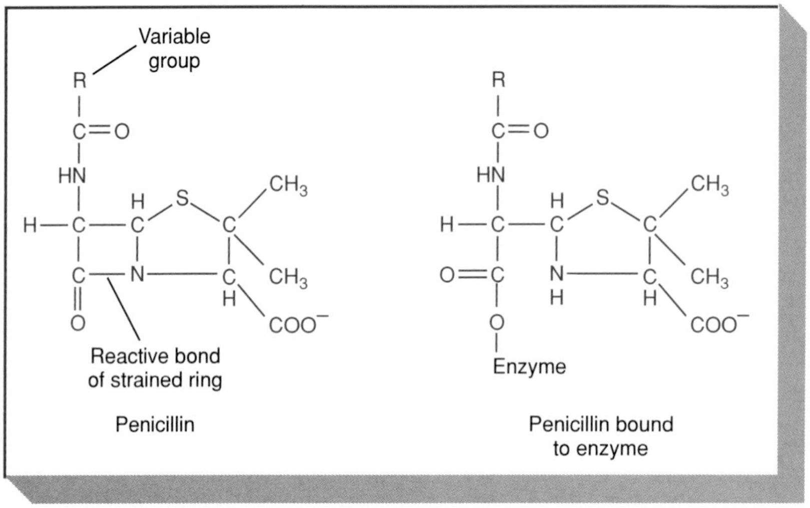

Figure 10.13
Penicillin binds to the active site of an enzyme that catalyzes reactions in the formation of cell walls of certain bacteria. While bound at the active site, penicillin reacts to form a covalent bond to the enzyme and render it unable to catalyze additional reactions.

Penicillin is an enzyme inhibitor that acts like a Trojan horse. It binds to the enzyme that completes the construction of cell walls for certain bacteria. Because the penicillin molecule resembles a portion of the cell wall, it is bound to the active site of the enzyme. While bound, it reacts with the enzyme to form a new covalent bond as indicated in figure 10.13. (Penicillin has a highly reactive four-membered ring that is cleaved by a reactive group on the enzyme.) The modified enzyme can no longer act as a catalyst. The cell walls remain incomplete and leaky, so cell division remains incomplete also. Penicillin slows the growth of invading bacteria and allows the body's defenses the time it needs to mount a counterattack.

Questions—*Chapter 10*

1. During the first ten min. of counting, the rate of decay of a radioisotope dropped from 1,600 cpm to 800 cpm. How much time will be required for the decay rate to drop from 800 cpm to 400 cpm? From 400 cpm to 100 cpm?

2. The atmospheric testing of nuclear weapons stopped in 1963. A radioactive isotope of strontium ($^{90}_{38}Sr$) released in the tests decays with a half-life of 29 years. In what year will the level of radioactive decay of $^{90}_{38}Sr$ released in 1963 drop to one-fourth its original value? To one-eighth the original value?

3. Why is it necessary to measure the quantities of both $^{238}_{92}U$ and $^{206}_{82}Pb$ and not just the quantity of $^{238}_{92}U$ to determine the age of a rock?

4. Potassium-40 is a long-lived radioactive isotope that decays to argon-40 by emitting a beta particle. The age of potassium-containing rocks can be determined by measuring the ratio of argon to potassium-40. If at a later time, a rock was heated to become molten, some argon might escape. How could measurements of the argon-to-potassium-40 ratio in rocks provide evidence for a large meteor impact?

5. The relative rate of decay of carbon 14 in CO_2 prepared from a charcoal artifact is one-eighth that of atmospheric CO_2. What is the approximate age of the artifact?

6. Criticize the following explanation for carbon-14 dating. While a tree is alive, it incorporates carbon 14 from the atmosphere. When the tree dies, the carbon 14 begins to decay. By comparing the fraction of carbon 14 in the wood to the fraction of C 14 in the atmosphere, the age of the tree can be calculated.

7. By comparing the fraction of carbon 14 in the inner layers of long-lived bristlecone pines to the fraction in outer layers and by determining the age of the tree by counting tree rings, it has been found that the fraction of carbon 14 in the atmosphere has not been constant. Why might the amount of carbon 14 in the atmosphere vary with incoming solar radiation?

8. Table salt and rock salt are samples of sodium chloride that differ in crystal size. Which dissolves faster in water? Why?

9. Would the rate of reaction of two gases be greater at a higher temperature if the frequency of collision of molecules of the gases was kept constant by increasing the volume of the gas? Give a reason for your answer.

10. In many gas storage cans, there is a small hole away from the spout. Does the presence of an open vent "catalyze" the emptying of gasoline from the can? Why or why not?

11. Platinum catalysts are often prepared by reducing PtO_2 powder. Why might the catalyst prepared by this reaction be far superior to one prepared by grinding platinum metal?

$$PtO_2 + 2\,H_2 \rightarrow Pt + 2\,H_2O$$

12. A petroleum refining company advertises that they use platinum in the production of their gasoline. Should you pay a premium price for the gasoline? What is the likely role of the platinum?

13. Ammonia for use in fertilizer is produced by the reaction of hydrogen and nitrogen at high temperature and pressure:

$$N_2 + 3\,H_2 \rightarrow 2\,NH_3 + heat$$

How do the following changes in conditions affect the rate of formation of ammonia?
   (a) Increase in pressure
   (b) Increase in temperature
   (c) Introduction of a catalyst while holding temperature and pressure constant

14. The reaction $A + B + C \rightarrow D$ follows the following rate equation:

$$Rate = k[A][B][C]$$

Argue which of the following mechanisms might be more likely, based on the analogy of the relative frequencies of automobile collisions involving two and three cars.
   (a) A simultaneous collision of molecules of A, B, and C to give D.
   (b) A two-way collision of molecules of A and B to give an intermediate that then collides with a molecule of C to give D.

15. Although methyl alcohol (CH_3OH) does not react with aqueous sodium chloride, it does react with aqueous hydrochloric acid to form CH_3Cl. Does the fact that hydronium ions speed the reaction let you know whether they act as a catalyst or a reactant? (To write an equation for this reaction, you need to consider the second product formed.)

16. Why are reptiles less active and more slow-moving in cold weather?

17. What forces might be involved in the binding of sulfanilamide to the enzyme involved in folic acid synthesis?

18. Do the antibiotics sulfa and penicillin kill germs or do they only slow the rate of reproduction of invading bacteria? Justify your answer.

CHAPTER

11

Energy, Entropy, and Chemical Change

Water runs downhill and fuels burn. Both the dripping of a measured volume of water and the burning of a measured length of a candle have been used to measure the passage of hours during the night. We can observe the passing of seasons by noticing the blossoms of early spring, the greens of summer, and the bright hues of autumn. We observe and experience the direction of change in time.

What accounts for the direction of chemical change? Scientists believe that processes of biology, physics, geology, and chemistry are governed by the same laws. To examine chemical changes, we need to consider a part of science called *thermodynamics* that deals with changes in energy and order.

■ Heat, Work, and Energy

Early in the nineteenth century, the American Benjamin Thompson (1753–1814) studied the quantitative relationship between *heat* and *work*. He measured the heat produced during the drilling of the bores of brass cannons (water was used to cool the hot metal). He found that the heat produced by the turning of a drill was proportional to the work done in turning the drill. Heat and work are two forms of energy (see Fig. 11.1). Thompson formulated the law of conservation of energy. This relationship, also known as the *first law of thermodynamics,* states that the energy of a system can only be increased by heating it or by doing work on it. Conversely, the energy of a system can only be lowered when the system loses heat or does work.

$$E_{\text{final}} - E_{\text{initial}} = \Delta E = q + w$$

Change in energy = heat added to system + work done on system

When we use an electric current from a lead storage battery to start an automobile or the burning of a fuel in a combustion engine to power it, we look to chemical reactions as a source of energy. The process may be efficient, converting much of the stored energy to work, or it may be inefficient with less work done and more heat produced. Whether a process is efficient or inefficient, the net energy change is the same. The first law of thermodynamics is the statement that energy is conserved in all processes, although the form of the energy may be changed.

Figure 11.1
Potential energy can be converted into heat or it can be converted into work. Heat and work are two forms of energy.

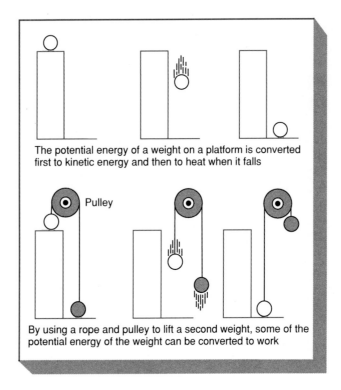

The potential energy of a weight on a platform is converted first to kinetic energy and then to heat when it falls

By using a rope and pulley to lift a second weight, some of the potential energy of the weight can be converted to work

Reactions that produce heat are called *exothermic reactions,* and reactions that absorb heat from their surroundings are called *endothermic reactions.* The burning of natural gas is an exothermic reaction, and the formation of nitric oxide from nitrogen and oxygen is an endothermic reaction:

$$CH_4 + O_2 \rightarrow CO_2 + 2\,H_2O + heat \qquad \text{exothermic}$$
$$heat + N_2 + O_2 \rightarrow 2\,NO \qquad \text{endothermic}$$

The energy of a reaction may appear partly as heat and partly as work, or it may appear entirely as heat. Consider the voltaic cell shown in figure 9.2. If the reaction of zinc with a copper salt is harnessed to drive an electric motor, then some of the energy from the chemical reaction is converted to work. If, however, the zinc metal and a solution of a copper salt are simply mixed, the redox reaction occurs directly on the zinc surface where copper plates out and zinc is oxidized. No work is produced, and the energy of the reaction is converted entirely to heat.

■ Measuring Heats of Reactions

Under certain restricted conditions, the heat liberated or absorbed by a chemical reaction depends only on the initial and final states of a system. (In the case of a reaction, the initial state describes the reactants and the final state, the products.) When a reaction that goes to completion occurs in a container with rigid walls to keep the volume constant, and no device is used to do work, then no work is done. Under these restricted conditions, the change in energy can easily be determined, for it is equal to the heat of the reaction

$$\Delta E = q \text{ when } w \text{ is zero.}$$

Chemists measure heats of reactions using a calorimeter shown in figure 11.2. For example, the heat of combustion of an organic compound is measured by burning it

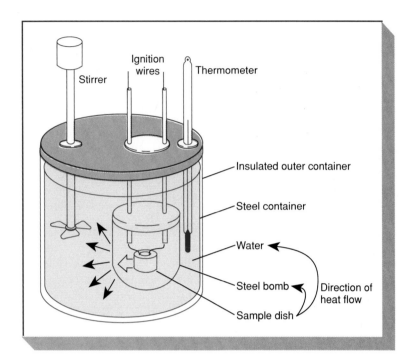

Figure 11.2
A bomb calorimeter is used to measure the heat of combustion of many compounds. The heat of the reaction flows into the bomb and then into the surrounding water. By measuring the temperature increase, the heat of reaction can be determined.

in an atmosphere of oxygen in a bomb calorimeter. Heat from the reaction is absorbed by the calorimeter and the surrounding water bath. The heat of combustion of the compound can be calculated from the temperature increase of the calorimeter and bath and their heat capacities.

When dieticians speak of the caloric content of a food, they refer to the heat of combustion of that food. For example, the combustion of 1 mole of the sugar glucose yields 673.0 kcal:

$$C_6H_{12}O_6 + 6\ O_2 \rightarrow 6\ CO_2 + 6\ H_2O + 673.0\ \text{kcal/mole}$$

The energy content of glucose is the same whether it is burned in the body or in a calorimeter. Dieticians, however, use different units. A dietician says that glucose contains 3.74 Cal/g but a calorie unit for food in the diet is 1 kcal of energy. In the body, that energy may be converted to heat or a combination of heat and work, or it may be stored chemically as fat.

■ A Closer Look at Work and Heat

Work and heat are two forms of energy in transit. Both the acceleration of a rocket and the thawing of snow involve the transfer of energy. Work is directional; we can lift weights or throw a ball. Electrical energy is carried by a current of electrons, and light energy is carried by a beam of photons. Heat is a less organized form of energy. The transfer of heat is a transfer of random molecular or atomic motion. The conversion of work into heat involves a loss of organization since energy with direction is converted to random molecular motion. Because of friction in mechanical systems and resistance in electrical circuits, some energy is lost as heat in every energy transfer.

Heat can be converted to either mechanical or electrical work. For example, heat is used to produce steam, and steam is used to power an engine or to turn a

Figure 11.3
Disorder increases as ice changes to liquid water and as liquid water changes to water vapor; more hydrogen bonds are broken and molecules begin to move past one another and then become separate.

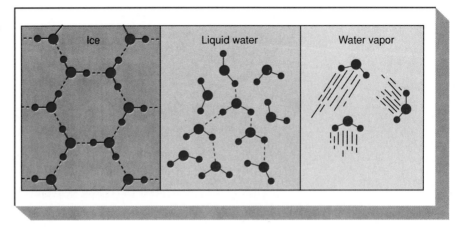

generator. The processes, however, are not efficient. The energy available from the expansion of steam depends upon the difference in temperature between the boiler and the exhaust. Usually only slightly more than thirty percent of the caloric value of a fuel can be captured and used to do electrical work.

■ Entropy—Disorder or Randomness

Some spontaneous processes involve no net energy change. Reversals of these spontaneous changes are never observed. For example, gases mix but they do not spontaneously separate. No person is known to have suffocated because oxygen and nitrogen in the air separated and he or she stepped into a pocket of nitrogen. Hot and cold objects in contact come to the same temperature. No one has seen a warm cup of coffee begin to freeze at the rim and boil in the center.

Neither the separation of gases nor the simultaneous generation of high and low temperatures would violate the law of conservation of energy; nonetheless, these events do not occur. Energy considerations alone do not explain why gases mix and why objects in contact come to the same temperature. To account for the direction of spontaneous change, an additional concept is needed.

Entropy is a quantitative measure of the disorder of a state. A highly ordered state such as a crystalline solid has a low entropy. A highly disordered state such as a gas has a high entropy. When ice melts or liquid water turns to steam, there is an increase in entropy for the system. In the solid state, water molecules are highly ordered—networks of hydrogen bonds orient each water molecule to its neighbors. In the liquid state, water molecules remain close to one another and hydrogen bonded, but there is greater freedom for molecules to move past one another or to tumble. In the gaseous state, molecules are much further apart and are not at all oriented to one another. Of the three *phases* of matter (solid, liquid, and gas), the solid has the lowest entropy and the vapor the highest (see Fig. 11.3).

■ Aside

Dynamite and the Nobel Prize

An explosive is a textbook case of applied thermodynamics. The shock from a detonator initiates a rapid exothermic reaction. As the reaction proceeds, the liberated heat causes the remaining explosive to rapidly rise in temperature, which, in turn, accelerates the rate of reaction. The hot

Figure 11.4
The Nobel Prize medal for chemistry.
© UPI/Bettmann.

gaseous products formed in the reaction exert great pressure and expand, doing work against their surroundings. With the blast, there is a large increase in entropy for both the gaseous products and for the surroundings.

$$2\,C_3H_6N_3O_6 \rightarrow 6\,CO + 3\,N_2 + 6\,H_2O + energy$$
Nitroglycerin

Alfred Nobel (1833–1896), the man who instituted and endowed the Nobel prizes in science and the arts, was the inventor of dynamite. He found ways to stabilize the explosive nitroglycerin and make it safer to use. Later, he developed and patented still other formulations for explosives and the money from these patents funds the Nobel prizes (see Fig. 11.4).

Explosives are used to quarry rock and to construct dams and mountain roads. They are also used in bombs and shells. Nobel's inventions have been used in both constructive and destructive ways. They are among the many technologies that have increased our power for both good and evil.

■ The Second Law of Thermodynamics

Organized energy such as the expansion of a gas, the thrust of a piston, and the turn of a wheel can do work. However, in the process, some of the energy is converted to heat, a more random form of energy. In a very specific example, consider a taxicab in Vancouver, British Columbia, where taxis have been modified to burn natural gas (methane) produced in the western provinces. The cab burns fuel climbing a hill. Heat from the combustion expands gaseous products of the reaction to drive pistons in the cylinders of the engine. The drive train of the car converts the rise and fall of pistons to the revolution of wheels, which propels the car, and the carbon dioxide and water vapor from combustion go out the exhaust pipe.

Can the process be reversed? If the taxi were to roll backwards down an incline, could the rolling of the wheels turn the engine in reverse to draw in the exhaust gases, to heat the mixture of carbon dioxide and water vapor and convert them to methane and oxygen? Clearly the answer is no. The exhaust gases have mixed with the atmosphere, and the heat of combustion has been dissipated to the surroundings. Neither the separation and concentration of the exhaust gases nor a temperature rise within the cylinder will occur spontaneously, in contrast to the spontaneous burning of methane.

Matter can be recycled; energy cannot. This seemingly simple statement underscores the fact that there are natural limits to any human intervention in the world. It is a restatement of the *second law of thermodynamics,* a basic underpinning of all sciences. According to the second law of thermodynamics, only those processes for which the total entropy increases or remains constant are observed to occur. The total entropy change is the entropy change of both the system and its surroundings:

$$\Delta S_{total} = \Delta S_{system} + \Delta S_{surroundings} > 0$$

That ice melts in surroundings of 20° C and not at −20° C is a consequence of the second law. Although there is an increase in entropy for the system at either temperature, the total change of entropy for the process is positive at the higher temperature and negative at the lower temperature. However, the calculation of entropies for a system or for its surroundings is beyond the scope of this course.

The second law of thermodynamics is a profound statement about the direction of change observed in nature. It begins with simple observations. When a hot object is in contact with a cold object, they spontaneously exchange heat to come to the same temperature. The reverse process in which one of the objects becomes hot at the expense of the other becoming cold never occurs—it is not a spontaneous process. Heat, like water, does not run uphill. In the direction of spontaneous change, mountains are leveled and valleys are filled.

■ **Aside**

Life and the Second Law

Let us reconsider the example of a taxi powered by the combustion of natural gas. We could construct a process for reforming the methane and oxygen from the exhaust gases by operating it in a giant greenhouse. In the reactions of photosynthesis, plants produce sugars and oxygen starting from carbon dioxide and water. Let cows grazing in the greenhouse eat these plants, so that the sugars are, in part, converted to methane. Atoms of carbon, hydrogen, and oxygen have indeed been recycled. Does not this example illustrate that it is possible to recycle both matter and energy in violation of the second law?

Our example does not violate the second law of thermodynamics because the greenhouse was enclosed so that matter did not enter or leave, energy in the form of light entered the greenhouse and drove the reactions of photosynthesis. Nuclear reactions in the sun provided the energy of the sunlight. Those nuclear fuels cannot be renewed.

How does life exist in a universe moving toward greater randomness? Life processes continuously create order in the synthesis of proteins and nucleic acids, in the organized movements of muscle, and in the processing of information in the brain. How is the low entropy of living organisms possible when the universe is moving toward greater disorder?

Plants make complex sugars from carbon dioxide and water. The removal of gaseous CO_2 and liquid H_2O from the environment to be stored as sugars in a fruit is a tremendous increase in order. However, the second law asserts that the entropy of the plant and its surroundings must be increasing. For green plants, the surroundings include the sun since radiant energy from the sun drives photosynthesis. The sun is aging, burning up its nuclear fuels, and radiating energy into space. The second law is not violated.

For an animal that digests food and eliminates waste, the system and its surroundings do not extend so very far. Food molecules are intermediate in order. Animals can maintain and increase order by exporting disorder. The carbon dioxide and water vapor that are exhaled have the high entropy associated with gases. By exhaling gases of high entropy into the surroundings, an animal maintains its organization. Upon death, the processes that maintain low entropy cease, "Ashes to ashes; dust to dust."

■ Equilibria

We are familiar with many *equilibrium* phenomena. For example, when a small amount of table salt is stirred in water, it dissolves. Still more salt can be dissolved until the solution becomes saturated, and no more salt will dissolve. The solid salt is then at equilibrium with the salt in solution (see Fig. 11.5).

$$NaCl_{(s)} \rightleftarrows Na^+_{(aq)} + Cl^-_{(aq)}$$

At the molecular level, processes at equilibrium are dynamic. The rates of the two processes become equal at equilibrium. In a saturated salt solution, ions in a salt lattice continue to go into solution, but an equal number of ions leave the solution for the crystal.

The salt used above may have been obtained by the evaporation of seawater. As seawater evaporates in the sun, the concentration of salt increases until the solution becomes saturated. Upon further evaporation of water, salt crystals form, but the concentration of dissolved salt remains constant, and the same equilibrium exists between the solid and the ions in solution.

In another familiar example, ice and water can remain in equilibrium at 0° C. If excess ice is added to warmer water, some of the ice melts and the water cools to 0° C. If excess water at 0° C is in contact with ice below 0° C, some of the water freezes and the ice warms to 0° C.

Figure 11.5
The composition of a saturated solution of sodium chloride is the same whether the solution is prepared by dissolving salt or by concentrating a dilute salt solution.

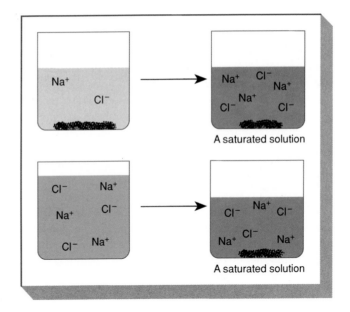

Chemical reactions can also proceed to equilibrium. At equilibrium, the rates of the forward and reverse reactions are equal, and no net change of concentrations occurs. Equilibrium represents the balance of competing dynamic processes. (Rates of reaction may be very slow. Discussions of equilibria always assume that there is a process for getting to equilibrium.)

The acid-base reaction of two water molecules to form hydronium ions and hydroxide ions is an example of chemical equilibrium:

$$H_2O + H_2O \rightleftarrows H_3O^+ + OH^-$$

The product of the concentration of hydronium ions times the concentration of hydroxide ions is a constant K_w in all aqueous solutions:

$$K_w = [H_3O^+] \times [OH^-] = 1 \times 10^{-14}$$

When an acid is added to water, the concentration of H_3O^+ in water goes up and that of OH^- goes down. Conversely, when a base is added to water, the concentration of OH^- in water goes up and that of H_3O^+ goes down.

■ Reactions Depend on Energy and Entropy Changes

The position of equilibrium, and indeed the direction of a reaction, can depend on temperature and pressure. For example, when limestone is heated to a high temperature, it is decomposed to form lime and carbon dioxide:

$$heat + CaCO_3 \rightarrow CaO + CO_2 \quad \text{at } 1{,}000° C$$

When lime is used to mark lines on an athletic field, it reacts with carbon dioxide from the air to reform calcium carbonate:

$$CaO + CO_2 \rightarrow CaCO_3 + heat \quad \text{at } 25° C$$

The position of equilibrium for a system consisting of $CaCO_3$, CaO, and CO_2 changes with temperature. At low temperatures, the exothermic reaction to form

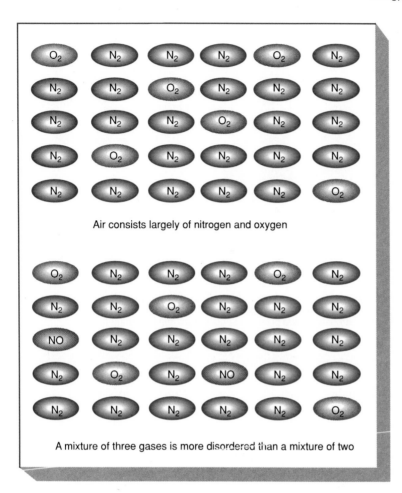

Figure 11.6
An increase in entropy accompanies the formation of a small amount of nitric oxide from nitrogen and oxygen.

$CaCO_3$ is favored. As the temperature is increased from 25 to 1,000° C, the direction of the net reaction is reversed. At high temperatures entropy considerations become dominant so that the reaction forming the highly disordered gas CO_2 as one of the products is favored.

■ Manipulating Chemical Equilibria— Le Châtelier's Principle

Both energy and entropy influence the position of equilibrium in a system. For example, the reaction of nitrogen with oxygen to form nitric oxide takes place only at high temperatures.

$$\text{Heat} + N_2 + O_2 \rightarrow 2\ NO$$

The reaction is endothermic, for nitrogen molecules and oxygen molecules have stronger bonds than those found in molecules of nitrogen oxides. However, a mixture containing only nitrogen and oxygen molecules is more ordered and is at a lower entropy than is a mixture of all three gases (see Fig. 11.6). An increase in entropy accompanies the reaction to form some NO. Hence, the formation of NO becomes more favorable as the temperature is increased.

Figure 11.7
Le Châtelier's principle can readily be demonstrated by showing the effect of temperature change on the amount of nitrogen dioxide in equilibrium with its dimer.

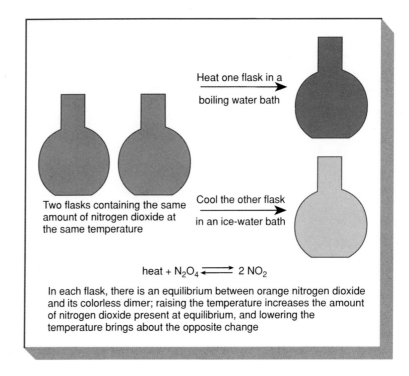

Two flasks containing the same amount of nitrogen dioxide at the same temperature

Heat one flask in a boiling water bath

Cool the other flask in an ice-water bath

$$\text{heat} + N_2O_4 \rightleftharpoons 2\,NO_2$$

In each flask, there is an equilibrium between orange nitrogen dioxide and its colorless dimer; raising the temperature increases the amount of nitrogen dioxide present at equilibrium, and lowering the temperature brings about the opposite change

When conditions such as temperature or pressure are changed, the composition of a system at equilibrium changes. Henri Le Châtelier (1850–1936), a French scientist, formulated a rule, *Le Châtelier's principle*, which can be applied to predict the direction of chemical change. This principle states that when conditions are changed, the position of equilibrium shifts to resist external changes. For example, an increase in temperature favors the reaction that absorbs heat. By converting some heat to chemical potential energy, this equilibrium shift resists a temperature rise (see Fig. 11.7).

The formation of nitrogen dioxide from N_2 and O_2 is favored by increases in both pressure and temperature.

$$\text{Heat} + N_2 + 2\,O_2 \rightarrow 2\,NO_2$$

To resist a pressure increase, the volume decreases as 3 moles of reactant gases forms 2 moles of product gases. A shift toward more nitrogen dioxide at equilibrium would also resist a temperature increase, for its formation is endothermic.

■ Manipulating Equilibria to Reduce Harmful Auto Emissions

Both nitrogen oxides and unburned hydrocarbons from fuel contribute heavily to photochemical air pollution, or smog. The reduction of nitrogen oxide and hydrocarbon emissions from automobiles has been mandated by federal law. One approach to reducing these emissions involved changes in engine design based, in part, on Le Châtelier's principle.

Although gases in an automobile cylinder do not reach equilibrium, the change in concentrations accompanying reaction is in the direction of equilibrium. As gases leave the cylinders, they rapidly cool and expand, stopping reactions and trapping nitrogen oxides that were formed at the elevated temperatures and pressures. To decrease nitrogen oxide emissions, the compression ratio was lowered at the expense of engine power. Lowering the compression ratio lowered the combustion temperatures and decreased the formation of nitrogen oxides.

Unfortunately, these changes in engine design resulted in less complete combustion of fuel. To further reduce emissions, exhaust gases are passed over a catalyst. The catalyst promotes the further oxidation of unburned hydrocarbons to form carbon dioxide and water. These catalysts also shift the amounts of nitrogen oxides in exhaust gases toward quantities in equilibrium at lower temperatures so they help limit NO and NO_2 emissions.

■ Energy, Entropy, and Resources

Thermodynamics provides an overview of mineral and energy resources. Whether it is petroleum or iron ore, a resource can be defined as a useful material found at high concentration (low entropy). Resources vary in the amount of energy needed for their recovery. For example, the greater the amount of dirt and rock covering a coal seam, the greater the cost to remove the overburden and mine the coal.

As we exhaust supplies of our most highly concentrated ores, we will need to use less-concentrated, lower-grade ores. The amount and the cost of energy needed to recover lower-grade resources will necessarily increase. (Occasionally, the development of new, more efficient recovery processes will offset these cost increases.) In the future, the cost of energy will be an increasingly large part of the total cost of producing many materials. Since energy resources are limited, the conservation and efficient use of energy will become increasingly important to our economic well-being.

We need to assess both the economic and the environmental impact of industrial processes because the processing and use of resources greatly affects our environment. Chemical wastes are a necessary part of the recovery of all minerals, but the amounts and types of wastes depend on the resource and on the particular processes used. The production and use of energy generates waste as well, since heat is a by-product of all energy production and use. Both the short-term cost of pollution control and the long-term cost of the failure to control pollution need to be taken into account.

■ Questions—*Chapter 11*

1. A person perspires more during exercise than when at rest. How does this observation demonstrate whether or not the chemical energy in food is converted completely into muscular work?
2. The heat produced when a piece of zinc is dropped into a solution of cupric sulfate is (less than, the same as, more than) the heat produced when Zn reacts with Cu^{2+} in a galvanic cell. Justify your answer.
3. The heat of combustion of glucose is 673.0 kcal/mole. Calculate the calories in 1.00 g of the sugar glucose.

4. If the flow of water in a waterfall were spontaneously reversed, would it violate the first law of thermodynamics? The second law?

5. Is the increase in entropy for the vaporization of 1 mole of water (less than, the same as, more than) the increase in entropy for the vaporization of 1 mole of a nonpolar liquid having the same boiling point? (The final volume of each gas would be the same; consider differences in the structure of the liquids.)

6. Why do the clothes on a line dry, even though the temperature is below the boiling point of water?

7. Why does a gram of water vapor have a higher entropy than a gram of liquid water?

8. Some eighteenth century scientists thought that air was a compound because the relative amounts of oxygen and nitrogen did not vary. They thought the lighter nitrogen should be on top of the heavier oxygen if these elements were not chemically combined. Why do nitrogen and oxygen mix, and why is the composition of air nearly constant on the surface of the earth?

9. It is much more difficult to determine the half-life of a radioactive sample when the number of decays occurring is small. Why is it easier to determine this average quantity with larger samples or shorter half-lives?

10. Some natural gas contains a relatively high amount of helium. What is the probable source of the helium? People concerned with future supplies of helium advocate separating helium from natural gas and storing it in abandoned mines. Why would it be much easier to recover helium from natural gas than from the atmosphere?

11. Why does it cost more to obtain iron from a low-grade ore than from ore with a higher iron content?

12. Gunpowder is a mixture of saltpeter (KNO_3), sulfur, and charcoal. What gases are formed by the oxidation of sulfur and carbon? The oxidation of carbon and sulfur is exothermic. Why is the production of heat important for the action of gunpowder?

13. Ammonia for use as a fertilizer is made by the reaction of hydrogen and nitrogen at high temperature and pressure.

$$N_2 + 3 H_2 \rightarrow 2 NH_3 + \text{heat}$$

   (a) How does an increase in the temperature affect the amount of ammonia produced at equilibrium?

   (b) How does an increase in the pressure affect the amount of ammonia produced at equilibrium?

14. In a capped bottle containing a soft drink, the carbon dioxide in the vapor and the carbon dioxide in the liquid are in equilibrium. Apply Le Châtelier's principle to account for the loss of carbon dioxide from the soft drink upon standing in an open glass.

15. Greater amounts of NO and NO_2 are formed when fuel is burned in an excess of air. Apply Le Châtelier's principle to show that nitrogen oxide emissions can be reduced by decreasing the amount of air so that there is just enough oxygen to completely burn a fuel.

■ Interlude

Chemistry, Symbol, and Language—A Personal Reflection

At the time when I first began this text, I was troubled by the thought that chemistry appeared to be the modern science most lacking in a rich philosophical history. My readings in both physics and biology offered me no consolation, for they seemed only to confirm my fears. Writers in physics return again and again to a rich history of discourse concerning the nature of the universe, and writers in biology take up questions concerning the nature of life and the evolution of man. In comparison, the history of chemical discourse appeared to be rooted in the particulars of systems being studied, and notably lacking in grand themes. Furthermore, I was not reassured by my earlier musings on the subject. I had often joked that while many physicists appeared to be thin in body and ethereal in manner, most chemists seemed drawn to the sensual things in life. They loved good food, married attractive spouses, and spoke lovingly of crystals, colors, and smells.

Just as many chemists are drawn to worldly things, chemistry seems to be a worldly science. It is intrinsically joined to technology. Chemistry offers both profound insights into the material world and mundane recipes for making this product or modifying that one. For example, Lavoisier, widely acknowledged to be the father of modern chemistry, developed procedures to improve the gunpowder that was later used by Napoleon's armies. Throughout its history, chemistry retained close ties to technology. Because it is the study of the properties and interconversions of material substances, chemistry has played key roles in the production of metals for industry, pigments for artists, papers and ink for printing, textiles and dyes for clothing, pesticides and herbicides for agriculture, pharmaceuticals for medicine, and chemicals for preserving food and purifying water.

Recently, I have been struck by the realization that chemistry, although a science having only shallow philosophical roots, shares a long and deep association with both religion and magic. Indeed, it is a most catholic science. The colors, symbols, shapes, and smells that marked the sensual appeal of chemistry in some ways echo the changing colors of vestments during the church year, the iconography of saints, the often striking architecture of churches, and the rich odors of incense. Chemistry shares symbols and substance with many religions. Both the medicine man and the pharmacist offer chemical potions for healing.

Why had it taken me so long to arrive at the perception that our chemical heritage is more tied to religion and magic than to philosophy? Was this not the knowledge that I had as a child? A book of "chemical magic" came with my first chemistry set. I recall my delight in watching a glass of water turn to a burgundy colored solution when I added a drop of indicator and a small amount of base. Like tens of thousands of other child-like chemists, I had transformed "water into wine."

That child-like insight disappeared when I was educated to be a chemist. Along with the lessons of science there came an awareness that science could and should address questions labeled "How?" but not those labeled "Why?" Questions of ultimate purpose were placed outside the realm of science. This separation was assumed rather than being developed explicitly as a part of the history of chemistry. The instructors and texts in introductory chemistry courses, then as now, gave lip service to the mention of atoms by early Greek philosophers and then dove into modern chemical concepts. Furthermore, modern chemistry had emerged and had become respectable as a science only after much of the mysticism present in alchemy had been repudiated. By totally excluding alchemy from the chemistry curriculum, instructors seemed to imply that the heritage of alchemy was an embarrassment to chemistry in much the same way that the heritage of astrology is to physics.

There is a second fundamental reason that scientists abandoned the metaphoric language of alchemy. In order to answer questions by appeals to observations, scientific questions are framed using operational definitions. While much has been gained by a reliance on operational definitions in the practice of science, something is lost when chemistry is cut off from its historical roots in alchemy. The science of chemistry presents a description of the modern material world using words and symbols that it shares in common with the metaphoric languages of religion and magic. Chemists trained in the modern scientific tradition offer a description of the world that is entirely rational. They represent water by the formula H_2O, and they analyze that water to identify the impurities present in a given sample. But the words and symbols they use evoke older, powerful meanings—that water is sacred to life and plays a sacramental role in baptism.

For many people, an understanding of the world of technology is informed, in part, with the metaphorical images taken from religion and fairy tale. At the marriage feast at Cana, Jesus turned water, a symbol of life, into wine, a symbol of celebration. Is it not also a miracle to make barren fields to bloom using fertilizers or to convert petroleum into clothing? Have we not been promised that life can be made a celebration through the miracles of technology? In the fairy tale, Sleeping Beauty, a wicked queen poisons an apple to kill her rival. Sleeping Beauty takes a bite of the apple and falls into a coma. A wandering prince kisses the sleeping beauty to break the spell. Fairy-tale technology is defeated, and the prince and princess live happily ever after. Don't some people say that our environmental problems are the due to the evils of technology practiced by wicked corporations? Don't some people long for a modern-day prince or princess who will save an innocent society?

Like many other chemists, I am troubled that the word "chemistry" has acquired a negative connotation for many people. The growth of modern technology has given rise to a host of seemingly new problems. As people survey the destruction to the environment associated with the excesses of industrialization, they confront new questions. Can the degradation of the environment be slowed or stopped? What technologies need to be altered and in what ways? I am concerned that people may be doomed to frustration and defeat as they pursue unreachable goals that reflect a reality present only in magical worlds. Yet I also am concerned that too many policies may be formulated only by the consideration of risks on a cost-benefit basis. As a chemist, what responsibility do I bear and what kind of response am I to make to these issues?

I conclude that I cannot convey the insights of chemistry without also addressing issues of technology and the environment. As a chemist, I know that there are risks and benefits associated with applications of chemistry. Early laboratory instructions for the practice of chemistry described procedures for "safely" tasting poisonous substances. Phosgene, one of the poison gases used in World War I, is an important intermediate in the production of agricultural chemicals. The same drugs used to decrease the risk of organ rejection in transplant surgery increase the risk that a patient will develop cancer at a later time. These connections can be seen as inherent in the relationships between chemical structures.

How can people enter into a dialogue concerning visions for the future that require action now? With what language can they talk about the world of everyday experience when that world and that experience are rapidly changing? The language of bread and water, of earth and sky, of health and disease is overlaid with both magical and sacramental meanings. The language of chemistry, the language of science, is rational. It is important to connect or to reconnect the language of the sacred and magical description of the world so important to human experience with the language of science used in chemistry. It is important to join the visions of prophets and storytellers to the visions of scientists.

This reflection and this text are offered in the hope that they might enrich the dialogue between scientists and nonscientists, between advocates for change and those defending the status quo. For the nonscientist, I hope to provide insight into the chemistry of the world about us. For the scientist, I hope to present familiar ideas in new and stimulating contexts. For all, I hope to convey the pleasures of seeing new connections between things, the pleasures that are the stuff of science.

12

Energy for the Twenty-First Century

The greatly increased use of energy that accompanied the Industrial Revolution brought about major changes in the way people lived. In preindustrial societies the density of the population was low and the use of energy was limited. Feed for domestic animals was used on the farm where it was grown. Trees on the property were cut for the owner's fuel and lumber. Small towns might be located at a site where a waterfall powered a mill; people who worked in the mill would live in the nearby houses. With the Industrial Revolution there began a rapid growth in the use of energy. Steam engines replaced water wheels and windmills; tractors and automobiles replaced mules and horses.

New sources of energy were developed to meet the needs of industrialization. The Industrial Revolution was first fueled by the burning of wood and threatened to denude the forested countryside of England. The development of the technology to mine coal helped avert this early environmental crisis; coal replaced wood as the fuel for industrial growth. In the twentieth century, petroleum products have replaced coal as the principal source of fuel for producing energy.

Changes in the organization of society accompanied the use of increased amounts of energy. The generation and distribution of electricity to homes and businesses, and the construction of pipelines to distribute natural gas enabled the consumption of energy to be at a distance from its production. The use of energy increased, and its production became concentrated as industry and cities grew. Transportation systems were developed to let workers commute and to facilitate the wide distribution of mass-produced goods. Modern industrial society came into being (see Tab. 12.1).

■ Energy Use Affects All Aspects of Modern Life

The cost of energy greatly affects our standard of living. Energy costs may be readily apparent, as are those of transportation and home heating, or the costs may be hidden. The price of food rises with increased costs of energy to make fertilizers, to run farm machinery, and to process and transport food. The price of clothing varies with the price of the petroleum-based chemicals used to make synthetic fibers. Energy costs influence all areas of modern life.

Table 12.1

Greatly increased use of energy accompanies industrialization.

Sources: U.S. Congress, Office of Technology Assessment, *Energy in Developing Countries*, OTA-E-486 (Washington, D.C.: U.S. Government Printing Office, January 1991).

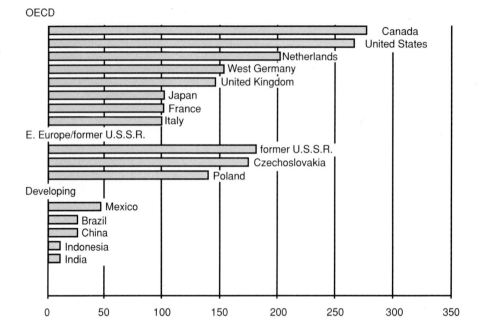

Million Btu's per person per year

Changes in the price of oil can have tremendous impacts on the economies of both developed and developing nations. An Arab oil embargo following the 1973 Yom Kippur war between Israel and the Arab countries led to immediate shortages in gasoline and heating oil. A series of rapid price increases led to the quadrupling of gasoline prices in less than a year. In 1986, a rapid drop in the price of oil stimulated the economies of industrialized nations; at the same time, it wreaked havoc with economies of poor, oil-producing nations such as Mexico. One result of such a drop in oil prices is that the prospect of default on payments by a major debtor nation such as Brazil threatens to disrupt the world banking system.

The nuclear power industry has suffered instabilities of a different sort. Increases in construction costs and high interest rates have left the nuclear power industry in shambles. No new nuclear plants have been ordered in the United States since 1978, the completion dates for some nuclear plants have been delayed, and the construction of others has been abandoned.

Because industrial production and economic stability are intertwined, there is widespread belief that the availability, cost, and distribution of energy will dominate American and world politics in coming decades. Attention has repeatedly been focused on the Middle East, where over fifty percent of known petroleum reserves are located. The 1991 war in the Persian Gulf stopped the immediate threat of an Iraqi seizure of the oil fields of Kuwait, but the political situation in the area remains volatile. Should energy availability and cost become unpredictable once again, conflicts between groups within nations, between supplier and user nations, and between rich and poor nations will increase.

At the same time, groups concerned about the waste of finite resources, the safe disposal of nuclear wastes, the effects of acid rain on the environment, and the stability of climates throughout the world have become active in the political processes of western nations. People are concerned with the profound environmental impacts that can accompany our use of energy.

In this chapter, sources of energy and the associated environmental concerns will be explored. The phrase, "There is no free lunch," summarizes an underlying theme of the chapter. One cannot produce and consume energy without doing some damage to the environment. Governments, corporations, communities, and individuals make choices that greatly affect environmental quality. Responsible action requires informed and thoughtful decisions.

■ Petroleum, Coal, and Natural Gas Are Fossil Fuels

Radiant energy from the sun captured by plants and by photosynthetic bacteria is stored in carbon compounds and in atmospheric oxygen. Photosynthesis, using light from the sun, continually produces sugars from carbon dioxide and water. When plants and animals die and decay, these reduced carbon compounds may be preserved and modified by geological processes involving high temperatures and pressures for long periods of time. The resulting fossil fuels are petroleum, coal, and natural gas.

Coal originated in forests which were submerged and then buried. Peat, which is mined from bogs and burned for fuel, is an intermediate product in the formation of coal from wood. Coals contain fossils of ancient trees, confirming their biological origin. Coal, like petroleum, consists of carbon, hydrogen, and frequently nitrogen and sulfur compounds.

Compounds in coal have a lower ratio of hydrogen to carbon than those in oil. Many of the compounds in coal have double bonds and rings. Their plate-like structures enable efficient stacking, giving coal a solid structure. Because it is a solid, coal is more difficult to extract from the earth and more expensive to transport than petroleum. In the 1980s, very large electricity-generating steam plants were the chief users of coal in the United States.

Petroleum is believed to have originated in microorganisms living in the sea. Because it is liquid, petroleum does not preserve imprints that furnish evidence of fossil origins. For that evidence, scientists searched for and found molecular fossils, compounds whose presence serves as suggestive evidence for a biological origin. These molecules contain repeating isoprene C_5H_8 units, like those found in vitamin A, rubber, turpentine, chlorophyll, cholesterol, and a variety of other naturally occurring compounds. The presence of these isoprene compounds is evidence that petroleum originated in living organisms.

$$(-CH_2\overset{\overset{\textstyle CH_3}{|}}{C}=CHCH_2-)$$

Compounds in petroleum have molecular weight ranges smaller than those found in coal. Petroleums are classified as sweet or sour, light or heavy. Sweet petroleums are low in sulfur; sour petroleums contain several percent sulfur. Light petroleums are free flowing liquids; heavy petroleums are viscous liquids that are difficult to pump and refine.

Natural gas is a by-product of the geological processes that produce petroleum and coal. Methane, the principal component of natural gas, is responsible for the fires and explosions that sometimes occur in underground coal mines. When geological formations are favorable, natural gas may be trapped in domes under impermeable rock formations. Because natural gas is low in sulfur and nitrogen, it is an environmentally cleaner fuel than either coal or oil.

■ Supplies of Fossil Fuels Are Limited

Concern for future supplies of fossil fuels is not new. Both gloomy and optimistic projections of supply and demand abound. These projections make very different assumptions about the growth rate of population, per capita energy use, and quantities of oil, gas or coal that remain to be discovered. However, the supplies of fossil fuel are finite and will run out someday. Choices made and implemented in the near future will affect the amount of time available to develop alternate energy sources.

Supplies of easily available oil and natural gas may be exhausted early in the twenty-first century. The known reserves of petroleum have remained nearly constant for many years because oil was being discovered almost as fast as it was being used. However, the average size of the newly discovered oil fields has been growing smaller and smaller. The future availability of natural gas is more certain, for new discoveries have led to increased estimates of reserves.

The quantity of petroleum in reserves is tied to the price of oil. When the price of oil rises, it may become attractive to reopen spent oil fields and to extract much of the original oil that remains. New petroleum recovery techniques that include the injection of high-temperature steam or detergents serve to free residual oil from the rocks and sands to which it clings. These tertiary petroleum extraction techniques add to the cost of producing oil, so they are not used while the price remains relatively low.

Supplies of coal are much greater than those of oil and gas. It is estimated that there is enough coal in the United States to supply this country's energy needs for 500 years. However, mining that coal can have serious environmental impact. In eastern mountain areas, strip mining has led to the erosion of slopes and the pollution of streams. In the west, abundant coal is found in areas where water is already in short supply. The water needed to mine and wash coal is also needed for agricultural irrigation and urban uses. The cost of coal varies with shipping costs and with the costs of the treatments to lessen adverse environmental impacts.

■ Acid Rains Present a Threat to Our Environment

It is not only the exhaustion of finite resources that is of concern. The use of large amounts of fossil fuels contributes to the production of acid rain and it may, in time, seriously disrupt the climate of the world. (The contribution of CO_2 emissions from the combustion of fossil fuels to global warning was discussed in Chap. 7.)

An unwanted side effect of burning large amounts of fossil fuels is the ejection of large amounts of nitrogen oxides and sulfur oxides high into the atmosphere as shown in figure 12.1. (The high-temperature reaction of nitrogen and oxygen to produce NO and NO_2 was discussed in Chap. 11. The oxidation of sulfur, present in many untreated fuels, first to form SO_2 and then to form SO_3, was discussed in Chap. 9.) Further reactions of these gases with water and with oxygen in the air lead to the formation of nitric and sulfuric acids. Rain carries these strong acids back to the earth, often hundreds of miles downwind of their source.

Higher smokestacks, once thought to minimize the effects of pollution by diluting these gaseous by-products of power production, have only served to increase their spread. In fact, the use of high stacks may have aggravated problems, for the gases remain aloft longer and have more time for reactions that produce strong acids. Because sulfur oxides and nitrogen oxides dissolve in water, they are concentrated in the tiny water droplets of clouds. Reactions occurring in clouds may hasten the formation of strong acids.

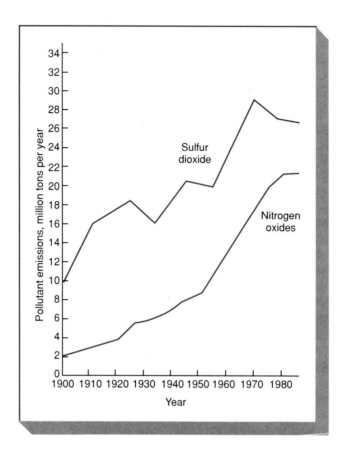

Figure 12.1
Sulfur dioxide and nitrogen oxide emission trends—national totals.
Adapted from *Acid Rain and Transported Pollutants*, Office of Technology Assessment, Congress of the United States.

Damage due to acid rain was first noted in Scandinavian lakes and forests, and in the northeastern portions of the United States and Canada. Since then, the effects have spread in ever-widening rings (see Fig. 12.2). The effects are irregular—fish may be killed in one lake, while a nearby lake continues to support a full spectrum of aquatic life. In some lakes, basic rocks, such as limestone, neutralize the acids:

$$H_3O^+ + CaCO_3 \rightarrow Ca^{+2} + HCO_3^-$$

Since the granites of the Adirondacks and the Canadian Shield do not neutralize the acids, these regions are particularly susceptible to damages.

While the initial concern about the effects of acid rain was focused on lakes and streams, attention has turned toward its effects on forests. In the higher elevations of New England, growth rates for trees have dropped dramatically during the last twenty years. Concern for the loss of vegetation and increased erosion is growing. Far greater damage to trees is found in the Black Forest region of Germany.

While the adverse effects on northern lakes and forests are apparent, the measurement of acid rain presents difficulties. The pH of "pure" rainwater is below 7. Rain is already naturally acidic, for it contains dissolved CO_2, and a solution of carbon dioxide in water is acidic. It is the presence of small amounts of the strong acids, HNO_3 and H_2SO_4, that is thought to be responsible for damages. Measured pH values may vary since the pH of unbuffered solutions is sensitive to small amounts of impurities and to changes in the amount of dissolved CO_2. It may be more reliable to concentrate samples and to measure the anions, NO_3^- and SO_4^{-2}, than it is to measure the pH (see Fig. 12.3). Indeed, proposed goals for the control of acid rain are stated in terms of the reduction of sulfur emissions.

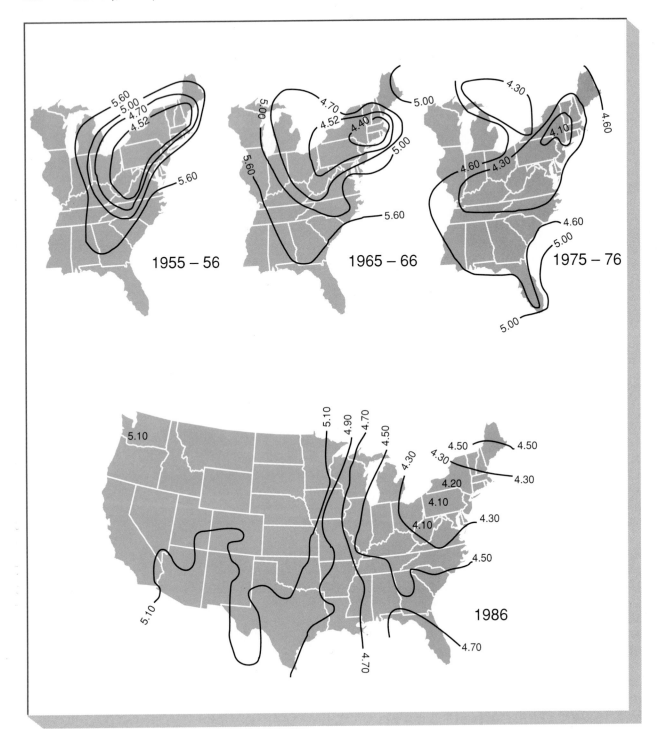

Figure 12.2
Precipitation acidity. Areas affected
by acid rain have grown in the last
four decades.

Sources: Department of Energy, *Acid Rain
Information Book: Final Report,* DOE/EP-0018
(Springfield, VA: National Technical Information
Service, 1981), pp. 3–44; National
Atmospheric Deposition Program, *NADP/NTN
National Data Summary: Precipitation Chemistry
in the United States* (Fort Collins, CO:
NADP/NTN, 1986, p. 26.

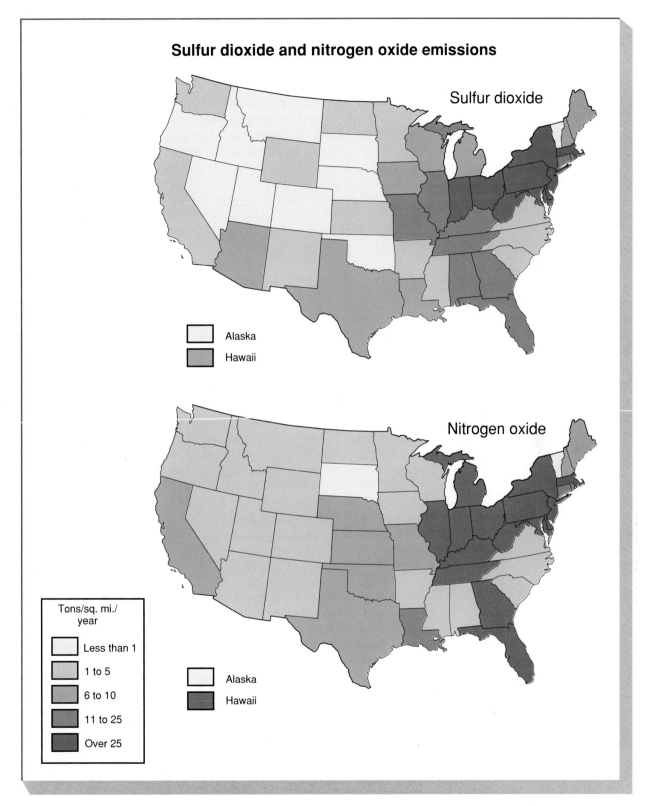

Figure 12.3
Average density of sulfur-dioxide
and nitrogen-oxide emissions.
Source: EPA, National Emissions Data System,
1987.

■ **Aside**

What About Acid Rain Causes Biological Damage?

What causes the mortality of fish in northern lakes and the deaths of trees at higher elevations? Scientists trying to answer these questions offer no simple answers. It is not yet certain which pollutants are the primary culprits, or how they act. Our lack of knowledge about cause-and-effect relationships leading to biological damage raises questions for control strategies.

It is not simply the acidity of rain that is destructive. Coal has been used in great amounts for many decades, yet the damages associated with acid rain have appeared recently. Some evidence points to nitrogen oxides and ozone as major contributing factors. Other evidence suggests that occasional events involving high acidity may be more important than average levels of acidity.

The complexity is illustrated by the damage to forests at higher elevations. It is found that high levels of acids are present in clouds; trees at high elevations are frequently bathed in clouds. But what causes the damage? Do the acids and ozone damage leaves, causing nutrients to be leached? Or does the nitric acid overfertilize trees so that they do not adequately prepare for winter? We do not know. The damage to the forests in Germany, where automobile emissions were not controlled before 1984, is worse than that found in the United States; an observation that supports some role for nitrogen oxides and ozone in acid rain action.

Similarly, evidence for the causes of fish kills is mixed. It has been found that the pH of high-elevation streams is lower in the spring than at other times of the year. Perhaps acid deposition accumulates in snow and ice, so that a surge of acidity accompanies the snow melt. When in their life cycle are fish most vulnerable? The timing of acidity peaks may be important.

Complete answers to questions of cause may be slow in coming. In the absence of this knowledge, the effects of various control strategies on emissions cannot be known with certainty.

■ Strategies to Reduce Acid Rain

The control of acid rain is less a technical problem than it is a political and an economic one. Because much of the emissions come from large utilities and industries, control is possible. A low-cost approach to achieving modest reductions in SO_2 emissions involves the burning of coal containing less sulfur. Simply washing coal reduces the amounts of impurities high in sulfur such as iron pyrite, FeS_2. To switch from burning higher-sulfur coals mined in the Midwest and northern Appalachians to lower-sulfur coals mined in the west or southern Appalachians would further reduce sulfur emissions.

The technology to reduce further the emission of oxides of nitrogen and sulfur exists, but it is more expensive. The installation of scrubbers in smoke stacks to dissolve acidic gases so they can be neutralized is one approach. In another approach, known as fluidized bed combustion, pulverized coal is mixed with crushed limestone and then burned. The basic limestone neutralizes much of the acidic gases even as the coal is burned:

$$SO_{2(g)} + CaCO_{3(s)} \rightarrow CaSO_{3(s)} + CO_{2(g)}$$

Passing the exhaust gases through scrubbers then removes much of the remaining sulfur and nitrogen oxides.

The capital costs of emissions control are high. At a national level, who should pay for the control of acid rain? Should utilities in the Northeast and Midwest pay the full costs? If so, will the industries in those regions be at a competitive disadvantage? To complicate the issue further, coals found in the Midwest are high in sulfur, while those in the West are low. If we shift the sources of our fuels, there are economic dislocations; if we do not, there are larger capital costs for pollution control.

Acid rain is both a national and international problem. Can its control be approached at just a national level? Environmental costs of pollution are paid by those downwind from the power producers and consumers. Many Canadians believe that the United States' failure to take steps to control sulfur emissions poses the greatest threat to good relations between the two countries. In Europe where boundaries are closer, no single nation can hope to control acid rain.

The United States government has set targets for the reduction of sulfur and nitrogen oxides during the coming decade. By the year 2000, the total emissions of sulfur oxides are to be reduced 10 million tons per year below 1980 levels. In addition, nitrogen oxide control devices must be installed on both existing and future generating plants. European countries have set still more stringent reduction goals.

■ Nuclear Fission Is a Source of Energy

Forces holding protons and neutrons together in atomic nuclei are far greater than those binding electrons to nuclei in atoms. In the fission of isotopes of uranium, the product nuclei have less mass and are more tightly bound than the reacting nuclei. Large quantities of energy are released as the kinetic energy of the particles produced by fission, and as high-energy photons called gamma rays.

$$\ce{^{235}_{92}U} + \ce{^{1}_{0}n} \rightarrow \ce{^{92}_{36}Kr} + \ce{^{141}_{56}Ba} + 3\ce{^{1}_{0}n}$$

$$\ce{^{235}_{92}U} + \ce{^{1}_{0}n} \rightarrow \ce{^{141}_{55}Cs} + \ce{^{93}_{37}Rb} + 2\ce{^{1}_{0}n}$$

The rate of energy production depends on the rate of fission reactions. If an average of one or more neutrons from each fission is captured by another uranium 235 isotope, a chain reaction can occur; if fewer neutrons are captured, a chain reaction is not sustained (see Fig. 12.4). If the amount of uranium 235 is above the critical mass, a chain reaction occurs and large amounts of energy are released. (In an atomic bomb two subcritical masses of uranium 235 are blown together by a conventional explosive. Then, in a small fraction of a second, the energy of the fission reactions is released before the force of the explosion separates the remaining uranium 235 and stops the chain reaction.)

In a nuclear reactor (see Fig. 12.5) the nuclear fuel is less concentrated than that used in the manufacture of weapons. (Power grade uranium fuel cannot explode because it is far from being concentrated enough to do so.) The rate of fission is governed by control rods containing cadmium and boron, two elements that absorb neutrons efficiently:

$$\ce{^{1}_{0}n} + \ce{^{113}_{48}Cd} \rightarrow \ce{^{114}_{48}Cd}$$

$$\ce{^{1}_{0}n} + \ce{^{10}_{5}B} \rightarrow \ce{^{11}_{5}B}$$

The fraction of uranium 235 decreases as the nuclear fuel is consumed. By withdrawing the control rods gradually, the rate of heat production by the reactor can be maintained. Should nuclear fission begin to occur too rapidly, the control rods can be thrust in to quench the reaction by reducing the flux of neutrons.

Figure 12.4
The rate of nuclear fission depends on the efficiency of neutron capture. If the mass of fissionable material is below a critical mass, too many neutrons are lost and a chain reaction is not sustained. If the mass of fissionable material is far above the critical mass, a highly branched chain reaction can occur, leading to an explosion. By controlling the mass of fissionable material and the flux of neutrons, a controlled, sustained rate of fission can be maintained.

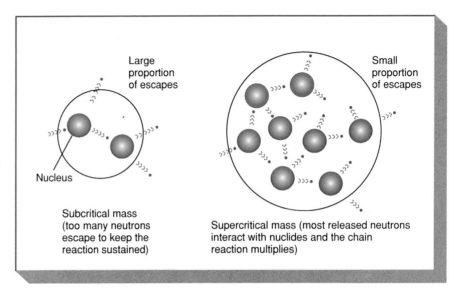

Large proportion of escapes

Small proportion of escapes

Nucleus

Subcritical mass (too many neutrons escape to keep the reaction sustained)

Supercritical mass (most released neutrons interact with nuclides and the chain reaction multiplies)

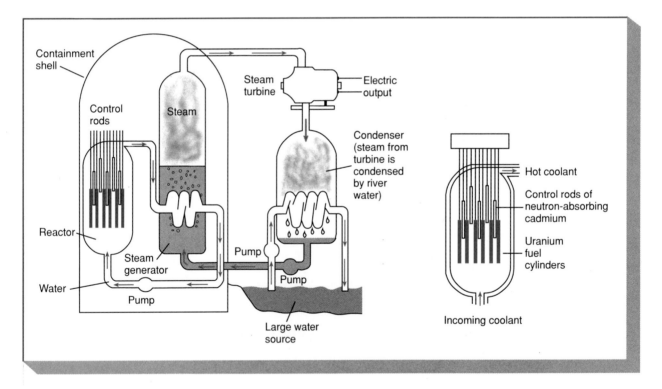

Containment shell

Control rods

Steam

Steam turbine

Electric output

Condenser (steam from turbine is condensed by river water)

Reactor

Steam generator

Pump

Pump

Water

Pump

Large water source

Hot coolant

Control rods of neutron-absorbing cadmium

Uranium fuel cylinders

Incoming coolant

Figure 12.5
In a nuclear reactor, the energy released by fission heats water to form steam. The steam is then used to generate electricity in the same way as in fossil-fuel-operated power plants.

In addition, a reactor contains a moderator, usually water or graphite, a form of carbon. Slow neutrons are captured more effectively in collisions with uranium 235 than are fast neutrons; collisions with water molecules or carbon atoms slow down the high-speed neutrons ejected at fission.

A heat exchanger is used to transfer thermal energy from the reactor core to a secondary loop for use in the generation of electricity. As in the case of the fossil fuel generation of electricity, only a part of the heat energy produced by the nuclear reactions can be converted into electricity.

The naturally occurring amount of uranium 235 is low; it comprises only 0.7 percent of uranium. For weapons use, or for the production of power, uranium is enriched to increase the fraction of uranium 235. It is difficult and expensive to separate isotopes of uranium. For example, during World War II, uranium enrichment facilities were built at Oak Ridge, Tennessee, and Paducah, Kentucky. Small differences in the rates of diffusion of uranium fluoride gas (UF_6) that differed in mass were used to produce weapons grade uranium. Since the enrichment in one step is small, the enrichment process had to be repeated thousands of times. If the use of uranium 235 for the production of power grows, the supply of this isotope will be depleted early in the next century.

■ Radioactive Wastes Pose Technical and Political Problems

The development of nuclear power is at a standstill in the United States; there are too many uncertainties about the future of nuclear power for utilities to commit large amounts of capital to it. The American public perceives nuclear power as more dangerous than most experts do. Public fears center around the disposal of nuclear wastes and the risks associated with accidents like those at Three Mile Island and Chernobyl. Environmental impact hearings, changes in regulations, and construction delays have added years to the completion of some plants and completely halted construction of others. In contrast, the state-run nuclear power industry in France receives broad support. There, the selection of a site for a nuclear plant and the construction of the plant can be completed with a minimum of delay.

Nuclear power plants produce radioactive wastes because the products of the nuclear fission reactions occurring in a reactor are themselves radioactive. In addition, radioactivity is induced in many of the components of the reactor that are exposed to a high flux of neutrons. Radioactive wastes differ greatly in their half-lives. High levels of radioactivity of short-lived isotopes present a relatively simple, short-term problem, for these materials can be stored until their radiation subsides (after 10 half-lives, radiation from any isotope is about one-thousandth of its original level).

Spent fuel rods from nuclear reactors contains fissionable plutonium. Plutonium 239, produced by the reactions of neutrons with uranium 238, has a half-life of 24,000 years. This half-life is short enough that the level of radioactive decay is high, but is long enough to present a long-term disposal problem. Plutonium is an alpha emitter; its compounds would be highly toxic if ingested or inhaled. The recovered plutonium may serve as a fuel in a nuclear reactor or it may be used to construct nuclear weapons. Since plutonium differs from uranium in its chemistry, it is far easier to separate plutonium from uranium than it is to separate the isotope uranium 235 from uranium 238.

However, the recovery and recycling of plutonium presents risks. One concern is that of possible nuclear proliferation if the plutonium falls into the wrong hands. Shipping spent fuel rods risks accidents and spills as well as the chance that terrorists could hijack a shipment and use the stolen plutonium to construct a bomb. Nuclear powers have recovered plutonium from fuel rods from specially designed reactors, have transported that plutonium, and have used it to construct nuclear weapons. The United States has chosen not to recycle fuels from nuclear power plants, but France has chosen a different policy and is recycling its own fuels as well as those from other nations.

To dispose of high-level radioactive wastes, a satisfactory means of containing them needs to be found. Proposals for the permanent disposal of radioactive wastes call for the wastes to be concentrated, encapsulated in glass or ceramic forms, and buried in geological formations that are impermeable to water. Some environmentalists are concerned that we cannot be certain that these wastes would not reenter the environment at a future time. In the meantime, spent fuels are accumulating at reactor sites and some operating reactors may be closed when the limit of allowed on-site storage of radioactive materials is reached.

The Nuclear Waste Disposal Act of 1982 provided strategies for the disposal of radioactive wastes. Low-level wastes such as those generated by hospitals were to be buried in specially designated landfills. Permanent disposal sites for high-level wastes are to be chosen and put into operation in the 1990s. However, state governments and the executive and legislative branches of the federal government have been reluctant to make the final difficult choices that will lead to permanent disposal of high-level wastes. The absence of a clear disposal policy makes it impossible to project costs for future nuclear power production. It is necessary to solve the problem of waste disposal if nuclear power is to contribute to meeting the future energy needs of the United States.

■ Breeder Reactors Would Extend the Future of Nuclear Power

Nuclear reactors designed to produce both power and fissionable isotopes are called breeder reactors. Breeder reactors have the potential to extend the supply of nuclear fuel more than a hundredfold, for they produce more fissionable atoms than they consume. As the reactor operates, uranium 238 is converted into fissionable plutonium 239. The plutonium produced by breeder reactors could be separated and used to fuel conventional reactors:

$$\begin{align}
{}^{1}_{0}n + {}^{238}_{92}U &\rightarrow {}^{239}_{92}U \\[6pt]
{}^{239}_{92}U &\rightarrow {}^{0}_{-1}e + {}^{239}_{93}Np \\[6pt]
{}^{239}_{93}Np &\rightarrow {}^{0}_{-1}e + {}^{239}_{94}Pu
\end{align}$$

The design of a breeder reactor differs greatly from that of a conventional reactor. The core of a breeder reactor is small and the operating temperatures are high. Liquid sodium is used as the coolant. Breeder reactors are inherently safer in operation than conventional reactors. Should fission speed up and begin to go out of control, the liquid sodium would cool by convection (should the pumps fail) without nearing its boiling point. In tests using experimental breeder reactors, it has been shown that the core expands as temperatures rise, lowering the neutron flux and slowing nuclear fission.

The future of breeder technology is uncertain. In 1984, Congress discontinued the funding for the Clinch River Breeder Reactor, a reactor intended to demonstrate the commercial potential of this technology. Breeder options are being pursued in Europe but not in the United States, so the uncertainties about the cost of producing power using this technology remain. The choice to pursue conventional nuclear technology rather than breeder technology is, in part, a historical accident.

The Navy was averse to developing breeder reactors to power nuclear submarines, because sodium reacts violently with water. The peacetime nuclear industry followed the Navy's lead in reactor design. However, reactor features desirable on a submarine and those desirable for the peacetime generation of power need not be the same.

■ The Nuclear Accidents at Three Mile Island and Chernobyl

Two major nuclear accidents have eroded the public's confidence in nuclear power and shaken the complacency of the nuclear power industry. In 1979, an accident at the Three Mile Island nuclear reactor in Pennsylvania disabled the reactor and melted a part of the core. Some radioactive gases were vented, but the fuel and the majority of the radioactive species did not escape the containment vessel. In 1986, a far more serious nuclear accident occurred near Chernobyl in the former Soviet Union. There a large release of radioactivity accompanied the reactor meltdown.

The two accidents did have striking similarities—in each case, a reactor with inherent design weaknesses was operating near its limits as a part of a test procedure. In each case, operators had turned off safety devices contrary to operating procedures. In each case, the accident was due to human error by inadequately trained operators working with too little supervision. In each case, the accident has led to a review of design, of operating procedures, and of training, In each case, the accident has led to heightened public concerns about nuclear power.

In the Three Mile Island accident, not enough cooling water reached the reactor and the water that was already there vaporized to steam. The heat from the nuclear reactions melted the zirconium alloy fuel rods in part of the core. In addition, the hot metal reacted with the steam to form metal oxides and hydrogen gas. A buildup of pressure was relieved, in part, by explosions due to the reaction of hydrogen and oxygen. However, when the cooling water was lost, neutrons were no longer slowed by collisions with water molecules, and the efficiency of neutron capture dropped, slowing nuclear fission.

At Chernobyl, in an experiment designed to take place at low reactor power, the power dropped more than was planned. To correct this condition, operators removed additional control rods causing the power to come up too fast to be controlled. In this case, there was not a strong containment vessel and a part of the fuel and its fission products were vented into the atmosphere. In the Chernobyl reactor, graphite was used to slow neutrons, so the loss of cooling water did not slow fission to nearly the extent that it had at Three Mile Island. In addition, the graphite burned for several days in a conventional blaze that carried radioactive materials high into the atmosphere (as part of the firefighting procedure, boron carbide was dropped on the blaze to absorb neutrons and slow the continuing fission).

■ Nuclear Fusion May Someday Provide Energy

The fusion of hydrogen nuclei (the deuterium isotope) to form helium releases far more energy than does the fission of the nuclei of heavy elements.

$$_1^2H + {}_1^2H \rightarrow {}_2^4He + energy$$

However, this reaction is hard to bring about. As positive hydrogen nuclei approach one another, the repulsive forces are great. In a hydrogen bomb, for example, the explosion of a fission bomb is used to implode the fuel and hold it together long enough for fusion to occur. It is this hydrogen fusion that fuels the Sun and other stars. In stars, the combination of strong gravitational forces and high temperatures cause nuclear collisions to be sufficiently frequent and violent to sustain fusion despite the repulsive forces present.

For the production of power, a means to contain the nuclear fuel needs to be found since the required temperatures exceed the melting point of conventional materials. Two approaches are currently being taken. One is to contain the ionized hydrogen gas (plasma) magnetically. The other is to use lasers to heat lithium hydride to temperatures high enough to cause the lithium nucleus to fuse with the hydrogen nucleus:

$$^{6}_{3}\text{Li} + ^{2}_{1}\text{H} \rightarrow 2\ ^{4}_{2}\text{He} + \text{energy}$$

The research on fusion continues, but it is not yet known whether this technology will ever contribute to future energy needs.

■ Solar Energy and Other Alternative Energy Sources

Solar energy is the largest underdeveloped source of energy available. The energy of the sun can be used to heat water and homes or it can be used to generate electrical power. However, the amount of solar energy striking the earth's surface varies with the time of day, location, climate, and season. Although solar power can help to meet our energy needs in some ways, in some places, it is not a panacea.

The use of solar energy for direct heating purposes is called passive solar energy. The use of electrical energy, particularly resistance heating, to heat homes or water is wasteful. A portion of the available energy is degraded to heat in the production of electricity, in the transmission of electricity, and in the conversion of electricity back into heat. There is a great potential for saving energy by using the sun for heating purposes in many locations, and the technology is already available.

The conversion of sunlight into electrical energy, sometimes called photovoltaic energy, is a technology still under development. In one application, solar-powered, photovoltaic devices have been used to provide power to orbiting satellites. The manufacture of solar voltaic materials is still expensive, but the cost has dropped considerably during the past decade.

In another approach to harnessing solar energy, sunlight is used to drive the electrolysis of water to generate hydrogen at one electrode and oxygen at the other. The hydrogen serves to store up energy. The hydrogen is then oxidized by oxygen in fuel cells to generate electricity. This approach to solar energy would make it available when there is a need for electrical energy, and not just when the sun is shining, although the large-scale use of hydrogen would present difficulties and dangers.

There remains a variety of undeveloped energy sources that can be used to help meet future needs. Energy in motion can do work, so the movement of wind and of the tides can be harnessed for energy production. At certain geological hot spots, the heat from geothermal wells can be used to produce energy. To the extent that these resources are developed, supplies of finite resources can be conserved.

■ Energy Efficiency and Conservation

Low-cost energy is a thing of the past. We live in a society in which energy is used to grow our foods, clothe our bodies, warm and cool our homes and workplaces, and to transport people and goods. A simpler life may be desirable for a few, but it cannot sustain civilization as we know it. At the same time, we are faced with dwindling resources and the destruction of environments caused by energy production by-products.

"Conservation" is a bad word to some individuals, who associate it with a call to reject an industrial society. Energy efficiency may be synonymous with conservation for many others. Energy is too valuable to waste. The careful use of energy in the present can extend energy supplies for the future, and reduce the demands on the environment. At the very least, it buys the world time. But third world countries see industrialization and increased energy use as the way to escape from their poverty. Whether or not western societies increase energy use or efficiency, there will, in all probability, be greater energy demands in the world.

What will be the energy legacy that we leave to the twenty-first century? This chapter pointed out the environmental concerns associated with each of the major sources of energy. These risks need to be evaluated and our options chosen carefully. We have a short time in which to plan for the future, and a part of that planning should be to keep a variety of options open. We need to develop nuclear power, solar power, and alternative energy sources, or by default we will find ourselves relying exclusively on coal. The choices may not be easy, and the efforts to keep open the threatened options for research and development—particularly those of nuclear power—may not be politically popular, but to close our options now may be dangerously short-sighted.

■ *Reflection*

The Environment and the National Interest

As the era of the Cold War came to a close, people became aware of the environmental degradation that had occurred in Eastern Europe. State-run economies had pushed industrial production at the expense of the environment and they had used primitive technologies that caused widespread, serious damage to air and water quality.

At about the same time, the American public was informed that there was serious radioactive contamination at a number of sites in this country. Facilities producing nuclear weapons for the military had not been held to the same standards as those pursuing peaceful uses of atomic energy. The cost of the cleanup of these sites is expected to be hundreds of billions of dollars. As the cleanup of past radioactive contamination is begun, a welcome new problem is emerging. How should we dispose of the nuclear material in warheads as we move to lower the number of nuclear weapons?

Does our national interest depend only on our military and economic strength? Does it also include our nation's environmental interests? These are questions with important policy implications. Should American aid to developing countries be tied to their environmental practices? To their efforts to control their population growth? Or should that aid be tied only to their trade policies and their potential military cooperation?

---

1. During the 1970s and early 1980s, the price of natural gas was regulated at a low level. What are the short-term and long-term effects of such a regulatory policy on the availability of natural gas?

2. Why is natural gas considered a cleaner fuel than petroleum?

3. Which fossil fuel is most nearly free from noncombustible contaminants when it is extracted from the earth? Which is least free?

4. Why does oil low in sulfur command a premium price on the market?

5. Because of interruptions to the supply of petroleum from the Middle East, western nations have sought to reduce their dependence on foreign oil. During the early 1980s, America urged that allies not furnish equipment for the completion of a pipeline to deliver natural gas from Siberia to western Europe. Why did European nations believe that it was not in their national interests to support the boycott?

6. In the early development of the automobile, low-sulfur oil from Pennsylvania was used as a lubricant. Why would oil high in sulfur have an adverse effect on engine life?

7. Statues of marble (a metamorphic rock that is chiefly $CaCO_3$) have shown dramatically increased rates of destruction since automobiles have come into wider use. Why? Write the equation for the reaction responsible for the damage.

8. Is there the same need to control auto emissions in Montana and Wyoming as in California? Should the regulation of emission controls be undertaken by the states or by the federal government? Why?

9. Ohio is a state that produces little energy but uses much. How would congressional delegations from Ohio and Vermont differ on the need for control of acid rain and on the formula for cost allotment? Should the regulation of emissions be undertaken by the states or by the federal government? Why?

10. How do increases in the price of oil affect the funding of research directed toward the development of new sources of energy? As the price of oil dropped during the mid-1980s, what happened to the level of funding for developing alternative energy sources?

11. In mountain regions of the eastern United States, trees at higher elevations are stunted in growth more than the same species at lower elevations. Why might acid fog (present in low-lying clouds) pose a wider threat to the environment than acid rain?

12. Proposals to use electric cars for driving in cities cite reduced pollution as a major benefit. In what ways would this be true? In what ways would it be false?

13. Both nitrogen oxides and sulfur oxides contribute to air pollution. Acid rain west of the Mississippi River has more nitrogen oxides than that east of the Mississippi. Describe how regional differences in power production and the use of automobiles might account for this variation in the source of acid rain.

14. Why is a nuclear power plant better able to respond to changes in electrical demand than a coal-fired steam plant?

15. Should decommissioned nuclear submarines be scuttled in deep parts of the ocean as they have been in Russia? Why or why not?

16. What scientific reason was there for the several year delay in cleaning up the damaged nuclear reactor at the Three Mile Island?

17. Why do the tailings from plants that separate compounds of uranium from its ore present a disposal problem?

18. Third world countries purchasing nuclear reactors from developed countries are concerned with fuel reprocessing capabilities. Why are developed countries hesitant to share this technology?

19. Should environmental groups have the right to seek injunctions to halt construction of new power plants for environmental reasons? What is the effect of a one year delay on the construction costs? (Add interest and inflation.) What would be the effect of the loss of this privilege on the right of the public to seek policy reviews?

20. A nuclear fission chain reaction is believed to have occurred hundreds of thousands of years ago in a uranium-containing deposit in Africa. Which of the following lines of evidence would not be needed to support this hypothesis? Give a reason for your choice.
    (a) Lower than normal ratio of uranium 235 to uranium 238
    (b) Occurrence of plutonium
    (c) Observation of products of fission

21. What are the advantages of a breeder reactor over a conventional nuclear reactor?

22. Should the development of commercial breeder reactors be subsidized by the federal government? Why or why not?

23. If nuclear fission were to produce power commercially in the future, plants would need to be located in regions of high population density since the plants would produce very large amounts of electricity. (Electrical power is lost in transport over long distances.) Is fusion a political possibility in the United States? Give reasons.

13

Down to Earth Chemistry

At first glance, a glass chandelier, a ceramic vase, a concrete dam, and a freshly plowed field do not appear to have much in common. Yet each is fashioned by nature or by humans from rocks and/or weathered rocks. Potters fashion ceramics from clay, and farmers grow wheat in clay soils. People modify chemical structures by heating limestone with clay to make cement and by heating limestone with silica sand to produce glass.

There are similarities in the structures of ceramics, of concrete and glass, and of soils; they are all built up from a small number of chemical building blocks. There is a straightforward relationship between the structures and the properties of these familiar materials. In this chapter, we explore the everyday chemistry of glass, cement, ceramics, and agriculture.

■ Silicates Have a Variety of Structures

Silicon and oxygen are the two most abundant elements in the crust of the earth. Minerals called *silicates* contain these elements in varying ratios, combined with a wide range of other elements. The variety of silicates arises in part from the various ways that the tetrahedral silicate units, SiO_4^{4-}, can combine as shown in figure 13.1. Silicate tetrahedra can join to form dimers, trimers, linear chains, two-dimensional sheets, and three-dimensional networks. When tetrahedra are joined in a chain, two oxygen atoms of each tetrahedral unit are shared with neighboring silicon atoms. In a sheet, three of the oxygen atoms bonded to each silicon atom serve as bridges. In sand and quartz, each silicon atom is bonded to four other silicon atoms through these oxygen bridges.

The electrical charges on silicate species vary. A bridging oxygen atom connected to two silicon atoms carries no charge, but each oxygen atom bonded to a single silicon atom has a −1 charge. Two negative oxygen atoms and two neutral bridging oxygen atoms are bonded to each silicon atom in a chain. One negative oxygen atom and three neutral bridging oxygen atoms are bonded to each silicon atom in a sheet. All four oxygen atoms bonded to a silicon atom in the three-dimensional structures of quartz and sand form bridges and are, therefore, neutral.

In ionic silicate compounds, the negative charge on silicate anions is balanced by the positive charge on cations, but the identity of the cations can vary. For example, both the minerals asbestos $(CaMg(SiO_3)_2)$ and $LiAl(SiO_3)_2$ have long silicate chains, but in one of these minerals the negative charge is balanced by Ca^{2+} and Mg^{2+} ions, and, in the other, it is balanced by Li^+ and Al^{3+} ions.

Figure 13.1
Some of the wide variety of structures found in silicate minerals are illustrated. These structures can be considered to be built by combining tetrahedral silicate anions.

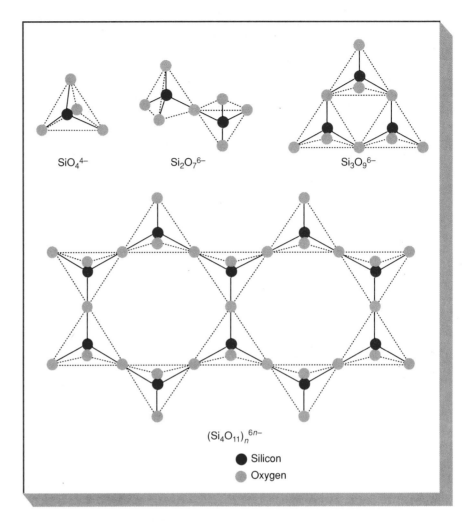

SiO_4^{4-} $Si_2O_7^{6-}$ $Si_3O_9^{6-}$

$(Si_4O_{11})_n^{6n-}$

● Silicon
● Oxygen

The physical appearances of some silicate minerals reflect their underlying molecular structure. For example, the mineral asbestos $(CaMg(SiO_3)_2)$ is fibrous, and its structure contains parallel silicate chains. Asbestos has been widely used for insulation, however, the small fibers in its dust contribute to the formation of lung cancers. When asbestos must be removed from schools and other public buildings, it is proving to be a slow and expensive process. A different structure is found in the mineral talc $(Mg_3(Si_2O_5)_2(OH)_2)$. Talc contains stacked two-dimensional silicate sheets. The layers can slide by one another, giving the mineral a slippery feeling (see Fig. 13.2).

■ Glass

To make common *soda-lime glass*, sand is heated with $CaCO_3$ (limestone) and Na_2CO_3 (soda ash). Heat decomposes the carbonates to give carbon dioxide and the oxides CaO and Na_2O:

$$Heat + CaCO_3 \rightarrow CaO + CO_2$$

$$Heat + Na_2CO_3 \rightarrow Na_2O + CO_2$$

(a)

(b)

Figure 13.2
The structures of some silicate minerals reflect their underlying molecular structure.

Left: Dr. Jeremy Burgess, Science Library, Photo Researchers. Right: J & L Weber/Peter Arnold, Inc.

Figure 13.3
In the formation of glass, basic O^{2-} ions attack Si atoms to convert the SiO_2 network to silicate chains.

Basic oxide ions furnish a pair of electrons to attack a silicon atom (a Lewis acid site) and to cut some, but not all, of the bridges of the SiO_2 network as shown in figure 13.3. The product of the reaction is a viscous liquid containing long silicate chains that vary in length and degree of branching. Positive sodium and calcium ions balance the negative charge of the stiff silicate chains (see Fig. 13.4).

When glass is heated to 600–700°C, it softens and becomes plastic. It can be blown into molds to make bottles, drawn into tubing, rolled into sheets, or fashioned into ornate figures. The silicate chains and cations in the glass are arranged somewhat randomly; they do not stack to form a crystal lattice. Upon cooling, the glass hardens and retains its shape.

Recipes for glass differ because the properties of glass are altered by the addition of different materials. Glass made with lead oxide (PbO) is used in crystalware since it refracts light more than ordinary glass and appears more brilliant. Manufacturers combine boron oxide (B_2O_3) with SiO_2 and Na_2CO_3 to make *borosilicate* glass that expands less than ordinary glass on heating. Borosilicate glass is used for the manufacture of cooking utensils because it better withstands rapid heating and cooling.

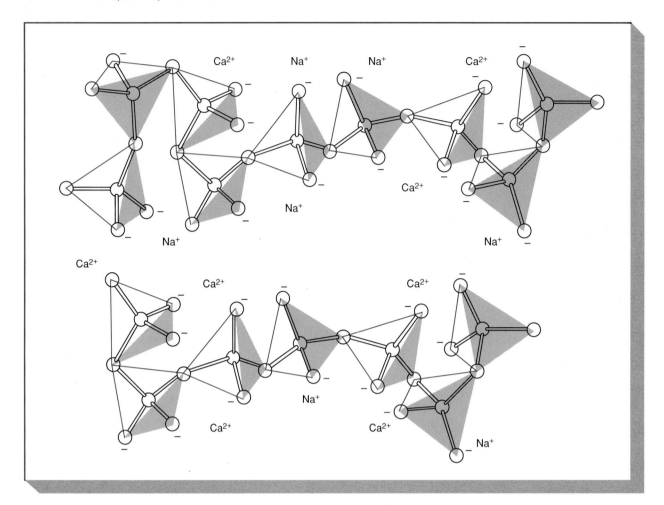

Figure 13.4
Glass consists of SiO_3^{2-} chains with cations to balance the negative charge.

■ Aside

Cement Is Dehydrated Rock

Cement is made by heating clay with limestone to 1500°C. The clay is dehydrated on heating and the limestone decomposes. Basic calcium oxide reacts with the Lewis acid sites of the clay to cut into the aluminosilicate structure. When water and sand are added to the dry cement, crystals of hydrated aluminosilicate minerals slowly form. The strength of cement increases as the crystals grow to form an interlocking network. Sand and gravel are added to the mix to make concrete. Reactions of sand with the basic components of cement help increase the strength at the interface between sand and cement crystals in concrete (see Fig. 13.5).

Plaster of Paris is also dehydrated rock. When water is mixed with powdered calcium sulfate hemihydrate, $(CaSO_4 \cdot 1/2H_2O)$ to form a slurry, crystals of hydrated calcium sulfate $(CaSO_4 \cdot 2H_2O)$ slowly form and the plaster hardens.

$$3\ H_2O + 2\ CaSO_4 \cdot 1/2H_2O \rightarrow 2\ CaSO_4 \cdot 2H_2O$$

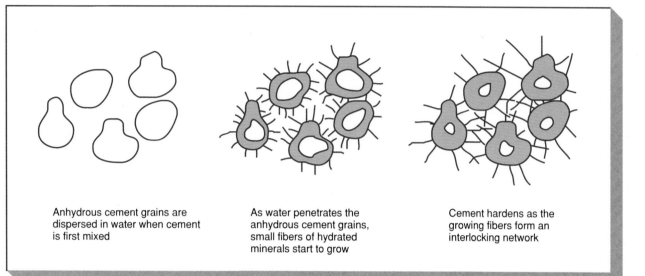

Anhydrous cement grains are dispersed in water when cement is first mixed

As water penetrates the anhydrous cement grains, small fibers of hydrated minerals start to grow

Cement hardens as the growing fibers form an interlocking network

Figure 13.5
Cements harden as water combines with the anhydrous minerals of cement grains to form a network interlocking hydrated mineral crystals.

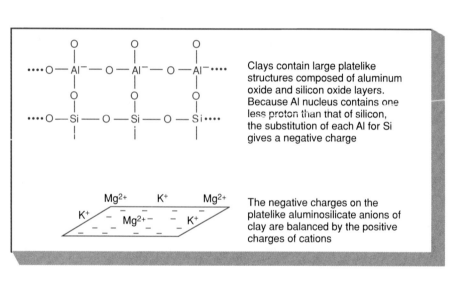

Clays contain large platelike structures composed of aluminum oxide and silicon oxide layers. Because Al nucleus contains one less proton than that of silicon, the substitution of each Al for Si gives a negative charge

The negative charges on the platelike aluminosilicate anions of clay are balanced by the positive charges of cations

Figure 13.6
Clays consist of large negative aluminosilicate anions with a variety of cations bound to the anions by the attraction of opposite charges.

■ Clays

Clays are *aluminosilicates*. Aluminum oxide, like silicon oxide, forms extended covalent structures. In clays, sheets of aluminum oxide are bonded to sheets of silicon oxide in sandwich-like layers. By substituting aluminum with a +3 kernel charge for silicon with a +4 kernel charge in an SiO_2 lattice or by substituting +2 magnesium for +3 aluminum in an Al_2O_3 lattice, charged anionic plates are formed. Cations are required to balance the negative charge of the polymeric anions in silicate minerals (see Fig. 13.6).

Clays bind water much more strongly than do silicate minerals. The cations that balance the electrical charge of anionic plates are solvated in tightly bound layers of water molecules. Because the water layers in clays can expand and contract, clays readily take up and release water.

■ Ceramics—From Clay to Glaze

Potters may prepare clay before shaping it by mixing clays from different sources. Ball clay, deposited from slow-moving streams, has fine particles that contribute plasticity to the mixture. Fire clay, deposited further upstream, has larger particles that contribute strength to ceramics and reduce shrinkage in its drying. Potters work the clay by adding water and by kneading it to align the aluminosilicate plates of clay particles.

Physical changes accompany the drying and firing of clay. When a clay pot is dried, weakly bound water evaporates and some shrinkage occurs. When a pot is fired, the remaining water is lost (a pot may shatter during heating if a trapped pocket of water turns to steam). As water is driven off, and the plates move closer together, forces between the bound cations and the anionic plates become greater. Finally the material is vitrified; water can no longer move between the bound plates, although the channels do remain, so the fired pot stays porous.

Glazes are used to color and sometimes seal ceramics. Many glazing compounds are carbonates. At firing, they are converted to oxides that react with the aluminosilicates of the clay to give it glassy surfaces. Other glazes simply impart a characteristic color to the surface.

The Indians of the American southwest controlled firing conditions in order to impart orange and black designs to their pottery. They used clays containing iron oxides that produce an orange Fe_2O_3 color when fired under oxidizing conditions. When a reducing atmosphere was introduced, for example, by putting leaves on the fire to give carbon monoxide, the orange Fe_2O_3 was reduced to black Fe_3O_4:

$$CO + 3\ Fe_2O_3 \rightarrow CO_2 + 2\ Fe_3O_4$$

Indians first fired the pottery under reducing conditions to obtain the black iron oxide. They then decorated the pots and covered part of the design with mud. When the pottery was fired under oxidizing conditions, the uncovered areas became orange and the unexposed surfaces remained black.

■ Aside

Colored Glass, Glazes, and Transition Metal Compounds

Glassmakers use small amounts of the compounds of transition metals, elements in which the $3d$ subshells are partially filled with electrons, to produce colored glass. Cobalt compounds produce deep blue cobalt glass, and iron compounds impart a brown color to glass beer bottles. Potters use glazes containing these and other transition metal compounds to color ceramics.

The ions of these transition metals are also colored in aqueous solutions where the origin of the colors have been studied. To absorb visible light, a transition metal ion must be surrounded by Lewis bases such as water molecules, ammonia molecules, or chloride ions. By changing the base around a given metal ion, chemists change the colors of the solution. For example, they can produce a lime-green chloride complex by adding hydrochloric acid to a pale blue solution of $CuSO_4$ in water. By adding ammonia to the same pale blue copper solution, they produce an intensely-colored deep blue complex of copper with ammonia.

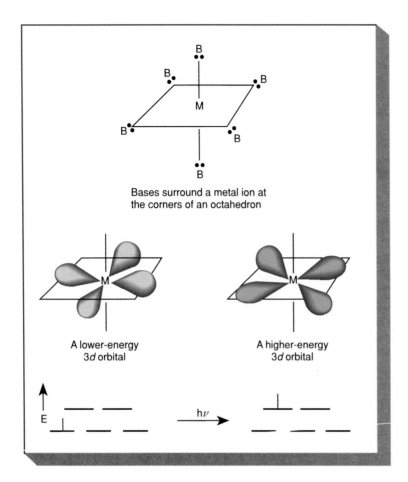

Bases surround a metal ion at
the corners of an octahedron

A lower-energy
3d orbital

A higher-energy
3d orbital

Figure 13.7
Transition-metal ions absorb light
when an electron is excited from a
lower-energy d orbital to a higher-
energy d orbital.

Compounds of transition metals are colored because they absorb some colors of the visible spectrum and transmit others. Consider an ion of a transition metal bound to six bases. The electrons binding the bases to the metal ions are at the corners of an octahedron. In such a complex, common for most of these ions, electrons in 3d orbitals differ in energy. The lower-energy d orbitals place electrons in regions between electrons of the bases. The higher-energy d orbitals place electrons closer to the electron pairs furnished by the bases. In these colored complexes, atoms of transition metals have one or more electrons in the lower-energy d orbitals, and they have one or more vacancies in the higher-energy d orbitals. Light with the right energy can excite an electron from a lower-energy d orbital to an open position in a higher-level d orbital (see Fig. 13.7).

■ Soils Release Nutrients by Ion Exchange

Consider a dilemma concerning potassium, essential for plant growth, that occurs with other nutrients in clay (and other) soils. Since simple potassium compounds are soluble, why hasn't all the potassium in soils been washed into the ocean? On the other hand, if potassium is tightly retained by soils, why is it available to growing plants?

Figure 13.8

Plants take up nutrient cations by ion exchange.

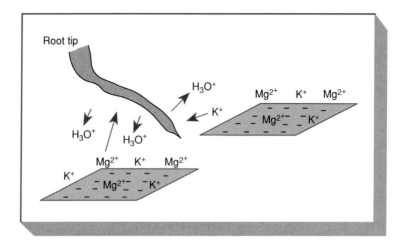

Mineral ions do not readily wash out of soil because the charge balance about a clay platelet or other charged soil particle must be maintained. Potassium, calcium, and magnesium ions needed by plants are bound to insoluble polymeric anions by strong electrostatic forces and are not free to migrate. Ion-binding sites also occur in organic soils, for organic material decays to produce anions of weak acids that act as cation-binding sites.

Plants take up mineral nutrients by *ion exchange* as shown in figure 13.8. Plants extrude hydronium ions from their roots, and they exchange the hydronium ions for nutrient cations in the soil. One H_3O^+ is extruded for each K^+ taken up, and two H_3O^+ are exchanged for each Mg^{2+}. When a plant dies and decays, the potassium and magnesium ions are released so they can again be captured by soil anions in exchange for H_3O^+.

Clay anions bind cations selectively. For example, they bind the +2 ions Mg^{2+} and Ca^{2+} more tightly than the +1 ions Na^+ and K^+. Among ions of the same charge, there is also selectivity; clays bind potassium ions more tightly than sodium ions. Clay particles washed out to sea selectively bind Mg^{2+} and K^+ from seawater rather than the more abundant Na^+, so soils deposited in river deltas retain high levels of nutrient ions.

This selective binding of potassium ions, rather than sodium ions, is important for agriculture. As water is used and reused for irrigation, the concentration of sodium chloride in the water builds up. The waters in the lower Colorado and Rio Grande rivers are somewhat salty, yet they are still used to irrigate crops. With good drainage, sodium ions do not extensively replace potassium ions. In poorly drained fields, the accumulation of salt is making once fertile fields unfit for plant growth.

The fertility of a soil depends, in part, on its ion exchange capacity. Fertile soils hold greater quantities of nutrient ions and release them to plants more readily. The availability of nutrients can depend on soil moisture; some clay soils hold cations more tightly as they dry and shrink. In these soils, growth occurs principally in the spring and early summer when the soil retains water from winter rains.

To grow in acid soils, plants must extrude hydronium ions into a more acidic environment. Because some ion-exchange sites are occupied by hydronium ions, fewer sites are occupied by nutrient ions. Treating a lawn or garden with lime CaO or ground limestone $CaCO_3$ adds needed calcium ions and neutralizes hydronium ions occupying sites. However, too much lime can cause iron, a nutrient needed in small amounts, to be tied up as insoluble $Fe(OH)_3$ and be unavailable to the plants.

■ Soil Degradation Follows the Burning of Tropical Forests

Soils found in tropical rain forests are chiefly composed of decaying organic vegetation. Much of the mineral content needed for plant growth is bound to ionic sites of these organic soils. Many indigenous peoples practice "slash and burn" agriculture. They clear a small plot, burn the organic material to free its mineral content, and then grow a variety of vegetables on the newly cleared land. After a few years, productivity decreases so they abandon old plots and move on to clear new ones. Regrowth of the forest renews the soil in small, abandoned clearings over a period of years.

In the Amazon basin and in some far eastern countries, far larger areas of rainforest are now being cleared for agriculture as indicated in figure 13.9. Crop yields are initially good since the fire releases nutrients tied up in decaying vegetation. However, the remaining soil lacks the ion-exchange capacity to retain mineral ions, so they soon wash out. The land slowly degrades, supporting livestock for a few years, and then hardens and becomes barren. As the vegetation cover shrinks, more soil is washed away, and downstream flooding becomes more frequent and severe.

The long-term degradation of the environment accompanying the effort to convert tropical rainforest to productive agricultural land far outweighs its temporary benefits. And unlike the effects of "slash and burn" farming, the effects of large-scale clearing and burning cannot be readily reversed.

■ Aside

Strontium 90 and the Test Ban Treaty

In the 1960s, public concern for the dangers posed by a single radioactive isotope, $^{90}_{38}Sr$, led to the adoption of a limited nuclear test ban treaty. Strontium 90 is one of the radioactive products of nuclear fission. It undergoes beta decay with a half-life of 29 years. Atmospheric testing of nuclear weapons at that time introduced radioactive material high into the atmosphere, and the resulting radioactive fallout was scattered over wide areas of the globe.

Since Sr^{2+}, like Mg^{2+} and Ca^{2+} in the same family of the periodic table, binds to clay, strontium 90 in radioactive fallout is retained by soils. By the process of ion exchange it is taken up by grasses. Dairy cows then consume large quantities of this contaminated grass, grain, and ensilage, and they concentrate strontium along with calcium in their milk (see Fig. 13.10).

Radioactive strontium in milk increases the risk of leukemia, a cancer of the blood. Strontium accompanies calcium in the formation of bones; as a result, the radioactivity of strontium 90 can be concentrated in the bones. The risk of leukemia is especially great for children since they drink more milk than most adults, and they are building new bone tissue.

Linus Pauling (b. 1901), a chemist who had received the Nobel Prize for his contributions to bonding theory and to the understanding of molecular architecture, led the drive for the end of atmospheric testing of nuclear weapons. A partial test ban treaty was concluded in 1963. Pauling received the Nobel Peace Prize and became the first chemist since Marie Curie to receive two Nobel prizes.

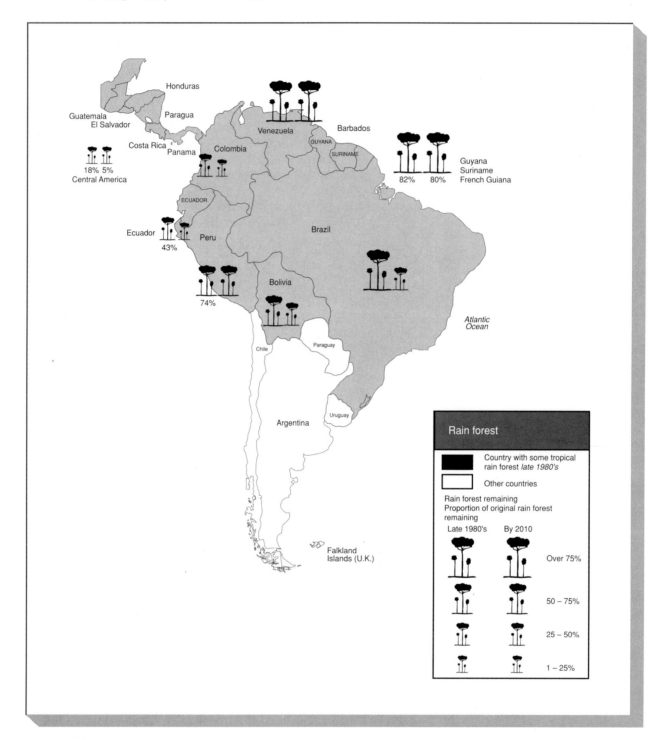

Figure 13.9
Loss of rainforest in South America.
Sources: Campbell & Hammond (1989); Myers & Houghton (1989). Data compiled by Friends of the Earth/Christine Lancaster.

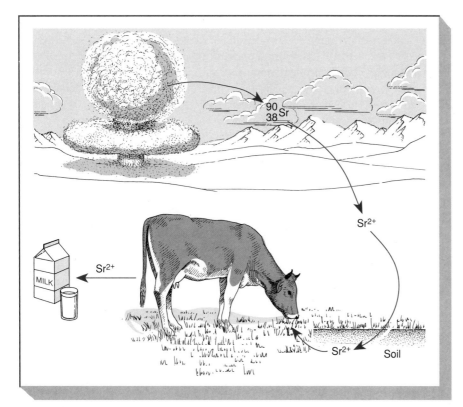

Figure 13.10
Concern for leukemia caused by radioactive strontium helped lead to the ban of atmospheric testing of nuclear weapons. Because strontium resembles calcium in its chemistry, it can accompany calcium in the food chain and be deposited in the bones of growing children. Radioactivity in bone tissue can produce leukemia because red blood cells are produced in bone marrow.

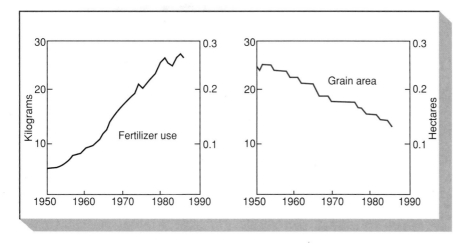

Figure 13.11
World fertilizer use and grain area per capita, 1950-86.

Reproduced from *State of the World*, 1987, A Worldwatch Institute Report on Progress Toward a Sustainable Society, Project Editor: Lester R. Brown. By permission of W. W. Norton & Company, Inc. Copyright © 1987 by the Worldwatch Institute.

■ Chemical Fertilizers and Nitrogen Fixation

In the United States, fewer and fewer farmers produce food for more and more people. This increased production is tied, in part, to applications of agricultural chemicals. As crops are harvested from fields, needed elements go with them; consequently, plant nutrients need to be replenished. Furthermore, many soils are deficient in one or more nutrients. County agricultural agents have helped farmers to analyze soils in order to determine and correct mineral deficiencies, primarily through the use of chemical fertilizers (see Fig. 13.11).

The most widely used chemical fertilizers contain nitrogen. Nitrogen is essential for making the proteins and nucleic acids of every cell, yet few compounds of nitrogen occur in abundance (although, there are some natural deposits of sodium and potassium nitrate found in arid regions of Chile that are mined for fertilizer). The incorporation of N_2 into compounds of nitrogen is called *nitrogen fixation*. Bacteria that grow symbiotically in nodules of the roots of legumes fix nitrogen from the air. The rotation of crops by planting legumes, such as soybeans and alfalfa, one season and crops with high-nitrogen demands, such as corn, another season has been a time-honored way to maintain soil fertility.

In 1913, Fritz Haber, in Germany, developed a chemical process for the fixation of nitrogen. In the *Haber process*, nitrogen and hydrogen react over an Fe–FeO catalyst to give ammonia. Even with the catalyst, a temperature of 500°C is required for a satisfactory rate of reaction.

$$N_2 + 3 H_2 \rightarrow 2 NH_3 + heat$$

In an application of Le Châtelier's principle, pressures as high as 800 atmospheres are used to increase the yield of ammonia. Upon cooling, the ammonia liquefies and is withdrawn. Unreacted nitrogen and hydrogen gas are recycled. By passing reactants over the catalyst bed, complete conversion of nitrogen and hydrogen to ammonia is attained.

A second German chemist, Wilhelm Ostwald, had earlier developed a chemical process to convert ammonia to nitric acid. In the *Ostwald process*, ammonia is oxidized to nitric acid over a platinum catalyst:

$$2 NH_3 + 4 O_2 \rightarrow 2 HNO_3 + 2 H_2O$$

Nitric acid and ammonia are used in the production of ammonium nitrate (NH_4NO_3) a fertilizer rich in nitrogen:

$$HNO_3 + NH_3 \rightarrow NH_4NO_3$$

Nitric acid is also used in the manufacture of explosives, and potassium nitrate is a component of gunpowder.

The chemical fixation of nitrogen has contributed to the destructiveness of war as well as to the productivity of agriculture. The British navy exercised its superiority during World War I by instituting a blockage to deprive Germany of the needed potassium nitrate that occurs naturally in Chile. Germany depended on the successful application of the Haber and Ostwald processes to produce munitions and to grow food. A strategy to combat wartime blockades was used to produce nitrates in the United States during the War of 1812 and in the South during the Civil War. It was based on the fact that nitrogen-fixing bacteria that live on decaying organic matter are abundant in some caves, and nitrates were leached from the cave dirt.

In what has been called the "green revolution," new hybrid varieties of rice and wheat have been developed and introduced in many countries. The use of these hybrids increased agricultural productivity; countries which once had to import food are now able to feed their growing populations. The green revolution has increased the demand for chemical fertilizers since these hybrids require higher levels of nutrients than can be supplied by traditional farming methods, using manure for fertilizer.

The green revolution helped to increase agricultural productivity during decades of rapid population growth. It provided time to begin to deal with the stress that population growth placed on food production, but it did not provide a permanent solution to agricultural productivity. Recently, the gains in agricultural productivity have slowed, and in 1991, the world-wide production of food actually decreased.

■ Fertilizers Containing Potassium and Phosphorus

The most abundant metal ions in living tissues are sodium, potassium, magnesium, and calcium. Potassium salts deposited from the evaporation of ancient seas are mined as a source of fertilizers. Limestone, and lime derived from the heating of limestone, are used as fertilizers to supplement calcium deficiencies.

Insoluble minerals containing calcium phosphate are mined to provide phosphate for fertilizer. The mineral hydroxyapatite ($Ca_5(PO_4)_3OH$) has the same composition as the hard part of bones and tooth enamel. A harder, less soluble mineral, fluoroapatite, in which fluoride ions have replaced the hydroxyl groups of apatite, is often used for phosphate production as well. (Fluoride-containing toothpaste helps increase the hardness of teeth by the same substitution.) In one process used to make the phosphate available to plants, these minerals are treated with sulfuric acid to make superphosphate fertilizer, a mixture of the more soluble compounds, $CaSO_4$ and $Ca(H_2PO_4)_2$.

■ Water Purity and Water Treatment

There are natural processes for removing biological wastes from water. Bacteria feed on organic wastes, using them for food and oxidizing them to CO_2 and water. In turn, these bacteria are infected and destroyed by viruses called bacteriophages. However, the capacity of these natural processes is limited, and the discharge of large amounts of biological and chemical wastes can overpower the ability of streams to regain purity. As populations have grown, so has the need to take a more active role in the treatment of both drinking water and sewage.

Drinking water is treated to kill harmful bacteria. Chlorine is added as a germicide in the United States, and ozone (O_3) is used for the same purpose in Europe. Turbid water may be treated with aluminum sulfate and a base to improve water clarity. A flocculent (fluffy) precipitate of aluminum hydroxide settles out and carries down suspended clay and other particulates.

Concern for water quality extends to naturally-occurring, large, underground reservoirs of water known as *aquifers* that are the source of well water for domestic or agricultural use. Nitrates from agricultural wastes contaminate well water in some rural areas making it unsuitable for use as drinking water. Spills from leaking or abandoned underground gasoline storage tanks contaminate well water in other areas. However, it is proving more difficult to treat underground contaminants than those on the surface, and so an increasing number of households are dependent on bottled water for drinking.

Water treatment processes help control the wastes added to streams. In sewage treatment plants, wastes are aerated and held to introduce additional oxygen and to allow time for the organic material to be digested. Many inorganic nutrients remain after sewage treatment (in some systems, effluents from sewage treatment are sprayed onto fields where the nutrients are retained and contribute to soil fertility).

The presence of inorganic nutrient ions, whether from municipal treatment plants or from agricultural runoff, can be a problem for water quality. The plant growth in many lakes and streams is normally limited by deficiencies in one or another nutrient, often nitrogen or phosphorus. As lakes and streams fill with a richer nutrient broth, blooms of algae spread over their surfaces. When these plants die, they sink and decompose. The algae produce oxygen by photosynthesis near the surface, but the oxygen is consumed by the decay process. With excessive plant growth and decay, deep waters become oxygen deficient and no longer support quality populations of fish. *Eutrophication* is the name given to the deterioration of water quality by excessive plant growth and decay.

Concerns for water quality and agricultural productivity will grow as population growth makes greater demands on the world's soil and water resources. Environmental progress has been made, and we have a good comprehension of the further actions needed to protect water quality and to promote soil conservation. Our current and future choices and actions in these areas will directly affect the world's future, and that of its inhabitants.

■ Questions—*Chapter 13*

1. The mineral wollastonite has the formula $Ca_3Si_3O_9$. What is the charge on the $Si_3O_9{}^{n-}$ anion? Is this anion linear with two oxygen atom bridges between silicon atoms or is it triangular with three bridging oxygen atoms?

2. What structural features of asbestos give rise to the formation of small dustlike particles? What are the advantages and disadvantages of policies that emphasize asbestos removal? Of those that leave asbestos in place but cover it with sealing compounds?

3. Caves are frequently found in limestone. Decaying organic matter produces acids in groundwater. Write the reaction for dissolving limestone in weakly acidic water.

4. Stalactites and stalagmites are formations sometimes found in caves. Apply Le Châtelier's principle to show that the loss of CO_2 to the air from water saturated with $Ca(HCO_3)_2$ leads to an increase in the concentration of $CO_3{}^{2-}$ and, hence, to the deposition of insoluble $CaCO_3$.

5. The properties of glass are dependent on the ratio of SiO_2 to Na_2CO_3. Why would each of the following changes in the standard recipe produce unsatisfactory glass?
    (a) a large decrease in the ration of SiO_2 to Na_2CO_3
    (b) a large increase in the ration of SiO_2 to Na_2CO_3

6. Glass is etched by solutions of hydrofluoric acid. Where do H_3O^+ ions attack silicate chains? Where do F^- ions attack?

7. Why is concrete often kept wet after pouring?

8. Why does a glaze containing $MgCO_3$ impart a glassy finish to a clay pot?

9. Which of the following compounds would be expected to give color to glass? What do these compounds that color glass have in common?
    (a) Fe_2O_3
    (b) PbO
    (c) $CoCO_3$
    (d) $MnCO_3$
    (e) B_2O_3

10. Shale is a sedimentary rock formed from compacted clay. Shale is relatively impervious to the flow of groundwater. Why?

11. Proposals for the long-term storage of nuclear wastes call for encasing radioactive wastes in glass or ceramic containers prior to burial. It is important that the container material withstand the heat generated by nuclear decay reactions and that it provide an additional barrier to groundwater. Why are glass and ceramic materials superior to metal or concrete for containment?

12. Like sodium salts, potassium salts are very soluble. Why are potassium ions available to plants? Why are they not primarily found in the ocean?

13. Why are mineral cations bound more tightly to a dry clay soil than to a water-saturated clay?

14. Can the deposit of acid compounds in watersheds be monitored by measuring the flow and pH of streams carrying runoff from them? List some ways that the pH of precipitation is modified as it moves through soil?

15. Why do soils differ greatly in their capacity to retain added fertilizer?

16. Why are clay soils on mountain slopes subject to mudslides? Would a single heavy rain or a series of frequent light rains be more apt to set the stage for a mudslide?

17. Clay sediments remain suspended in rivers for long times. These sediments quickly settle when the rivers flow into the ocean. Why would suspended clay particles settle more rapidly from saltwater than fresh water?

18. An isotope of cesium ($^{137}_{55}$Cs) is also a by-product of nuclear fission. It decays by emitting a beta particle. This isotope has a half-life of 30 years. What nutrient cation does Cs^+ most resemble? Is Cs^+ retained by soils and taken up by plants? Does $^{137}_{55}$Cs pose the same health risk as $^{90}_{38}$Sr? Why or why not?

14

Metals

From the dawn of civilization, the use of metals has played a prominent role in human culture. Early artisans fashioned jewelry of silver and gold and tools of copper. According to archaeologists, the introduction of bronze constitutes a significant turning point in the development of civilization. Bronze is an alloy made by combining the metals copper and tin. Because bronze is harder than either copper or tin, artisans could use it to fashion sharper knives and harder hammers. It is important to note that deposits of copper and tin are not found together so that the development of a civilization based on bronze tools depended on trade.

The production of iron and hard, tough steel made the growth of industry possible in the nineteenth century. Steam engines powered looms and drove the ships and trains that brought nations closer together. In America, farmers used steel plows to break the prairie sod, and laborers spanned the continent with steel rails as the West was settled. Cities grew upward as architects designed and built skyscrapers using steel beams for support.

Flight and the use of aluminum are twentieth-century developments. The fragile biplanes flown in World War I have been replaced by jets flying faster than the speed of sound and by airliners carrying hundreds of passengers across oceans and continents. Light, strong aluminum helped speed the pace of life and lower the costs of transportation.

Copper is a relatively rare metal, and aluminum is the most abundant metallic element present on the surface of the earth. How did it happen that copper was one of the first metals to be widely used and aluminum was one of the last? An understanding of the chemistry involved in the production of metals can increase our appreciation of history, economics, and geography.

■ Copper and Tin

Copper, a major component of both bronze and brass (an alloy of copper and zinc), has played a variety of important roles in technology. Because it is a relatively scarce metal and is a superior conductor of electricity, copper's value remains high even though other metals have replaced it for many applications. Copper, like the metals gold and silver, is sometimes found free in nature, but more often copper is found combined with sulfur in the CuS and $CuFeS_2$ ores found in mountainous regions, including the southern Appalachians in Tennessee, the Rocky Mountains in Montana, Utah, and Arizona, and the Chilean Andes.

Figure 14.1
In the processing of copper ore, CuS is first concentrated by flotation. Water wets the silicate minerals that comprise the major portion of the ore; these minerals sink to the bottom and are separated. Oil coats the surface of the CuS, and a suspension of this mineral is first carried to the surface in a froth and then skimmed.

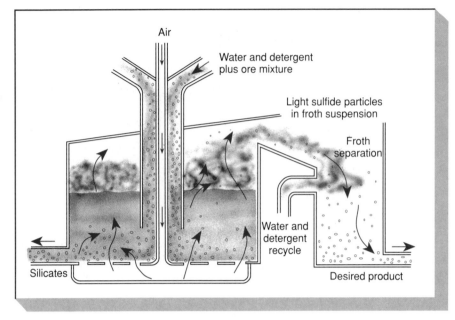

In a single step, copper compounds are separated from the sulfide ores and concentrated. In this process, called flotation, oil and water are added to finely crushed rock. Water wets the ionic silicate minerals that compose most of the rock, and these minerals sink. In contrast, water does not wet the surface of the less polar copper sulfide compounds found in the ore. When air is blown through the mixture, a froth containing oil and copper sulfide floats to the surface, where it is skimmed off (see Fig. 14.1).

Copper sulfide is then roasted in air to obtain copper. Oxygen has the seemingly paradoxical effect of both reducing copper to its metallic state and oxidizing sulfur to sulfur dioxide:

$$CuS + O_2 \rightarrow Cu + SO_2$$

In the case of $CuFeS_2$, the sulfur is oxidized by O_2 while the copper is reduced:

$$CuFeS_2 + 5/2\ O_2 \rightarrow Cu + FeO + 2\ SO_2$$

The release of sulfur dioxide from early copper smelters led to acid rain that devastated the nearby countryside. At Copper Hill, Tennessee, acid rain killed the forest cover and the heavy rains of the region gouged the steep mountain slopes. Years later, the smelting process was modified to recover the sulfur dioxide and to make sulfuric acid as a by-product of the production of copper. Because copper is refined in the sparsely populated mountain areas where it is mined, the public was not alerted to the destructive effects of acid rain.

Tin, like lead, is an unreactive metal in the same family of the periodic table as carbon. It is found as an oxide (SnO_2), which is readily reduced by carbon.

$$SnO_2 + C \rightarrow Sn + CO_2$$

Although it is not an abundant metal, tin has properties that have led to its valuable applications. People call the cans that have been widely used to preserve and market foods "tin cans." These tin cans contain only small amounts of the metal tin; they are usually made from steel that has a thin protective layer of the unreactive tin.

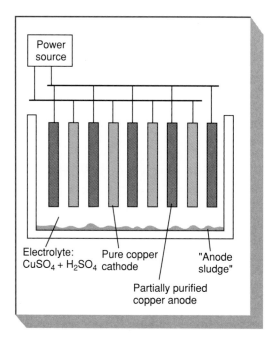

Figure 14.2
Electrolytic cell for refining copper.

(Figure labels: Power source; Electrolyte: $CuSO_4 + H_2SO_4$; Pure copper cathode; "Anode sludge"; Partially purified copper anode)

■ The Electrolytic Purification of Copper

Very pure copper is required for all electrical applications. Since copper from the smelter still contains small amounts of other metals, it is purified by electrolysis. The impure copper is used as an anode, therefore, it is oxidized and dissolved. Copper ions migrate to the cathode where they are reduced to give copper that is more than 99.9% pure.

The electrolysis removes small amounts of iron and silver as well as other metals. Iron is oxidized and goes into solution, but it does not plate out with the purified copper at the cathode since Fe^{2+} is harder to reduce than Cu^{2+}. Because it is harder to oxidize than copper, silver in the impure anode falls to the bottom of the cell as the copper in the anode is oxidized. The recovery of silver from the anode sludge helps pay the cost of the electrolytic purification of copper (see Fig. 14.2).

■ Aside

Silver Mining in the Amazon Basin

The country of Brazil is confronting difficult choices that arise from the conflict between miners who wish to extract some of the mineral wealth found in the eastern slopes of the Andes Mountains and the native peoples inhabiting the rainforests of the Amazon River basin. It is profitable to mine and extract silver even in remote regions of the earth since silver metal is valuable and easily transported by plane.

However, the technology used to extract silver from ores is destructive to the environment. Because such small quantities of silver are present in the crushed ores, it is necessary to extract the silver chemically, using solutions of cyanide salts. In the presence of cyanide, silver metal is oxidized by oxygen in the air and the compounds of silver and sulfur present in the ore dissolve. The cyanide ion CN^- binds Ag^+ tightly in a Lewis acid-base reaction:

$$O_2 + 4\, Ag + 8\, CN^- + 2\, H_2O \rightarrow 4\, Ag(CN)_2^- + 4\, OH^-$$

$$Ag_2S + 4\, CN^- + H_2O \rightarrow 2\, Ag(CN)_2^- + HS^- + OH^-$$

Metallic silver can then be recovered from the aqueous solution by reducing the silver with a more active metal such as zinc:

$$2\, Ag(CN)_2^- + Zn \rightarrow 2\, Ag + Zn(CN)_4^{2-}$$

The wastes from silver mining are endangering the environment and destroying the way of life of the aboriginal people. Escaping cyanide compounds poison downstream fish and aquatic plants. Natives who once depended on fishing for food are forced to become increasingly dependent on imported foods.

Is it possible to mine silver without allowing cyanide salts to escape? Unfortunately, the answer is no so long as the scale of operation and the investment of capital remain small. It is often easier to introduce quick-and-dirty technologies than to introduce technologies less destructive to the environment.

■ Lead

Lead, like copper, occurs as sulfide ores in mountain regions and is readily recovered from these ores. It is the most abundant heavy metal in the earth's crust. Lead is found most often in the +2 oxidation state (note that tin, in the same column of the periodic table as lead, occurs in the +4 oxidation state when combined with oxygen). Like copper, lead is readily recovered from its ores:

$$2\, PbS + 3\, O_2 \rightarrow 2\, PbO + 2\, SO_2$$

$$PbO + C \rightarrow Pb + CO$$

Lead has been used in a wide variety of applications. A plumber was one who worked with lead (Latin, *plumbum*), for all pipes used to be made of lead. Some lead compounds are both relatively insoluble and brightly colored: they have been used in glazes, paints, and colored inks. The softness and high density of lead made it desirable for use in shotgun shells; lead shot is "easy" on the gun barrel and "hard" on the duck or pheasant. Lead's low-melting point made it desirable for the casting of hot type for printing.

From the time that automobiles came into wide use until the late 1980s, a major use of lead has been in the production of gasoline additives. Tetraethyl lead, a compound with four ethyl $(CH_3CH_2\text{—})$ groups covalently bonded to lead, was added to gasoline to improve its antiknock characteristics. Upon heating in the engine cylinder, the long, weak covalent bonds between carbon and lead separate to give carbon species with odd numbers of electrons. These odd electron species react readily with molecular oxygen and help to promote smooth combustion (see Fig. 14.3).

The decomposition of tetraethyl lead produces metallic lead. To get the lead out of engines, compounds containing chlorine and bromine were also added to gasoline. An aerosol of PbBrCl and related compounds of lead is emitted in the

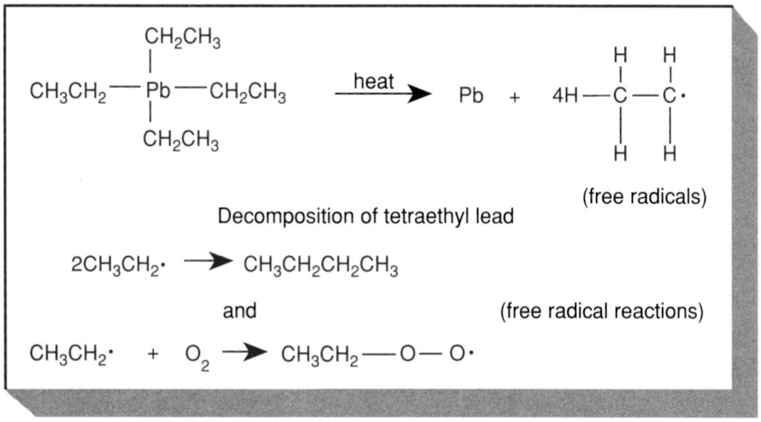

Figure 14.3
Tetraethyl lead has been used as a gasoline additive to improve the antiknock characteristics of the fuel. When heated, tetraethyl lead decomposes to form reactive free radicals that help promote smooth combustion.

exhaust. Because lead compounds poison the catalysts used to reduce hydrocarbon and nitrogen oxide emissions, lead additives are not permitted in gasoline for newer cars with emission control devices. Gasoline blenders gradually removed lead additives from all gasoline during the 1980s.

■ The Toxic Legacy of Lead

Unfortunately, the widespread use of lead has had adverse health effects. Lead, like many other heavy metals, is poisonous because lead binds to the sulfur atoms of proteins. Brain damage often results from long-term exposure to low levels of lead. Lead poisoning poses a particular problem for growing children since peeling lead-based paint and lead-containing dust are present in many older homes. Lead poisoning may have been a hazard as early as the time of the Roman Empire, when lead acetate, "sugar of lead," was sometimes used to sweeten wine. Lead was also used for plumbing and lead-containing glaze was used on their earthenware pots.

Because of its toxicity, lead has been replaced in many of its uses. Manufacturers of paints began removing lead from paint formulations during the 1950s. The use of hot lead type for printing has been replaced by newer technologies. Many states have outlawed the use of lead shot for hunting game birds.

It is a challenging task to remove the lead compounds from the environments containing high levels of them. Removing the lead used to solder joints between copper pipes in older domestic water systems can involve replacing the entire plumbing system. Efforts to remove lead paint can themselves pose risks. Dust containing paint particles can be more dangerous to people than if the paint had remained in place. It may be necessary to bag and remove paint scrapings and dust so that the lead-based paint might be buried as toxic waste.

■ Iron

Iron has seen limited use for many centuries. The Bible mentions the early use of iron by the Hittites and archaeologists have identified sites where iron was produced during Roman times. However, the properties of iron and steel are sensitive to impurities, and the early steels made from some iron ores were brittle. Since ores found in different locations contained different impurities, the production of iron and steel remained an art.

Figure 14.4
The blast furnace is used in the production of iron from its ore.

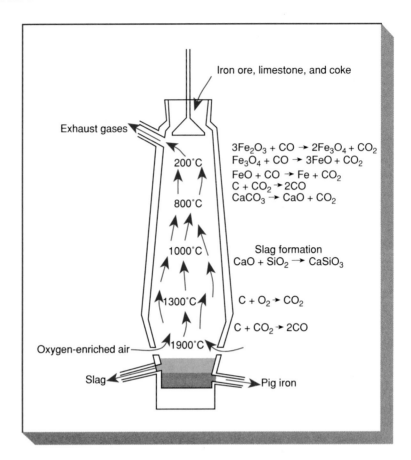

Improved methods of refining iron and making steel introduced in the nineteenth century led to a rapid increase in the production of iron. The ability to produce consistently high-grade iron and steel followed developments in chemical analysis and inventions that improved production processes. The availability of limestone and coal in western Pennsylvania to use in iron recovery and the ability to ship ore through the Great Lakes from mines in Minnesota contributed to the early growth of Pittsburgh as a center for steel production in the United States.

Iron is recovered from iron oxides, Fe_2O_3 and Fe_3O_4, that are found in sedimentary deposits. The ore is reduced in blast furnaces as illustrated in figure 14.4. A charge of iron ore, coke (C), and limestone ($CaCO_3$) is heated in the furnace, and air is introduced at the bottom. The partial combustion of the coke to carbon monoxide provides heat for the process. Carbon monoxide, not carbon, reduces the iron oxides in a series of chemical reactions. (Reactions between coke and iron ore are very slow, for there is only a small amount of contact at the surfaces of the two solids.)

The role of the limestone is to lower the melting temperature of silicate impurities in the ore and to remove acidic phosphates that adversely affect the quality of iron and steel. The slag formed in these reactions floats on the molten iron. Periodically, the furnace is tapped; slag and cast iron or pig iron are drawn off. Slag is widely used in the making of concrete; therefore, highway construction provides

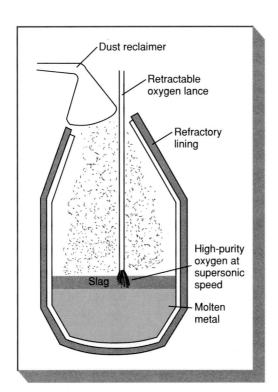

Dust reclaimer

Retractable
oxygen lance

Refractory
lining

High-purity
oxygen at
supersonic
speed

Slag

Molten
metal

Figure 14.5
To make steel, the carbon content of molten pig iron is reduced by a high-temperature reaction with oxygen. A flux of CaO is used to help remove phosphorus and silicon remaining in the iron.

a useful outlet for some of this by-product. The cast iron from the blast furnace contains about 94% iron, with carbon and traces of other elements comprising the remainder. Blast furnaces are operated continuously for months and years. Ore, coke, and limestone are added at the top, and iron and slag are withdrawn from the bottom.

■ Galvanized Iron and Steel

Iron is subject to corrosion; it reacts with oxygen, particularly when wet, to form rust, Fe_2O_3. Iron can be protected by paint or by coating it with a layer of zinc to produce galvanized iron (zinc is more readily oxidized than iron). In the first step of the oxidation of iron to form rust, Fe^{2+} is formed. When Fe^{2+} is formed in galvanized iron, it is reduced back to iron by the zinc before it can be further oxidized:

$$Fe^{2+} + Zn \rightarrow Fe + Zn^{2+}$$

Zinc reacts as a sacrificial electrode, protecting the exposed iron from oxidation.

Steel is made by burning the excess carbon from the iron as shown in figure 14.5. Modern steels differ in carbon content and properties. High-carbon steel can be tempered to hold a sharp edge while low-carbon steels are tougher. Alloy steels are made by introducing other metals. Steel containing manganese is very hard and tough; it is used in the blades of earth-moving machinery and in safes. Elastic, tough chrome-vanadium steel is used to make wrenches and engine valves. Stainless steel, an alloy containing chromium and nickel, does not rust because its surface is protected by a tightly adhering oxide layer.

Figure 14.6
The small +3 aluminum ion greatly increases the acidity of solvating water molecules.

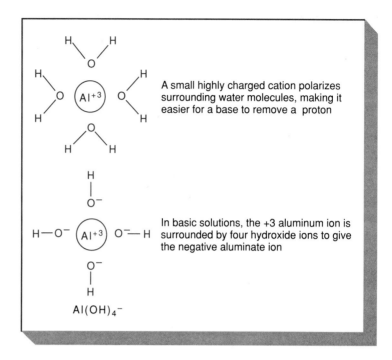

A small highly charged cation polarizes surrounding water molecules, making it easier for a base to remove a proton

In basic solutions, the +3 aluminum ion is surrounded by four hydroxide ions to give the negative aluminate ion

$Al(OH)_4^-$

■ Aluminum

Aluminum is a light, strong metal widely used in transportation and packaging. Its high strength-to-weight ratio makes it attractive for use in airplanes and beer and soft drink cans. Automobile manufacturers are replacing steel components of many cars with parts made of aluminum because the fuel economies of lightweight cars are increasingly attractive. Transportation savings help to offset initial costs for aluminum users. Although aluminum cans cost more to manufacture than steel cans or glass bottles, they are lighter and less bulky, therefore, they are less expensive to transport.

Aluminum occurs widely in granites and clays, but aluminum cannot be recovered economically from these minerals. Bauxite ($Al_2O_3 \cdot 2H_2O$) is a sedimentary ore used for the production of aluminum (commercial bauxite deposits are found in Arkansas and Jamaica). The aluminum in bauxite is separated by extraction from the iron and silica that occur with it. Al_2O_3 dissolves in strongly basic solutions, while Fe_2O_3 is insoluble. When base is added to bauxite, the aluminum oxide dissolves while the iron oxide remains behind:

$$Al_2O_3 + 2\ OH^- + 3\ H_2O \rightarrow 2\ Al(OH)_4^-$$

After the aluminum is leached from the bauxite, the resulting solution is heated to precipitate hydrated crystals of aluminum oxide from solution. The crystalline solid is collected and heated to form Al_2O_3, and the remaining basic solution is reused.

Aluminum oxide, unlike many other metal hydroxides, dissolves in strong base because the high positive charge on the Al^{3+} ion increases the acidity of the attached water molecules. The small +3 ion pulls the electrons of the attached water molecules toward the aluminum and away from the hydrogen atoms of the water. This electron shift polarizes the oxygen-hydrogen bond so that a hydroxide ion in solution can remove a proton to give the soluble $Al(OH)_4^-$ ion (see Fig. 14.6).

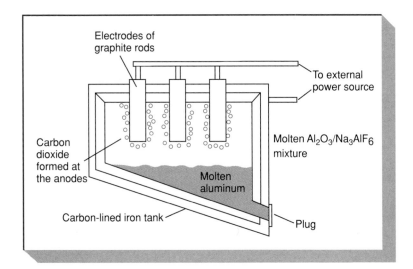

Figure 14.7
An electrolytic cell is used for the production of aluminum.

Aluminum-oxygen bonds are so strong that aluminum ore cannot be reduced using common chemical reducing agents. Aluminum is produced by electrolysis. To electrolyze Al_2O_3, a suitable solvent is required. Water is not a suitable solvent for the electrolysis since it is more easily reduced than Al^{3+} and hydrogen gas, rather than aluminum metal, would form at the cathode. Charles Hall in the United States and Paul Heroult in France found that Al_2O_3 dissolves in the fused mineral, cryolite (Na_3AlF_6) to give conducting solutions. Their independent discoveries, made in 1886, opened the way to low-cost aluminum production. A current of electricity is passed through a molten solution of Al_2O_3 in cryolite. Liquid aluminum, collected at the bottom of the cell, serves as one electrode and carbon as the other. The electrode reactions produce aluminum at one electrode and carbon dioxide at the other. The carbon rods are graphite and their use is a factor in the cost of refining aluminum since they are consumed in the electrolysis (see Fig. 14.7).

$$Al^{3+} + 3\ e^- \rightarrow Al \text{ reduction}$$

$$C + 2\ O^{2-} \rightarrow CO_2 + 4\ e^- \text{ oxidation}$$

A large amount of energy is required to refine aluminum since its electrolysis requires both a high voltage and a high current. In the United States, aluminum refineries were built in the Tennessee Valley and in the Pacific Northwest, areas that have abundant, low-cost electric power.

■ Aluminum Is Protected by an Oxide Coat

Aluminum is protected by a thin, strongly adhering film of aluminum oxide (Al_2O_3). If the film is scratched and the metal exposed, it reacts with oxygen in the air or with water on its surface to form a fresh surface of Al_2O_3. This layer of Al_2O_3 is impervious to oxygen so it protects the bulk of the metal beneath it. An aluminum beverage can lying by the roadside may take hundreds of years to oxidize completely.

The reaction of aluminum with oxygen can be easily demonstrated. If the aluminum is treated with a small amount of mercury (or a solution of a readily reduced compound of mercury) and a scratch is made, the presence of the mercury keeps the oxide coat from adhering. Aluminum oxide continues to form and the aluminum becomes deeply etched.

Figure 14.8
Large-scale recycling of metal waste (magnetic separation of iron and steel) from non-ferrous metals.
© Arthur R Hill/Visuals Unlimited.

Acids and bases react with the aluminum oxide coating to expose fresh metal. Acids attack oxygen atoms in Al_2O_3, and bases attack the aluminum atoms. The freshly exposed metal is then oxidized:

$$2\,Al + 6\,H_3O^+ \rightarrow 2\,Al^{3+} + 3\,H_2 + 6\,H_2O$$

$$2\,Al + 6\,H_2O + 2\,OH^- \rightarrow 2\,Al(OH)_4^- + 3\,H_2$$

For this reason, highly acidic foods requiring long cooking times should not be prepared in aluminum cookware.

■ Recycling Aluminum

Much of the energy used to produce aluminum remains stored in a discarded aluminum can or in an obsolete airplane. The recovery of aluminum from scrap requires only about five percent of the energy needed to produce aluminum from ore. Recycled aluminum is melted and the aluminum oxide is skimmed off. Energy is required to melt the metal, but the cost of reducing +3 aluminum is avoided.

For the recovery of aluminum to be effective, any iron must be removed. Iron impurities in aluminum would be weak spots where the oxide coating would not cling, and the oxidation of the metal could continue. Magnets are used to remove iron before the aluminum scrap is melted and recycled (see Fig. 14.8).

■ Aside

Metalloids and Semiconductors

Silicon and germanium lie below the nonmetal carbon and above the metals tin and lead in the periodic table. Silicon and germanium belong to a small group of elements classed as *metalloids*, elements with properties between those of metals and nonmetals as shown in figure 14.9. Metalloids conduct electricity, but do so poorly.

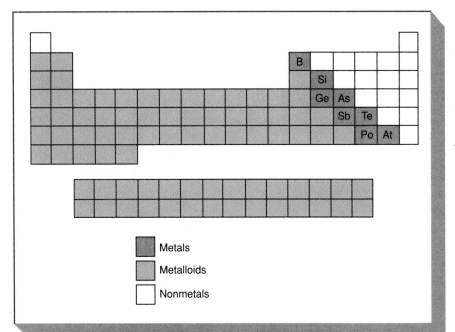

Figure 14.9
In the periodic table, metalloids form a narrow band dividing metals from nonmetals.

Metals

Metalloids

Nonmetals

In metals, conducting electrons are in orbitals that extend throughout the crystal. In the diamond structure of carbon, electrons are in localized bonds and are not free to migrate. The crystal structures of silicon and germanium resemble that of diamonds, but not all the electrons are localized. In crystals with the larger silicon or germanium atoms, electrons are not as tightly held, and a few are excited to conducting bands where the electrons are free to migrate as in a metal.

Silicon and germanium are *semiconductors.* Current is carried both by conducting electrons and by positive holes. A nearby electron can fill a hole left by electron excitation and thereby leave behind a new hole; this movement of a positive hole is an electrical current. Unlike metals, semiconductors become better conductors as the temperature increases, for the number of excited electrons increases with the increasing temperature (see Fig. 14.10).

Small but highly controlled amounts of impurities are introduced to modify the conductivity of semiconductors. If gallium with only three outer electrons is added, it produces a lattice with an excess of positive holes (a p-semiconductor). If arsenic with five outer electrons is added, it produces a lattice with an excess of conducting electrons (an n-semiconductor). Devices made of this "doped" silicon and germanium are widely used to construct circuits for computers.

Transistors are made from sandwiches of n- and p-conductors. These have electrical properties important for the miniaturization of complex circuits. For example, electrons can flow in only one direction through an np junction. This device (a rectifier) turns alternating current into direct current, allowing us to use a common electrical outlet in place of batteries.

Figure 14.10
Semiconductors carry current either by the movement of positive holes or by the movement of electrons in a conduction band. Electrical properties important for the miniaturizing of circuits can be prepared by modifying semiconductors by adding controlled amounts of other elements and by combining different semiconductors in a circuit.

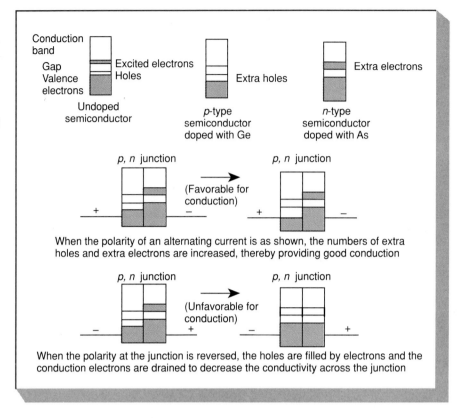

■ **Questions**—*Chapter 14*

1. The following elements have symbols derived from Latin names. Either the element or its compounds were known in ancient times. Which one of these elements was known only in compounds?
   (a) Ag ("argentum") silver
   (b) Au ("aurum") gold
   (c) Cu ("cuprum") copper
   (d) Na ("natron") sodium
   (e) Pb ("plumbum") lead

2. What is a chemical explanation for the finding of gold artifacts in Egyptian burial tombs even though gold is a rare element in the crust of the earth?

3. In the electrolytic purification of copper, why do iron impurities remain in solution? Why are silver impurities found in the anode sludge?

4. Copper, silver, and gold are in the same column of the periodic table. How does the ease of oxidation vary going down the table in this family? If oxidation was simply removing an electron from an isolated atom, which of these elements would be most easily oxidized? (Consider relative atomic size.)

5. Both copper and tin were liberated from ores near ancient campfires. Write the reaction for the formation of each metal. Which ore is reduced by carbon?

6. Why would a pulverized mixture of charcoal, ore, and limestone be unsuitable for use in a blast furnace?

7. Why are iron oxides reduced more rapidly by carbon monoxide than by solid carbon?

8. What is the role of $CaCO_3$ in the production of iron?
9. Scrap iron and steel are recycled to make more steel. Why would it be important to be able to analyze the metal content of scrap to control the properties of the steel? What complications are posed by the use of a wider variety of metals in auto construction?
10. Why is aluminum used in the construction of automobiles since it is weaker than iron and costs more to produce?
11. Magnesium is a lightweight metal with uses similar to aluminum. It is in the same family of the periodic table as the element calcium, which is readily oxidized by water or air. Why might metallic magnesium be resistant to oxidation?
12. Why does the solubility of aluminum compounds, leached from bauxite with base, decrease as CO_2 is absorbed from the air?
13. Why is it more economical to recycle aluminum than iron?
14. A solution of aluminum sulfate is acidic; its pH is less than 7. Write a reaction of a hydrated aluminum ion, $Al(H_2O)_6^{3+}$ with water to account for the decrease in pH.
15. Account for the observation that solutions of $FeCl_3$, $Fe(NO_3)_3$, and $Fe_2(SO_4)_3$ are acidic. How would the charge on Fe^{3+} affect the acidity of attached water molecules?
16. Carbon is used in the production of both aluminum and iron. Write the reactions of carbon in the production of each metal.
17. For many uses, aluminum and copper need to be free of impurities. Why is copper purified as a metal and aluminum as an ore? (Try to think of chemical differences that would account for this difference.)
18. A student writes that aluminum is more susceptible to oxidation than is iron. In what way is that answer correct? In what way is it wrong? Explain.
19. Why is it less attractive economically to collect and recycle glass than it is to collect and recycle aluminum? Consider the energy costs and savings for each material.

15

An Introduction to Organic Chemistry

Most known chemical compounds contain carbon and the branch of chemistry that involves the study of these carbon compounds is called organic chemistry. The first organic compounds were isolated from living organisms, and the name "organic" reflects an early belief that compounds from living organisms were different from other compounds due to the presence of a so-called "vital force." In 1828, Friedrich Wohler (1800–1882) synthesized urea, an organic compound first isolated from urine, from inorganic compounds. His synthesis helped bring an end to this vital-force theory in chemistry.

The variety of organic compounds is very great. Thousands of different compounds have been isolated from coal and oil alone. Volatile compounds serve as trail markers for ants and as sex attractants for boll weevils. Paints, medicines, insecticides, and textile fibers are products of the organic chemical industry. In the biological world, fats, carbohydrates, nucleic acids, enzymes, and hormones are all carbon-containing compounds.

This chapter provides an entry into organic chemistry. It introduces a small number of compounds drawn from the major classes of organic compounds. The reader is encouraged to begin to learn the names and structures of these representative compounds and of their extended families.

■ Alkanes

Compounds of only carbon and hydrogen are called *hydrocarbons*. Hydrocarbons containing only single bonds between the carbon atoms are called *alkanes*. Alkanes react with few other chemicals, and the reactions they do undergo, such as combustion, tend to be relatively unselective. Alkanes play a central role in the nomenclature of organic compounds since all other organic compounds are named as derivatives of the alkanes. The simplest alkanes are methane (the major constituent of natural gas), ethane, and propane (a compound used for heating and cooking in many rural areas).

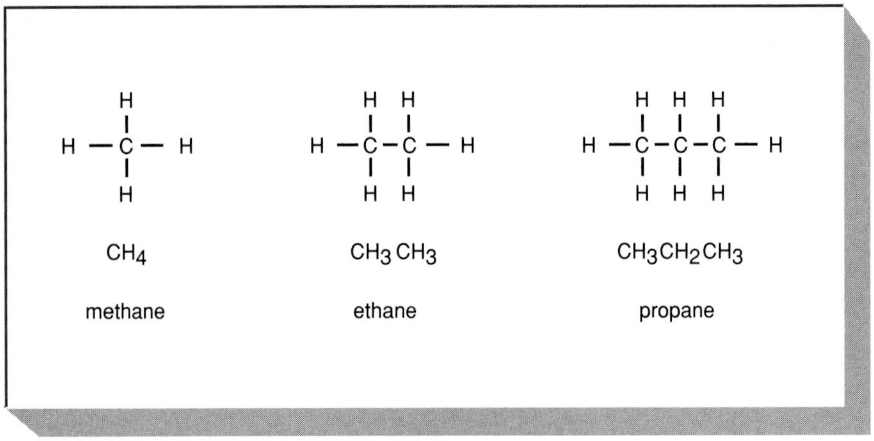

CH_4

methane

CH_3CH_3

ethane

$CH_3CH_2CH_3$

propane

Abbreviated or condensed structures are used to illustrate the organic structures. The number of hydrogen atoms attached to each carbon atom is indicated after the C. A student should be able to draw a Lewis structure if given a condensed structure. The use of condensed structures saves both paper and typesetting expense. In this text, the hydrocarbon portions will be frequently written as condensed structures.

Alkanes may be either branched or cyclic. The simplest case of branching is illustrated by the existence of two alkanes having the formula C_4H_{10}:

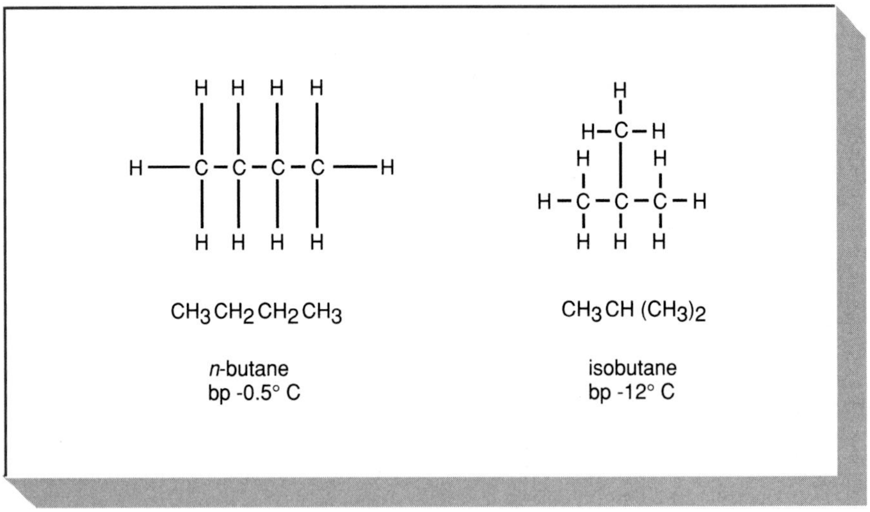

$CH_3CH_2CH_2CH_3$

n-butane
bp -0.5° C

$CH_3CH(CH_3)_2$

isobutane
bp -12° C

Compounds that have the same formula but differ in the way that the atoms are connected are called *isomers*. Isomers also differ in their chemical and physical properties. Butane and isobutane are sometimes called *carbon-skeleton isomers* because the carbon atoms alone show different structures.

Cyclohexane is an example of a cyclic alkane. In the popular convention for writing cyclic compounds, both carbon atoms and hydrogen atoms are omitted and just the skeletal structure is shown (see below). A carbon atom is present at every corner, and the number of hydrogen atoms bonded to a carbon atom is calculated by subtracting the number of carbon–carbon bonds from four (the total number of bonds to each carbon atom). In the case of cyclohexane, two carbon–carbon bonds are shown at each corner and two hydrogen atoms are attached to each carbon atom.

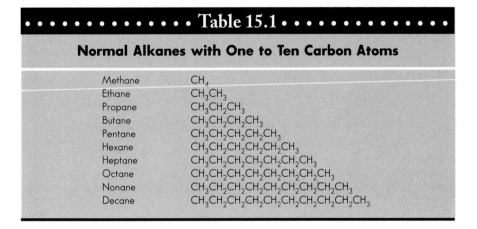

• • • • • • • • • • • • • Table 15.1 • • • • • • • • • • • • • • •

Normal Alkanes with One to Ten Carbon Atoms

Methane	CH_4
Ethane	CH_3CH_3
Propane	$CH_3CH_2CH_3$
Butane	$CH_3CH_2CH_2CH_3$
Pentane	$CH_3CH_2CH_2CH_2CH_3$
Hexane	$CH_3CH_2CH_2CH_2CH_2CH_3$
Heptane	$CH_3CH_2CH_2CH_2CH_2CH_2CH_3$
Octane	$CH_3CH_2CH_2CH_2CH_2CH_2CH_2CH_3$
Nonane	$CH_3CH_2CH_2CH_2CH_2CH_2CH_2CH_2CH_3$
Decane	$CH_3CH_2CH_2CH_2CH_2CH_2CH_2CH_2CH_2CH_3$

■ Naming Organic Compounds

As science has grown, the concern for information retrieval has also grown. Scientists have developed rules for generating systematic names that are clear and unambiguous. Rules are applied that allow one to go from a structure to a name or from a name to the correct structure. Because it is essential that scientists be able to communicate and that they have access to information in the literature, rules for naming compounds have international recognition.

In systematic nomenclature, organic compounds are named as derivatives of straight chain (normal) alkanes. (The naming of compounds with rings is often more complex.) The names of the first ten normal alkanes are given in table 15.1. Students should memorize these names and structures.

Common names continue to be used for both simple and very complex compounds. Examples of common names for the simple hydrocarbons of the isomers of pentane (C_5H_{12}) are shown. The prefixes *n-*, *iso,* and *neo* are used to designate variations in chain branching. Because the variety of arrangements of carbon chains becomes rapidly greater as the number of carbon atoms increases, the number and special meanings of the prefixes used in common names quickly become intractable. Chemists commonly use systematic names for branched compounds having greater complexity than the pentanes. (Systematic names can also have disadvantages: they may be long, awkward, and difficult to say.) For many compounds, the use of common names, like that of nicknames, persists:

Structure	Common name	Systematic name
$CH_3CH_2CH_2CH_2CH_3$	*n*-pentane	Pentane
$\begin{matrix} CH_3 \\ \diagdown \\ \diagup \\ CH_3 \end{matrix}$CHCH$_2CH_3$	Isopentane	2-methylbutane
$\begin{matrix} CH_3 \\ | \\ CH_3-C-CH_3 \\ | \\ CH_3 \end{matrix}$	Neopentane	2,2-dimethylpropane

The names of the isomeric pentanes provide a simple illustration of the systematic naming of branched hydrocarbons. The longest unbranched chain provides the root of the name (for example, pentane). Hydrocarbon branches are named by replacing the *ane* of the corresponding alkane (methane) by *yl* (methyl). Each branch replaces an H in the unbranched chain, so it is called a substituent. The carbon atoms of the longest chain are numbered (from 1 to 4 for butane), and the position of each alkyl substituent is indicated by a numerical prefix (for example, 2-).

■ Functional Group Isomers— An Illustration in Reasoning

Organic chemists reason by analogy. They place compounds with similar structures, which undergo similar reactions, in a single category. Then from a structure, a chemist can infer the physical properties and characteristic reactions of the compound. In the inverse direction, a chemist may infer the pieces of structure that are present in a new compound from its physical properties and the chemical reactions that it undergoes.

Let us sample the reasoning involved in organic chemistry. Two known compounds both have the formula C_2H_6O but they have very different properties. Under the name given each compound are listed some chemical and physical properties:

	Ethyl Alcohol	Dimethyl Ether
Dissolves in concentrated sulfuric acid	Yes	Yes
Reacts with Na to give H_2 and strong base	Yes	No
Boiling point	78° C	−25° C

Only two possible structures having the formula C_2H_6O can be written. Because each second-row atom has an octet of electrons about it, each carbon atom has four bonds and each oxygen atom has two bonds. A hydrogen atom forms one single bond to carbon or to oxygen because hydrogen can accommodate only two electrons in its outer shell (a hydrogen–hydrogen bond is present only in a hydrogen molecule, H_2).

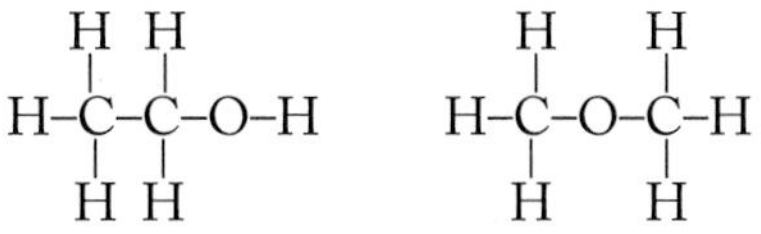

$$
\begin{array}{cc}
\mathrm{H\ \ H} & \mathrm{H\ \ \ \ H} \\
\mathrm{H{-}C{-}C{-}O{-}H} & \mathrm{H{-}C{-}O{-}C{-}H} \\
\mathrm{H\ \ H} & \mathrm{H\ \ \ \ H} \\
\mathrm{A} & \mathrm{B}
\end{array}
$$

To assign these structures to the alcohol and ether, we can reason by an analogy to water. Like water, each reacts as a base with concentrated sulfuric acid. The two compounds behave like water in their reactions with sulfuric acid because each compound has unshared pairs of electrons on its oxygen:

$$H_2SO_4 + H_2O \rightarrow H_3O^+ + HSO_4^-$$

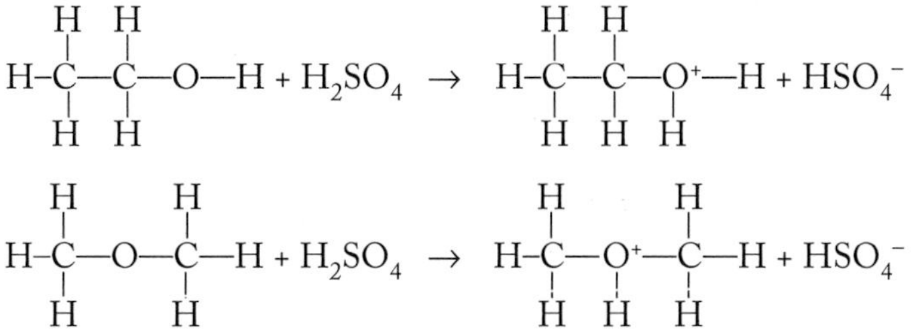

$$
\mathrm{H{-}C{-}C{-}O{-}H} + H_2SO_4 \rightarrow \mathrm{H{-}C{-}C{-}O^+{-}H} + HSO_4^-
$$

$$
\mathrm{H{-}C{-}O{-}C{-}H} + H_2SO_4 \rightarrow \mathrm{H{-}C{-}O^+{-}C{-}H} + HSO_4^-
$$

Water is reduced by sodium to form hydrogen gas and a solution of sodium hydroxide. This reduction is a reaction of the —OH bond. Only structure A has an —OH group. Like water, it reacts with sodium to give hydrogen gas and a basic solution. On the basis of the reaction with sodium, we can assign structure A to ethyl alcohol and structure B to dimethyl ether:

$$2\,H_2O + 2\,Na \rightarrow H_2 + 2\,Na^+ + 2\,OH^-$$

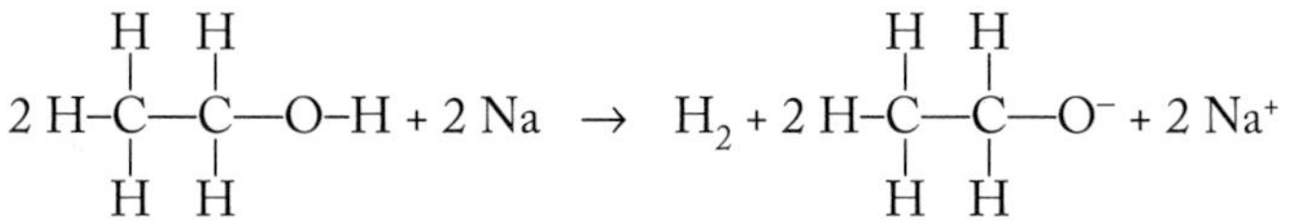

$$
2\,\mathrm{H{-}C{-}C{-}O{-}H} + 2\,Na \rightarrow H_2 + 2\,\mathrm{H{-}C{-}C{-}O^-} + 2\,Na^+
$$

A check on our assignment of structures is provided by a comparison of boiling points. Ethyl alcohol is expected to have a much higher boiling point than dimethyl ether. Because it has an —OH group like water, the alcohol has strong hydrogen bonds between molecules as shown in figure 15.1. With no —OH group, the ether has only weaker forces of attraction between its molecules.

Ethyl alcohol and dimethyl ether are isomers with different *functional groups*. Ethyl alcohol has the chemistry associated with the hydroxyl (—OH) or alcohol functional group, and dimethyl ether has the chemistry associated with the ether (—O—) functional group. Note that with many reagents, the portion of an organic molecule consisting only of C—C and C—H bonds remains unchanged.

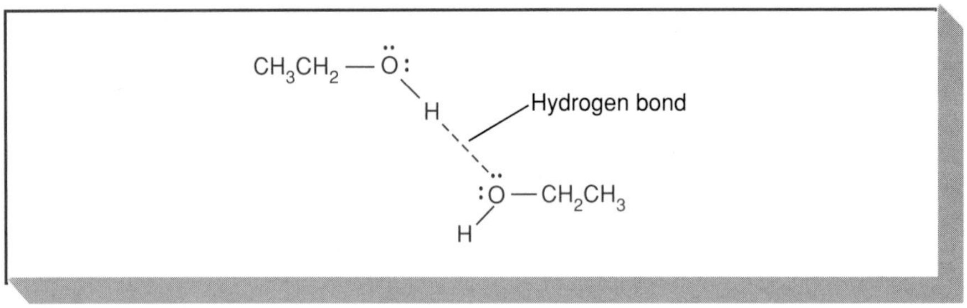

■ Naming Compounds with Functional Groups

Chemists name compounds with functional groups as the derivatives of alkanes. Names of compounds reflect both the functional group present and the parent hydrocarbon skeleton. The longest hydrocarbon chain containing the functional group provides the root that indicates the number of carbon atoms in the chain. The suffix of a name indicates the functional group present. For example, names of alcohols end in *-ol*. Methanol and ethanol are the systematic names of one- and two-carbon alcohols.

$$CH_3OH \qquad CH_3CH_2OH$$

methyl alcohol   ethyl alcohol

methanol   ethanol

When the position of a functional group on a hydrocarbon chain is not obvious, a numeral designates its position. The chain is numbered beginning from the end that will result in the smaller numeral for the position of the functional group. Using an example of three-carbon alcohols, we note that 1-propanol and 2-propanol are the systematic names for the two alcohols with the formula C_3H_8O. These alcohols are *positional isomers* because they have the same carbon skeleton and the same functional group, but they differ in the placement of the attachment of the —OH group.

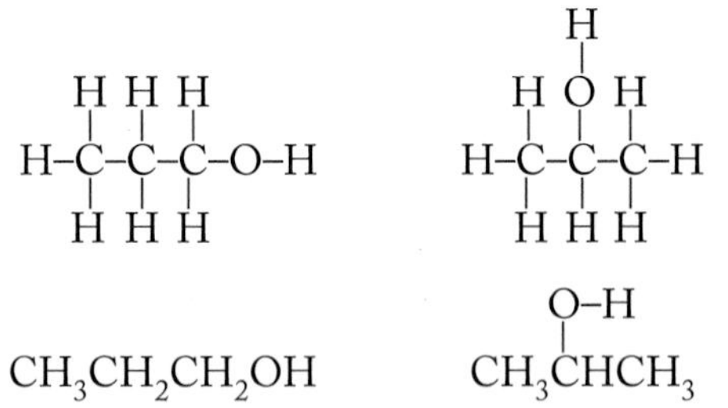

$$CH_3CH_2CH_2OH \qquad CH_3CHCH_3$$

n-propyl alcohol   isopropyl alcohol

1-propanol   2-propanol

Given a structure, a student should be able to propose a name, and from a given systematic name, the student can propose a structure. The following examples illustrate the application of the simple rules of nomenclature for more complex compounds.

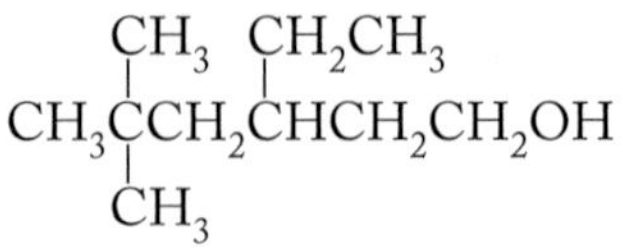

• • • • • • • • • • • • • • Table 15.2 • • • • • • • • • • • • • • •

Four-Carbon Alcohols

Alcohol	bp, °C	Common Name	Systematic Name
$CH_3CH_2CH_2CH_2OH$	118	n-Butyl alcohol	1-Butanol
$CH_3\overset{\displaystyle OH}{\overset{\mid}{C}}HCH_2CH_3$	100	sec-Butyl alcohol	2-Butanol
$CH_3\overset{\displaystyle CH_3}{\overset{\mid}{C}}HCH_2OH$	108	Isobutyl alcohol	2-Methyl-1-propanol
$CH_3\overset{\displaystyle OH}{\underset{\underset{\displaystyle CH_3}{\mid}}{\overset{\mid}{C}}}CH_3$	82	t-Butyl alcohol	2-Methyl-2-propanol

Since this compound has an —OH group, it is an alcohol. The longest chain containing the —OH group has six carbons, so it is a hexanol. The carbon atoms of the hexanol are numbered from that end which gives the lowest number (1) to the functional group. This substituted 1-hexanol has two —CH_3 (methyl) groups attached to C-5 and a —CH_2CH_3 (ethyl) group on C-3. The systematic name of the compound is 5,5-dimethyl-3-ethyl-1-hexanol. Note that each methyl group is given a position even though they are attached to the same carbon atom.

Write a structure for 4-ethyl-4-isopropyl-2-octanol. From the suffix -ol, the compound is identified as an alcohol. Oct- indicates eight. Numbering the carbon chain from left to right places the —OH group at C-2. An ethyl substituent has two carbons, and an isopropyl group is a branched 3-carbon substituent so that the correct structure would be:

$$CH_3\overset{\displaystyle OH}{\overset{\mid}{C}}HCH_2\overset{\displaystyle CH_2CH_3}{\underset{\underset{\displaystyle CH_3CHCH_3}{\mid}}{\overset{\mid}{C}}}CH_2CH_2CH_2CH_3$$

■ Extending the Variety of Isomers

The number of possible isomers with a given formula increases rapidly as the number of carbon atoms increases. For example, there are four alcohols having the formula $C_4H_{10}O$. The structures of these alcohols, their boiling points, and two names for each are given in table 15.2.

These four-carbon alcohols show different carbon skeletons as well as the different points of attachment for the —OH group. The first two alcohols are positional isomers that are derivatives of the straight chain hydrocarbon, n-butane. The other two alcohols are also positional isomers, but they have the same carbon skeleton as the branched hydrocarbon isobutane.

Compounds with the same functional group show both similarities and differences. For example, the four-carbon alcohols in table 15.2 are all hydrogen bonded, and they all react in a similar way with acids and bases. However, one of these alcohols, 2-methyl-2-propanol, fails to react with some reagents that oxidize the other butyl alcohols.

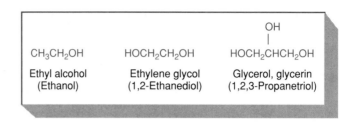

■ The Uses of Some Alcohols and an Ether

A wide variety of alcohols are readily available either from fermentation processes (this process is required by law for human consumption) or as products of the petrochemical industry. Many are used as either solvents or intermediates in the chemical industry. Three alcohols of major importance are ethanol, ethylene glycol, and glycerol, shown in figure 15.2.

Ethyl alcohol is used as a solvent and as a feed stock for other chemical reactions. Ethylene glycol (1,2-ethanediol) is used both in making textile fibers and as a coolant for automobile engines. Like water, it has a high heat capacity and can transfer heat from the engine to the radiator where it is exchanged with the surrounding air. With two —OH groups, it is a strongly hydrogen-bonded liquid with both a high boiling point and a low freezing point, properties desirable for a compound that must remain a liquid in a running engine on a hot day and in an idle engine on a freezing night. Glycerol or glycerin (1,2,3-propanetriol) is a by-product of the manufacturing of soap. It is used as a moisturizer in cosmetics, as a starting material for the manufacture of dynamite, and in making protective coatings for appliances.

Ethers are much less widely used than alcohols although diethyl ether played an important role as an early general anesthetic in the development of surgical procedure.

$$CH_3CH_2OCH_2CH_3$$
diethyl ether

The widespread use of this compound did much to reduce pain and to extend the range of surgery, but its use required extreme care, due to the highly explosive nature of the mixtures of ether vapor and air. Diethyl ether has been replaced as an anesthetic by less flammable compounds to reduce the danger of fire and explosion.

■ Alkenes and Alkynes

Alkenes are hydrocarbons that have carbon–carbon double bonds. The simplest alkene, ethene, or ethylene as it is commonly called, has the formula C_2H_4 and stimulates the ripening of many fruits. In the names of alkenes, the suffix *-ene* replaces the *-ane* of the corresponding alkane. The hydrocarbon chain is numbered to give the lowest integer to signify the position of the double bond as is shown in the case of 1-butene (see Fig. 15.3).

Alkenes exhibit still another type of isomerism when each carbon of the double bond is attached to two different groups. For example, there are two isomeric 2-butenes as shown in figure 15.4. In the *cis* isomer, the two hydrogen atoms are on the same side of the double bond, and in the *trans* isomer, they are on opposite sides. Isomers differing in the geometry of the attachment of groups about the double bonds are often called *geometric isomers*. They exist because there is no free

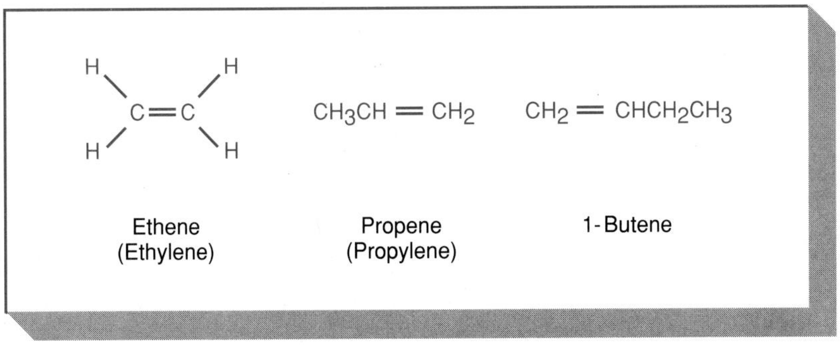

Figure 15.3
Alkenes are hydrocarbons that have carbon-carbon double bonds. The structures of three alkenes are illustrated together with systematic and common names (in parentheses).

rotation about double bonds (geometric isomers are not normally found in alkanes because there is free rotation about carbon–carbon single bonds).

Alkenes are more reactive than alkanes, and they undergo a number of characteristic reactions. For example, hydrogen adds to the double bond of an alkene to convert it to the corresponding alkane in a reaction catalyzed by finely divided platinum or palladium:

$$H_2 + -CH=CH- \rightarrow -CH_2CH_2-$$

Because a compound containing a carbon–carbon double bond can add hydrogen, it is called *unsaturated*. In contrast, a compound that has only single bonds is called *saturated*. For example, unsaturated fats and oils contain carbon–carbon double bonds, whereas saturated fats and oils do not.

Double bonds are also reactive with a variety of reagents. For example, the electrons of a double bond are available to be shared with an acid, or to react with an oxidizing agent. The production of ethyl alcohol and of ethylene glycol illustrate these reactions. Ethyl alcohol is the product of the acid-catalyzed addition of water to the double bond of ethylene:

$$CH_2=CH_2 + H_2O \rightarrow CH_3CH_2OH$$

In the production of ethylene glycol, the double bond is first oxidized by oxygen, then the three-membered ring of the resulting cyclic ether is opened to add water and to give a diol in a reaction catalyzed by acid:

$$CH_2=CH_2 + O_2 \rightarrow \overset{O}{\overset{/\ \backslash}{CH_2{-}CH_2}}$$

$$\overset{O}{\overset{/\ \backslash}{CH_2{-}CH_2}} + H_2O \rightarrow HOCH_2CH_2OH$$

Alkynes are hydrocarbons that have carbon–carbon triple bonds. Like alkenes, alkynes are unsaturated, and they undergo addition reactions similar to those of alkenes.

$$H-C\equiv C-H$$
ethyne or acetylene

The simplest alkyne, acetylene, was used as an illuminating gas in the headlights of early automobiles and is still used in lamps carried by some cave explorers. The reaction of water with calcium carbide is used to generate the acetylene; by controlling the drop rate of the water, we control the rate of the formation of gas:

$$CaC_2 + 2\ H_2O \rightarrow C_2H_2 + Ca(OH)_2$$

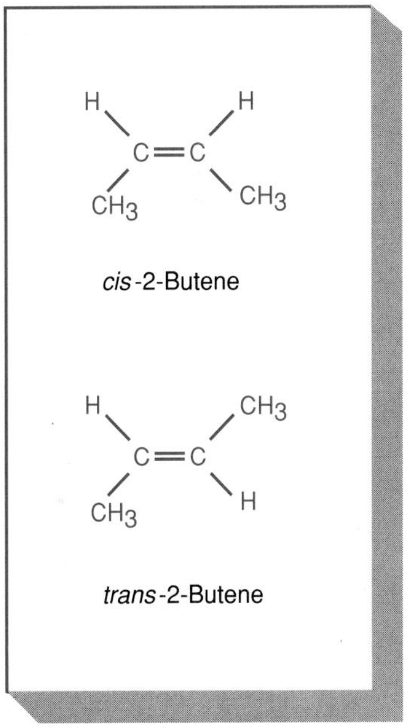

Figure 15.4
Geometric isomers are possible when each carbon atom of the double bond is attached to two different groups. In the case of the isomeric 2-butenes, the two hydrogen atoms are on the same side of the double bond in the *cis* isomer and on opposite sides of the double bond of the *trans* isomer. (Note that the two 2-butenes are positional isomers of 1-butene shown in Figure 15.3.)

Figure 15.5
Benzene and its derivatives are
called aromatic compounds.

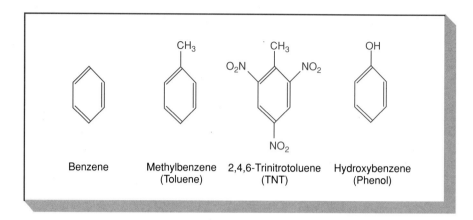

Benzene   Methylbenzene   2,4,6-Trinitrotoluene   Hydroxybenzene
                 (Toluene)   (TNT)   (Phenol)

Figure 15.6
The real structure of benzene is
intermediate between the two
structures shown. Chemists describe
benzene as being resonance-
stabilized and use the doubleheaded arrow to indicate resonance
rather than change or equilibrium.

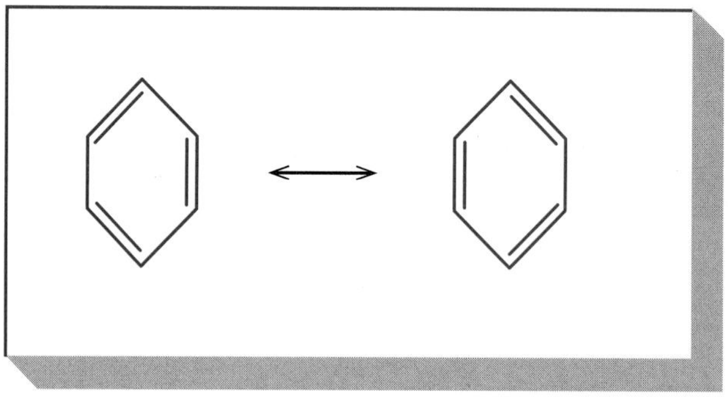

Acetylene burns in air with a bright sooty flame; the light is from the glowing carbon particles formed in the flame. But acetylene is burned with pure oxygen to give the extremely hot flame of an oxyacetylene torch used for the cutting and welding of metal.

■ Aromatic Hydrocarbons

Benzene (C_6H_6) and its derivatives belong to the class of compounds called *aromatic hydrocarbons*. One derivative, toluene, an aromatic compound used in making the explosive TNT, is methylbenzene. Another, phenol, a disinfectant also known as carbolic acid, is hydroxybenzene (see Fig. 15.5).

Benzene and its derivatives are generally less reactive than alkenes. The explanation for the low reactivity of benzene lies in the stability of the benzene ring. No single Lewis structure using octets of electrons can describe benzene because all six carbon–carbon bonds are equal in length. Chemists refer to benzene as being resonance stabilized, and they represent it as a hybrid of two fictional Lewis structures. The doubleheaded arrow is used to indicate that the real structure of benzene is intermediate between the two Lewis structures shown in figure 15.6. Compounds that are resonance stabilized are less reactive than would be predicted on the basis of a double bond in a Lewis structure, for their electrons are delocalized.

Reactions of aromatic hydrocarbons take a different course than reactions of alkenes and alkynes. Acids add to the double bond of alkenes, but, with benzene, however, the course of reaction is substitution. The preference for substitution, rather than addition, is related to the stability associated with the benzene ring, for

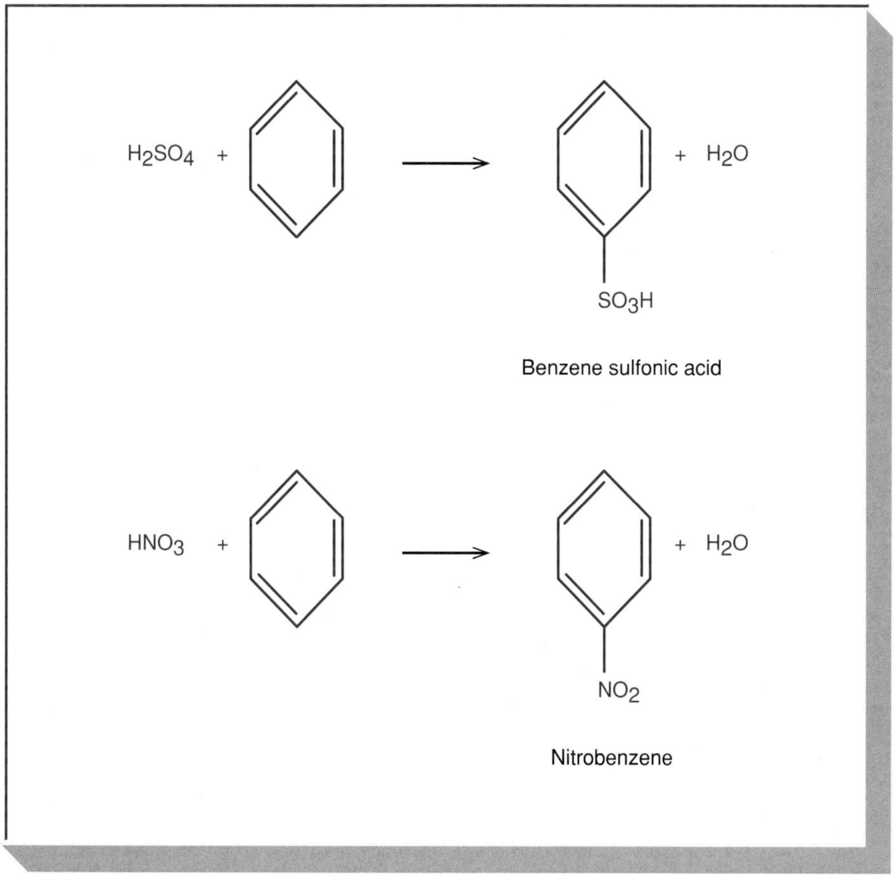

Figure 15.7
Benzene and other aromatic compounds undergo substitution reactions in which an incoming group replaces a hydrogen atom on the aromatic ring.

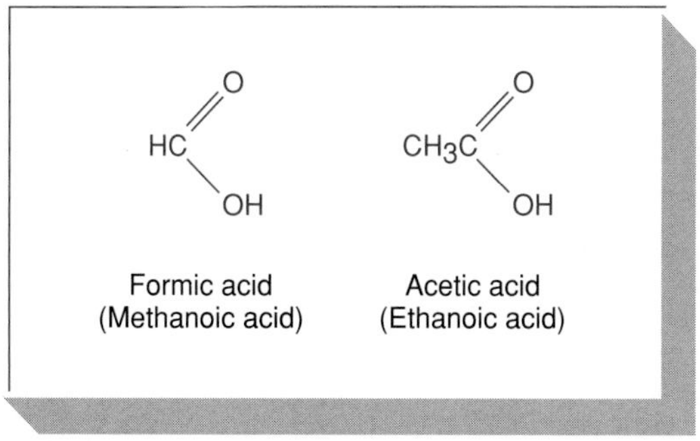

Benzene sulfonic acid

Nitrobenzene

Formic acid
(Methanoic acid)

Acetic acid
(Ethanoic acid)

Figure 15.8
Carboxylic acids are among the most important organic compounds. Among the organic acids are formic acids found in ants and acetic acid found in vinegar.

substituted benzenes are generally more stable than those that would have resulted from simple addition. Electrons are delocalized in substituted benzenes just as they are in benzene itself (see Fig. 15.7).

■ Carboxylic Acids

Carboxylic acids are among the earliest known and most important organic compounds. Among the organic acids are formic acid or methanoic acid, formerly isolated from ants, and acetic acid or ethanoic acid found in vinegar, shown in figure 15.8.

• • • • • • • • • • • • • • • Table 15.3 • • • • • • • • • • • • • • •

Structures of Some Carboxylic Acids

Acid	Common Name	Systematic Name
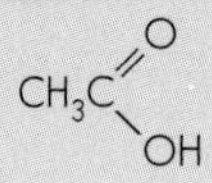 CH_3C (with $=O$ and OH)	Acetic acid	Ethanoic acid
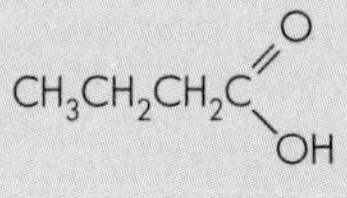 $CH_3CH_2CH_2C$ (with $=O$ and OH)	Butyric acid	Butanoic acid
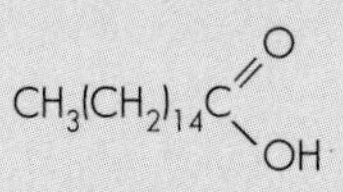 $CH_3(CH_2)_{14}C$ (with $=O$ and OH)	Palmitic acid	Hexadecanoic acid
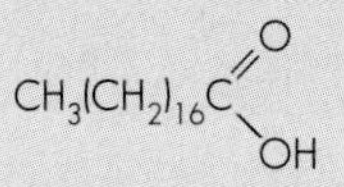 $CH_3(CH_2)_{16}C$ (with $=O$ and OH)	Stearic acid	Octadecanoic acid
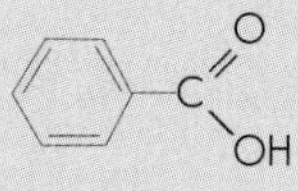 (benzene ring)$-C$ (with $=O$ and OH)	Benzoic acid	Benzoic acid
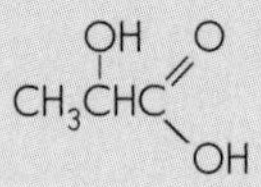 CH_3CHC (with OH, $=O$ and OH)	Lactic acid	
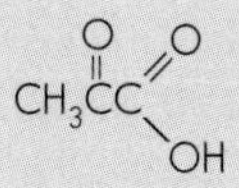 CH_3CC (with $=O$, $=O$ and OH)	Pyruvic acid	
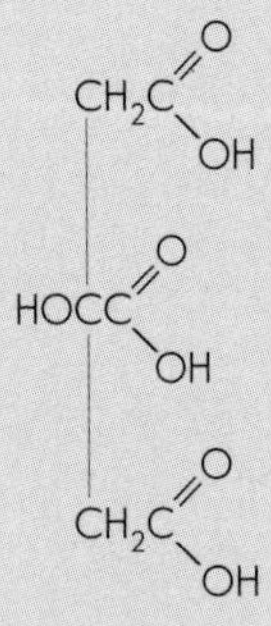 CH_2C (with $=O$ and OH) — $HOCC$ (with $=O$ and OH) — CH_2C (with $=O$ and OH)	Citric acid	

Common names of carboxylic acids remain in use while the systematic names end in the suffix *-oic acid*. The names and structures of some carboxylic acids are listed in table 15.3.

Acid-base chemistry facilitates the separation and purification of carboxylic acids. They can be separated from mixtures of organic compounds by extraction into aqueous sodium hydroxide followed by separation of the aqueous and organic layers. The weak carboxylic acids are converted by strong bases entirely to sodium salts (sodium carboxylates), which are far more soluble in water than the free acids. Upon acidification of the resulting basic solution with a strong acid, the carboxylate ion accepts a proton to reform the carboxylic acid. The free acid can be recovered from the aqueous solution by filtration or by extraction, depending on its solubility.

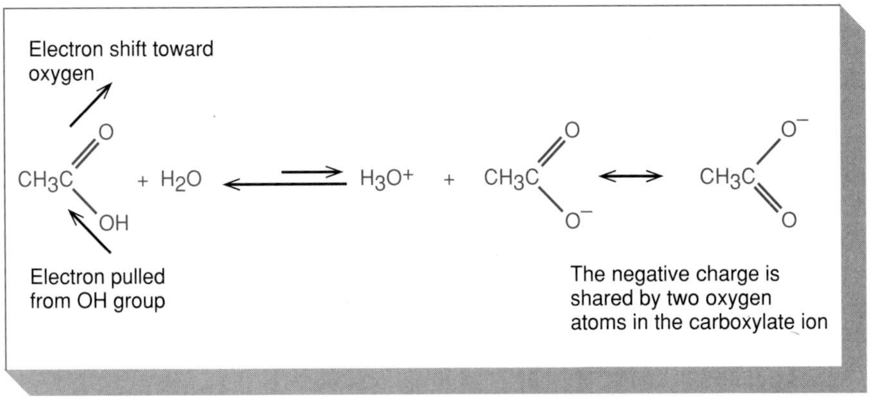

Figure 15.9
Carboxylic acids are more acidic than water or ethanol.

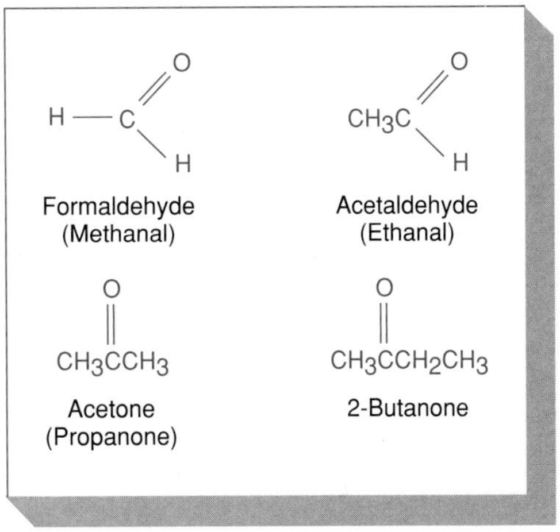

Electron shift toward oxygen

Electron pulled from OH group

$$CH_3C \overset{O}{\underset{OH}{\big\langle}} + H_2O \rightleftharpoons H_3O^+ + CH_3C \overset{O}{\underset{O^-}{\big\langle}} \longleftrightarrow CH_3C \overset{O^-}{\underset{O}{\big\langle}}$$

The negative charge is shared by two oxygen atoms in the carboxylate ion

Figure 15.10
Aldehydes and ketones are compounds that have a carbon-oxygen double bond called a carbonyl group. In an aldehyde, the carbonyl-carbon is attached to one or two hydrogens. In a ketone, the carbonyl-carbon is attached to two carbons.

$$H - C \overset{O}{\underset{H}{\big\langle}}$$

Formaldehyde
(Methanal)

$$CH_3C \overset{O}{\underset{H}{\big\langle}}$$

Acetaldehyde
(Ethanal)

$$CH_3\overset{O}{\overset{\|}{C}}CH_3$$

Acetone
(Propanone)

$$CH_3\overset{O}{\overset{\|}{C}}CH_2CH_3$$

2-Butanone

Why are carboxylic acids more acidic than water or alcohols? All have an —OH group, yet the —OH of a carboxylic acid is a better proton donor than the —OH of either alcohol or water. Both the polarity of the carbon–oxygen double bond of the acid and the resonance in the conjugate base contribute to the acidity. The polar C=O group pulls electron density from the —OH group making it more positive and better able to donate a proton than the —OH of an alcohol. In the carboxylate ion, the negative charge is shared equally by the two oxygen atoms. Because the negative charge is more diffuse, the carboxylate ion is a weaker base than the hydroxide ion (see Fig. 15.9).

■ Aldehydes and Ketones

Aldehydes and *ketones* are compounds that have a carbon–oxygen double bond. Formaldehyde is the simplest aldehyde, and acetone is the simplest ketone. In a ketone, the C=O or *carbonyl group* is flanked by carbon atoms. In an aldehyde, one or two hydrogen atoms are attached to the carbonyl group. Aldehydes and ketones are named by adding a suffix to the root of the name of the parent hydrocarbon. The characteristic suffix in the name of an aldehyde is *-al;* in ketones, it is *-one* (see Fig. 15.10).

The chemistry of the carbonyl group is common to both aldehydes and ketones. Bases, including water, add to the polar carbon–oxygen double bond of aldehydes and ketones as shown in figure 15.11. These Lewis acid-base addition reactions resemble the addition of water to the carbon–oxygen double bond of

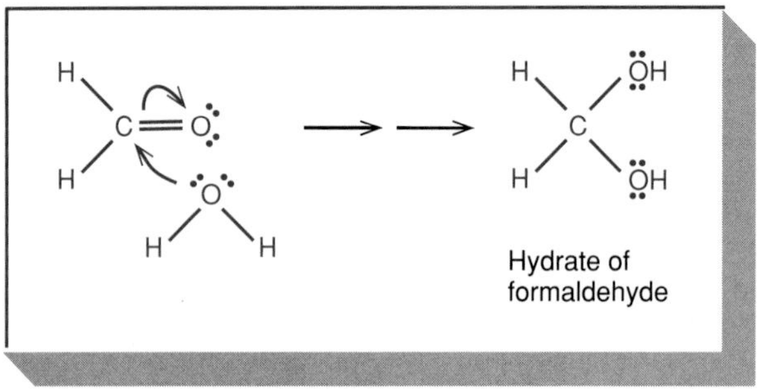

Figure 15.11
Bases add to the carbonyl-carbon atom of aldehydes and ketones. This Lewis acid-base chemistry is important in the chemistry of sugars and other carbohydrates.

Hydrate of
formaldehyde

carbon dioxide. The extent of the addition reaction varies. Formaldehyde in an aqueous solution is almost completely hydrated, whereas very little of the adduct is present in an aqueous solution of acetone.

The most important difference between these two functional groups is that aldehydes are more readily oxidized than ketones. Many mild oxidizing agents convert aldehydes to carboxylic acids under conditions that leave ketones unchanged. The chemistry of aldehydes and ketones, important in the structure of sugars and other carbohydrates, is discussed further in chapter 18.

■ Stereoisomers

The discovery of compounds that seemed to have the same structure, but different properties, posed an important puzzle to chemists. When lactic acid, produced by the fermentation of milk was compared to lactic acid isolated from muscles, the samples had different properties.

Two young chemists, Jacobus van't Hoff (1852–1911) of the Netherlands and Joseph Le Bel (1847–1930) of France, independently ascribed these differences in properties to differences in geometry. In 1874, more than thirty-five years before Rutherford proposed a nuclear structure for the atom, they postulated that the four bonds to a carbon atom are directed toward the corners of a tetrahedron. A tetrahedral shape for carbon accounted for the existence of mirror image isomers because the carbon atoms bonded to four different groups are asymmetric. The lactic acid from muscle consisted of a single isomer, and that from milk was shown to be a mixture of equal amounts of mirror image isomers.

Molecules that differ only in geometry are called *stereoisomers*. Stereoisomers that differ in the attachment of the four groups around an asymmetric carbon atom are nonsuperposable mirror images called *enantiomers*. Enantiomers can be thought of as right- and left-handed molecules. Consider the C—H bond of the central carbon atom of lactic acid to be a steering column, and the other three bonds to be spokes of the steering wheel. In the two enantiomers, the three groups correspond to opposite rotations of the wheel, just as the fingers of the right and left hands curve in opposite directions. These molecules are *chiral;* that is, they have "handedness." Their relationship resembles that of the right and left hands (see Fig. 15.12).

Enantiomers can be distinguished by their interactions with polarized light. Most physical properties such as melting point, solubility, and density are identical, but enantiomers rotate plane-polarized light in opposite directions. By passing light through a polarizing lens (similar to those used to reduce glare in sunglasses), then through a solution of the compound being tested, and finally, through a second polarizing lens, the rotation of polarized light can be studied. If a sample is chiral, the second lens will need to be turned or rotated in order to restore the maximum brightness. Enantiomers will rotate polarized light to an equal extent, but in opposite directions.

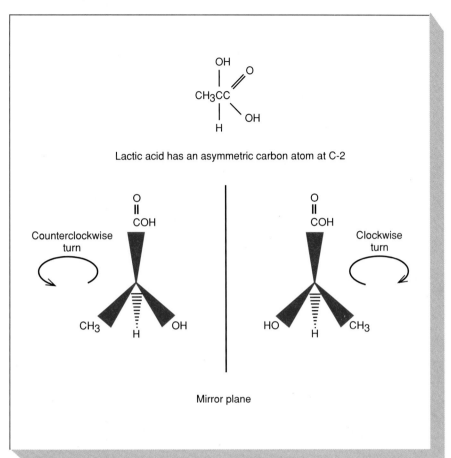

Figure 15.12
Enantiomers are mirror image isomers. The relationship is like that between left and right hands.

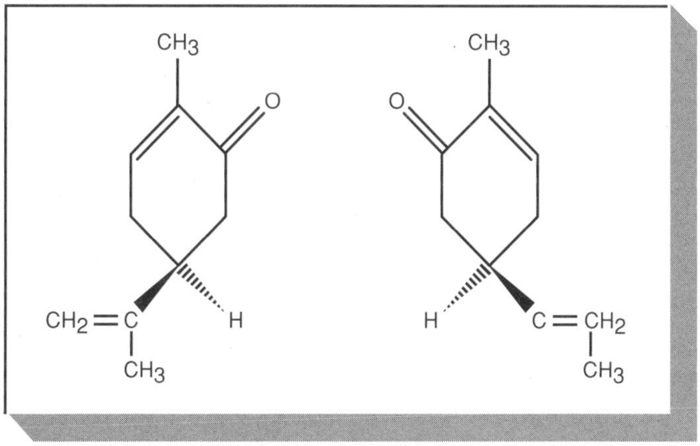

Lactic acid has an asymmetric carbon atom at C-2

Counterclockwise turn

Clockwise turn

Mirror plane

Figure 15.13
Enantiomeric carvones from spearmint and caraway differ in taste and odor.

The different odors and tastes of the carvones from dill or caraway seeds and from spearmint leaves illustrate that receptor sites can distinguish between mirror-image isomers. The carvone isolated from dill and caraway seeds is the enantiomer of that found in spearmint leaves as shown in figure 15.13. Molecules that carry chemical messages to proteins function as sense receptors for odor. A chemical messenger fits its receptor like a key fits a lock. Since proteins are chiral, signal

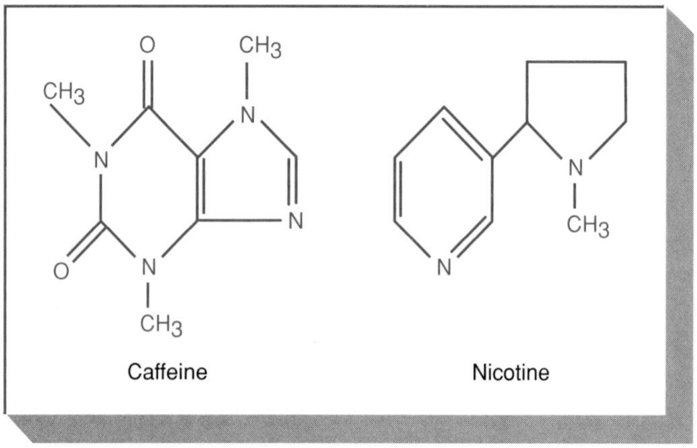

Figure 15.14
Alkaloids are organic bases isolated
from plants. Caffeine is found in
coffee and tea, and nicotine is found
in tobacco. Many alkaloids have
potent physiological effects.

Figure 15.14
Alkaloids are organic bases isolated from plants. Caffeine is found in coffee and tea, and nicotine is found in tobacco. Many alkaloids have potent physiological effects.

molecules having right- and left-handed shapes may not fit the same receptor protein, just as right and left hands do not fit the same glove. Each carvone triggers a different sensory sensation because it binds to a different receptor.

■ Amines

Amines are organic bases and just as alcohols can be considered to be derivatives of water, amines can be considered to be derivatives of ammonia. Amines may have one, two, or three alkyl groups substituted for the hydrogen atoms on ammonia. The basic character of amines depends upon the unshared pair of electrons on nitrogen:

$$CH_3CH_2NH_2 \qquad CH_3\overset{\overset{\displaystyle CH_3}{|}}{N}CH_3$$

ethyl amine   trimethyl amine

Many naturally occurring organic bases have powerful physiological actions. *Alkaloids* are physiologically active organic amines isolated from plants. For example, caffeine, nicotine, quinine, cocaine, and morphine are all alkaloids (see Fig. 15.14). People can readily separate these alkaloids from their plant sources by taking advantage of their acid-base chemistry. The ease of separation of alkaloids makes the control of the chemical processing of alkaloids used for illicit drugs very difficult. For example, cocaine is obtained from the coca plant by first extracting the dried leaves with kerosene and then reacting the base with an aqueous strong acid to separate it from the kerosene and from neutral organic compounds. After separation of the acid solution, the free base (present in the acidic solution as the salt of cocaine) can be reformed by the removal of an acidic proton in a reaction with the base baking soda ($NaHCO_3$).

■ Esters and Amides

Carboxylic acids react with alcohols to form products known as *esters*. In the reaction, a molecule of water is also formed. These reactions, in which two molecules combine by the elimination of a small molecule such as water, are called condensation reactions.

$$CH_3\overset{\overset{\displaystyle O}{||}}{C}OH + HOCH_2CH_2CH_3 \rightarrow CH_3\overset{\overset{\displaystyle O}{||}}{C}OCH_2CH_2CH_3 + H_2O$$

Acetic acid + *n*-propyl alcohol   → *n*-propyl acetate   + water
Ethanoic acid + 1-propanol   → 1-propyl ethanoate   + water

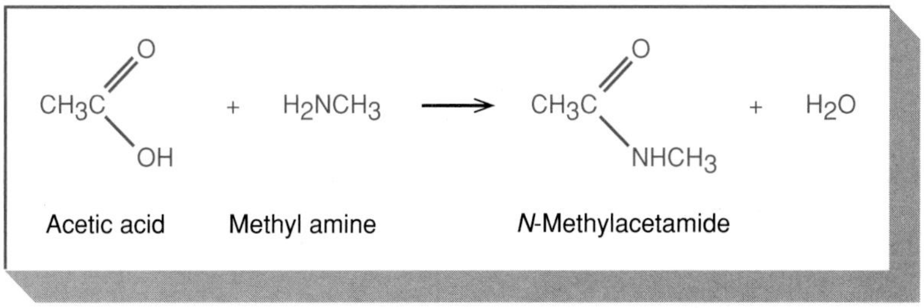

Figure 15.15
Amides are neutral compounds that can be formed by the condensation reaction of a carboxylic acid with ammonia or an amine.

Esters are named as derivatives of carboxylic acids. The alcohol portion of an ester is named first but with the suffix *-yl*. Then, the acid portion is named with the ending *-oic acid* changed to *-oate* in the ester. Esters have lower boiling points than either alcohols or acids of the same molecular weight because they are not hydrogen-bonded. Volatile, low-molecular-weight esters contribute to the sweet odor of many fruits.

Amides are neutral compounds that can be considered to be the products of a condensation reaction of an amine or ammonia with a carboxylic acid as shown in figure 15.15. Amides have structures analogous to those of esters, in which an amine moiety has replaced the alcohol moiety of the ester. The amide functional group plays a central role in the structure of proteins. Subsequent chapters explore examples of the importance of these and other organic compounds in the natural world and in man-made products.

■ Questions—*Chapter 15*

1. Which of the following functional group isomers has the higher boiling point? Why?

$$CH_3CH_2OCH_2CH_3 \qquad CH_3CH_2CH_2CH_2OH$$
diethyl ether   *n*-butyl alcohol

2. Draw the three isomeric ethers having the formula $C_4H_{10}O$.
3. Write structures for a compound fitting each of the following descriptions:
   (a) Branched alkane
   (b) Three-carbon alkene
   (c) Alkylbenzene
   (d) Three-carbon ether
4. Name the following compounds.
   (a) $CH_3CH_2OCH_2CH_3$
   (b) $CH_3CH_2CH_2CH_2CH_2OH$
   (c) $CH_3\overset{\overset{\displaystyle OH}{|}}{\underset{\underset{\displaystyle CH_3}{|}}{C}}CH_3$
   (d) $CH_3CH{=}CH_2$
   (e) $CH_3{-}\overset{\overset{\displaystyle OH}{|}}{\underset{\underset{\displaystyle H}{|}}{C}}{-}CH_2CH_2CH_3$

5. Each of the following compounds is an isomer of the 2-pentanol shown above. Match the correct isomeric relationship to each compound.
   (a) $CH_3CH_2OCH_2CH_2CH_3$

   (b)
   $$CH_3\underset{\underset{\displaystyle CH_3}{|}}{\overset{\overset{\displaystyle OH}{|}}{C}}CH_2CH_3$$

   (c)
   $$CH_3CH_2\underset{}{\overset{\overset{\displaystyle OH}{|}}{C}}HCH_2CH_3$$
   (1) Functional group isomer
   (2) Positional isomer
   (3) Carbon-skeleton isomer

6. Write structures for the following compounds.
   (a) Diisopropyl ether
   (b) 3-Pentanol
   (c) 2,3-Dimethylbutane
   (d) 1-Butene

7. Why are alcohols more soluble in water than hydrocarbons?

8. Why is the boiling point of ethylene glycol used in antifreeze higher than that of either ethyl alcohol or water?

9. Draw *cis* and *trans* isomers of 2-pentene, $CH_3CH=CHCH_2CH_3$.

10. Draw a complete Lewis structure of cyclohexene. What is the molecular formula of cyclohexene?

11. Draw resonance structures of ortho-xylene (1,2-dimethylbenzene). Why is the existence of a single isomer of this compound evidence for resonance? (If both the Lewis structures existed, would you expect to see isomers?)

12. Give reactions to account for the following observations:
    (a) When bromine is added to an alkene, cyclohexene, the bromine color fades.
    (b) When bromine is added to an aromatic hydrocarbon, benzene, the bromine color persists. If, however, a Lewis acid is present, the bromine color fades and the odor of hydrogen bromide can be noted.

13. Write structures for a compound fitting each of the following descriptions:
    (a) Two-carbon aldehyde
    (b) Four-carbon ketone
    (c) Three-carbon carboxylic acid
    (d) Two-carbon amine
    (e) Two-carbon amide

14. Write structures for the following compounds.
    (a) Ethanal
    (b) 2-Pentanone
    (c) Propanoic acid
    (d) Methyl butanoate
    (e) Dimethyl amine

15. Just as water reacts as a base to add to the carbon–oxygen double bond of formaldehyde, alcohols also add to this Lewis acid. Draw the structure of the product formed when methanol adds to formaldehyde.

16. Acetic acid, $C_2H_4O_2$, has the same density in the vapor state as molecules (dimers) having twice the molecular weight. What forces would exist in acetic acid dimers? Draw a structure for a dimer of acetic acid in the vapor state.

17. Account for the observation that trifluoroacetic acid (CF_3CO_2H) is a stronger acid than acetic acid.

18. What functional group is present in each of the following compounds?

(a) 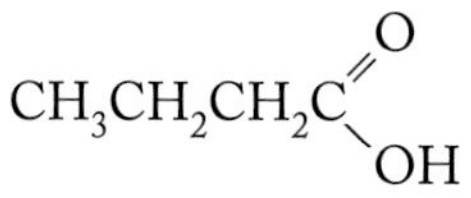 $CH_3CH_2CH_2C\overset{\displaystyle O}{\underset{\displaystyle OH}{}}$

(b) 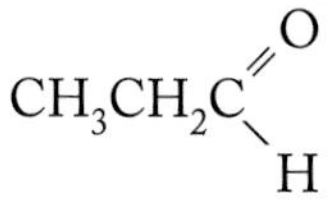 $CH_3CH_2C\overset{\displaystyle O}{\underset{\displaystyle H}{}}$

(c) 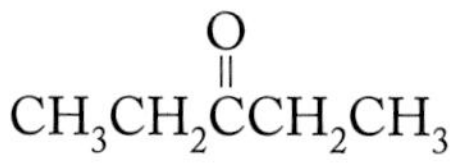 $CH_3CH_2\overset{\displaystyle O}{\underset{\displaystyle \|}{C}}CH_2CH_3$

(d)  $H\overset{\displaystyle O}{\underset{\displaystyle \|}{C}}OCH_2CH_3$

(e) $CH_3CH_2CH_2NH_2$

(f) 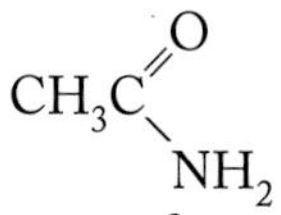 $CH_3C\overset{\displaystyle O}{\underset{\displaystyle NH_2}{}}$

19. Give a name for each of the compounds in question 18.

20. One of the butyl alcohols consists of a pair of enantiomers. Which butyl alcohol has a chiral center? Sketch the enantiomers of this butyl alcohol.

21. Draw the structures of the products of each of the following reactions.

(a) 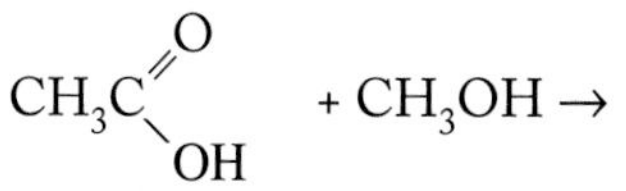 $CH_3C\overset{\displaystyle O}{\underset{\displaystyle OH}{}} + CH_3OH \rightarrow$

(b) 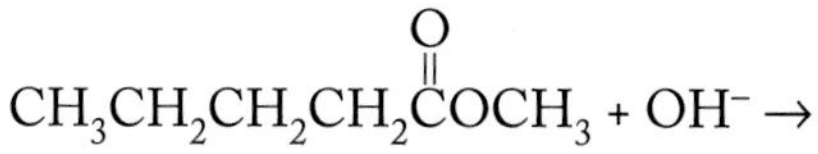 $CH_3CH_2CH_2CH_2\overset{\displaystyle O}{\underset{\displaystyle \|}{C}}OCH_3 + OH^- \rightarrow$

22. In the following pairs of compounds, one isomer has a single functional group and the other isomer has two functional groups. Name the functional groups present in each compound.

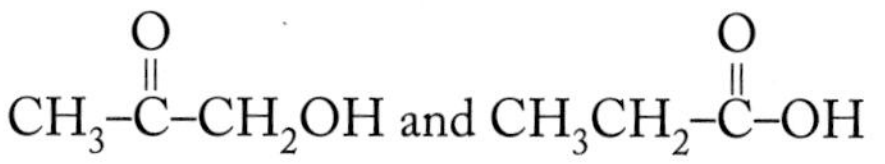

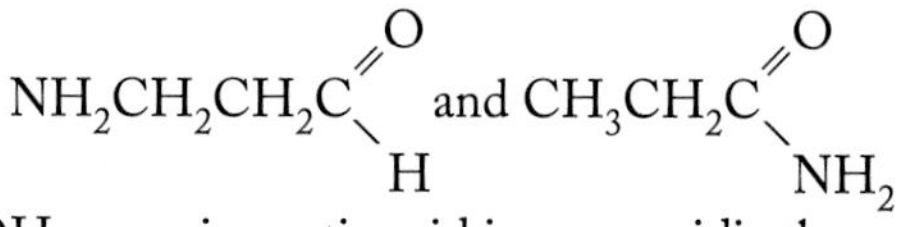

$$CH_3-\overset{\displaystyle O}{\underset{\displaystyle \|}{C}}-CH_2OH \text{ and } CH_3CH_2-\overset{\displaystyle O}{\underset{\displaystyle \|}{C}}-OH$$

$$NH_2CH_2CH_2C\overset{\displaystyle O}{\underset{\displaystyle H}{}} \text{ and } CH_3CH_2C\overset{\displaystyle O}{\underset{\displaystyle NH_2}{}}$$

23. Just as the —OH group in acetic acid is more acidic than water, the —NH_2 group in acetamide is less basic than ammonia. How does the C=O group make the nitrogen in the amide less basic?

C H A P T E R

16

Organic Compounds From Petroleum in Transportation and Agriculture

Petroleum refining provides the majority of the hydrocarbons used in the chemical industry as well as such products as gasoline and most heating oil. The petrochemical industry converts hydrocarbons produced from petroleum into a wide variety of high-volume organic compounds. Some are blended into fuels, some are used as solvents, and still others are used in the formation of plastics and other useful materials.

The quantities in production are large to take advantage of large-scale economies. Petroleum refining operations are tied to the marketplace. The demands for gasoline, diesel fuel, and heating oil dictate the amounts of these products being produced. During the summer, the demand for gasoline is greater; in winter, much more heating oil is needed. Refineries represent very substantial capital investments; therefore, operations are designed to be continuous because of the substantial large-scale benefits.

Several steps are involved in the refining of crude oil. Nitrogen and sulfur impurities need to be removed, and viscous mixtures of high-molecular-weight compounds need to be converted into more volatile lower-molecular-weight compounds. At all stages of refining, there are mixtures to separate. Crude separations are made by distillation, using columns like that diagrammed in figure 16.1. Low-boiling fractions are blended into gasoline and high-boiling fractions are subjected to chemical reactions to convert them into more useful products.

■ Chemical Reactions in Petroleum Refining

Catalytic cracking is a process by which high-molecular-weight alkanes are broken, or cracked, to give smaller fragments. High-boiling fractions (bp 250–500° C) are heated to about 500° C with aluminosilicate catalysts. The product stream from this process is rich in lower-molecular-weight branched compounds. Many of the compounds formed by catalytic cracking have carbon–carbon double bonds (see Fig. 16.2).

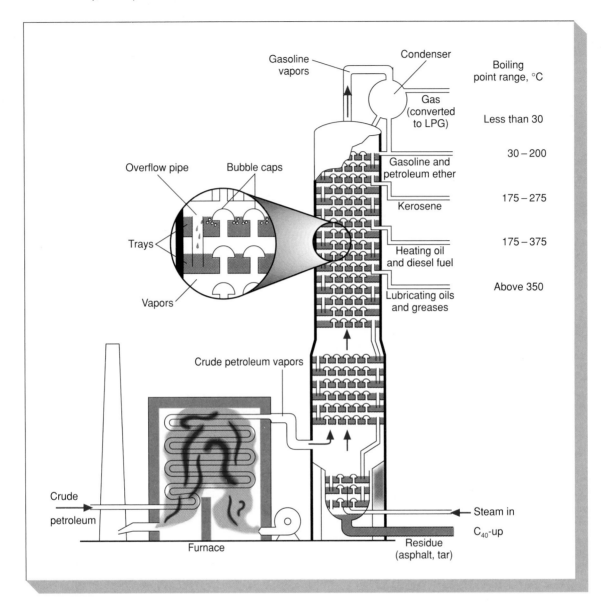

Boiling point range, °C

Gas (converted to LPG) — Less than 30

Gasoline and petroleum ether — 30 – 200

Kerosene — 175 – 275

Heating oil and diesel fuel — 175 – 375

Lubricating oils and greases — Above 350

Figure 16.1

Distillation is used to separate crude petroleum into fractions that differ in their boiling points. Fractions may be used directly or processed in subsequent refining steps.

In another refining process, low-molecular-weight alkanes and alkenes from catalytic cracking are combined, using a sulfuric acid catalyst to give branched chain alkanes. In this step, compounds too volatile for use in gasoline are converted to less volatile, high-octane products (see Fig. 16.3).

Catalytic reforming converts cyclic alkanes and some linear alkanes into aromatic hydrocarbons in reactions such as those shown in figure 16.4. Finely divided platinum catalyzes the removal of H_2. Some of the hydrogen gas that is a by-product of reforming reactions is used in other reactions that remove sulfur compounds from petroleum. Much of it is also used in the manufacture of ammonia for fertilizer.

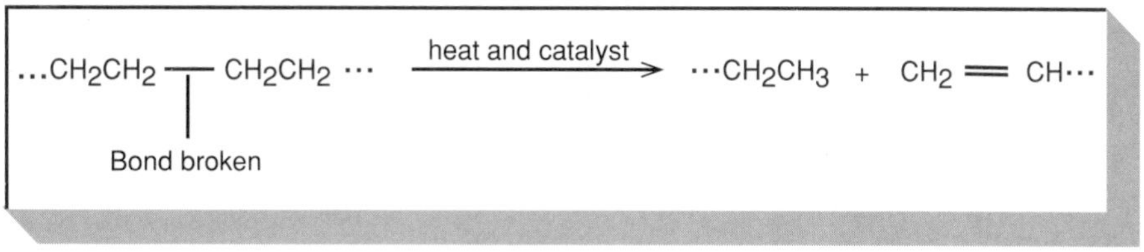

Figure 16.2
In catalytic cracking reactions, high-molecular-weight alkanes are converted mixtures of lower-molecular-weight alkanes and alkenes.

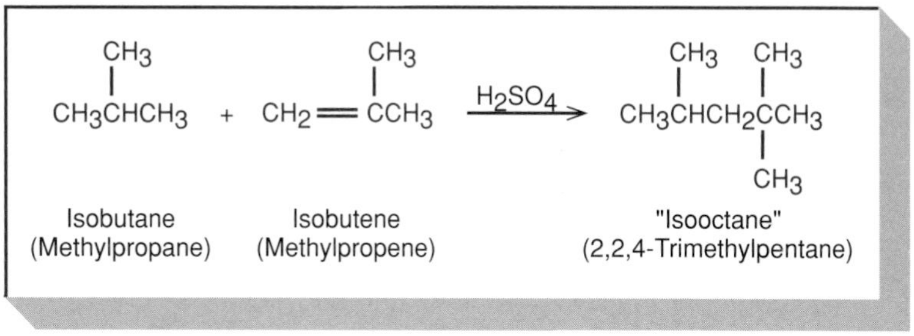

Figure 16.3
Small alkanes and alkenes from catalytic cracking are combined to give branched-chain alkanes.

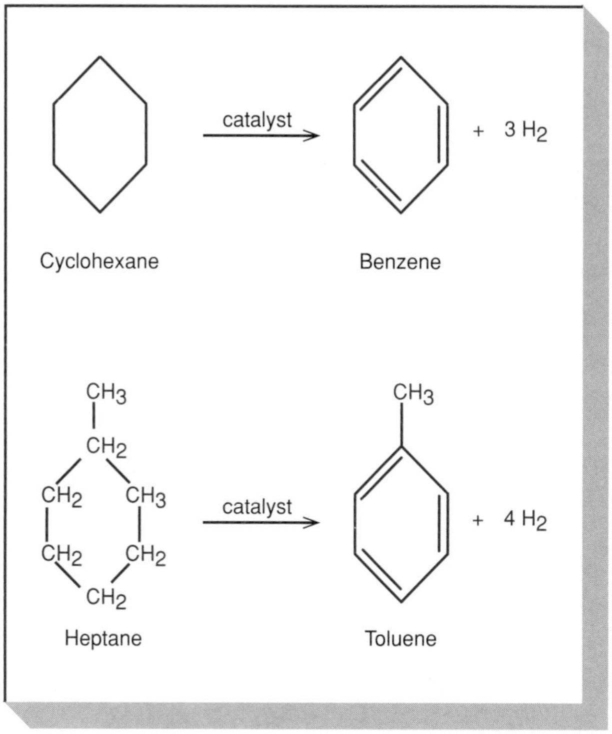

Figure 16.4
Aromatic hydrocarbons are formed from cycloalkanes and from some linear alkanes through catalytic reforming steps.

■ Blending Fuels for Engine Performance

Refiners blend fuels for the specific uses of automobiles. They vary the mixtures according to season and to region of the country. They increase the proportion of low-boiling hydrocarbons in fuels blended for winter use, and decrease that proportion for summer use.

The octane number of a gasoline is a measure of its performance in internal combustion engines. Higher-octane gasolines burn more smoothly and cause less engine knock. Standard compounds used for evaluating gasoline performance are isooctane (100 octane) and *n*-heptane (0 octane). Octane numbers from 0 to 100 can be measured by comparing the antiknock characteristics of a gasoline blend to a mixture of these standards. A fuel with an octane rating of 87 would give the same engine performance as a mixture of 87% isooctane and 13% *n*-heptane (some aromatic compounds in gasoline have octane ratings above 100).

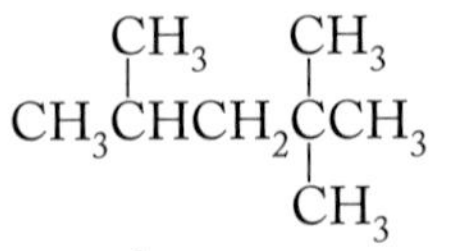

$$CH_3CHCH_2CCH_3 \qquad CH_3CH_2CH_2CH_2CH_2CH_2CH_3$$

isooctane $\qquad\qquad\qquad$ *n*-heptane

(2,2,4-trimethylpentane)

Diesel fuels differ from the gasolines used in most internal combustion engines. In a diesel engine, the mixture of air and fuel is compressed more and reaches a higher temperature before ignition so diesel fuel is less volatile than gasoline and has different burning characteristics. A cetane rating scale, in which cetane (value = 100) and 1-methylnaphthalene (value = 0) serve as standards, is used for diesel fuel:

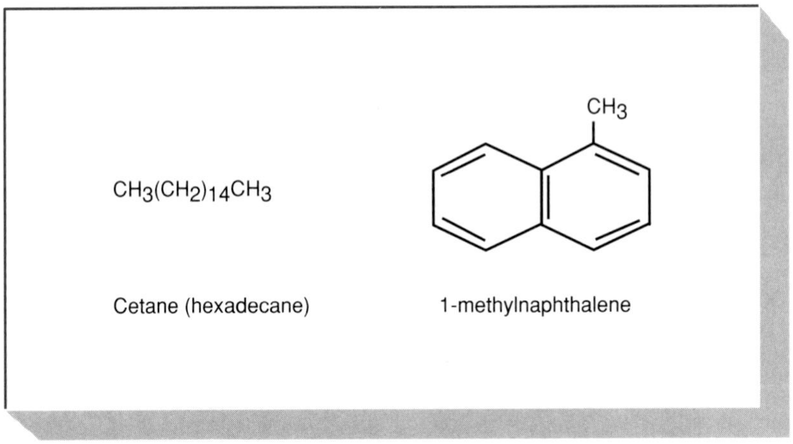

CH₃(CH₂)₁₄CH₃

Cetane (hexadecane)

1-methylnaphthalene

■ Changing Fuels to Reduce Air Pollution

To attain the standards for the improvement of air quality set in the Clean Air Act, significant changes in gasoline blends are required. These standards set permissible levels of carbon monoxide, hydrocarbons, nitrogen oxides, and ozone. Many metropolitan areas with large numbers of automobiles and unfavorable weather conditions are significantly out of compliance for several days each year.

Beginning in 1992, all gasoline sold in metropolitan regions not meeting carbon monoxide standards are required to contain 2.7% oxygen during the winter months. At cold temperatures, when engines operate less efficiently, some fuel may condense on cylinder walls prior to ignition so combustion is less complete, a condition that results in a higher fraction of CO in the exhaust gases. The addition of oxygen-containing compounds to gasoline produces a fuel mixture that burns more completely under extreme conditions and thus reduces the amount of carbon monoxide emissions.

Refiners now use ethanol, methanol, and methyl *t*-butyl ether to increase the oxygen content of fuels. Blends of methanol or ethanol and gasoline first came on the market in the mid-1980s. Because water can hydrogen bond to alcohols, it is far more soluble in alcohols than in the hydrocarbons found in gasoline, so small amounts of additives are used to stabilize the alcohol-gasoline mixtures and to prevent water from collecting and later separating in the fuel mixture.

Some cautionary thoughts are in order. Brazil made an extensive effort to produce ethanol for fuel by the fermentation of sugar cane in the belief that the greater use of alcohol in fuels would reduce its oil imports and ease its balance of payments difficulties. Cars were modified to use pure alcohols as fuels. Now, as farmers have shifted production to more profitable crops, Brazil faces a shortage of ethyl alcohol for fuel and possesses a fleet of cars that cannot run on gasoline.

The use of methyl alcohol in fuel blends presents different environmental problems than the use of ethanol. Methanol is far more toxic than ethanol, and, like ethanol, it is completely soluble in water. The widespread use of methanol in gasoline would be accompanied by increased risks that methanol from gasoline spills would enter the water supply.

Methyl *t*-butyl ether (shown in Fig. 16.5) was first added to gasoline to replace tetraethyl lead as an octane enhancer. It blends well with gasoline, and it produces more energy when burned than either ethanol or methanol. Even though the production of methyl *t*-butyl ether is growing faster than that of any other chemical, shortages of this compound may slow compliance with the provisions of the Clean Air Act.

Ozone is one of the most destructive components of the airborn mixture called smog because it damages membranes of the respiratory system, interfering with the body's ability to exchange oxygen and carbon dioxide in the lungs. Two components of auto exhaust, nitrogen oxides and hydrocarbons, contribute to ozone formation through a complex set of chemical reactions occurring in the atmosphere. The problem is particularly acute in the Los Angeles area where there is a high concentration of automobiles. The combination of thermal inversions and the blocking of air movement by the mountains to the east often allow the concentrations of pollutants to build up for days or weeks. Beginning in 1995, areas not attaining ozone standards will be required to sell gasoline containing at least 2% oxygen, no more than 1% benzene, and 25% aromatics (in total). Other components of smog produce a haze that causes lower visibility and detracts from the natural beauty of places as remote as the Grand Canyon.

California has led the effort to reduce smog and ozone. It has required automobile modifications that have greatly reduced the quantities of hydrocarbons escaping by either evaporation or incomplete combustion. It has enacted legislation to restrict hydrocarbon evaporations from gasoline in storage, from solvents used in paint, and from fuels used to light charcoal for a barbecue. However, despite decades of efforts, levels of atmospheric pollutants have continued to rise (see Fig. 16.6).

Some scientists now believe that an effort that emphasizes only the reduction of hydrocarbon emissions cannot succeed. They point out that trees and shrubs place far more hydrocarbons into the air than had been previously estimated, so that there are limits to the reductions that can be made with the best of efforts. They argue that the future emphasis should be on the more difficult task of achieving greater nitrogen oxide emissions reduction.

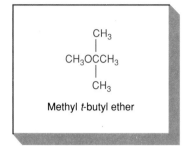

Figure 16.5
Methyl *t*-butyl ether is used as a gasoline additive.

Figure 16.6
This photo of Los Angeles, California, shows the day's built-up smog and hazing.
© Margaret McCarthy/Peter Arnold, Inc.

■ Petrochemicals—High-Volume Chemicals for a Variety of Uses

Alkenes and benzene derivatives obtained from petroleum are feedstocks for the petrochemical industry. Low-cost chemicals are used whenever possible so sulfuric acid, oxygen, water, chlorine, and sodium hydroxide are used in large quantities. As the use of chemicals produced from petroleum has grown, oil refining companies have moved into the production of chemicals, and large chemical companies have acquired oil producing companies.

The petrochemical industry provides the starting materials used to produce synthetic polymers. The oxidation of the xylenes, isomeric dimethylbenzenes, provides the starting materials for the production of a textile fiber, plastic beverage bottles, and coatings for appliances (see Fig. 16.7). The production of styrene, a compound used in making plastic materials and synthetic rubber, utilizes both ethylene and benzene in three reactions as shown in figure 16.8. In the first, hydrogen

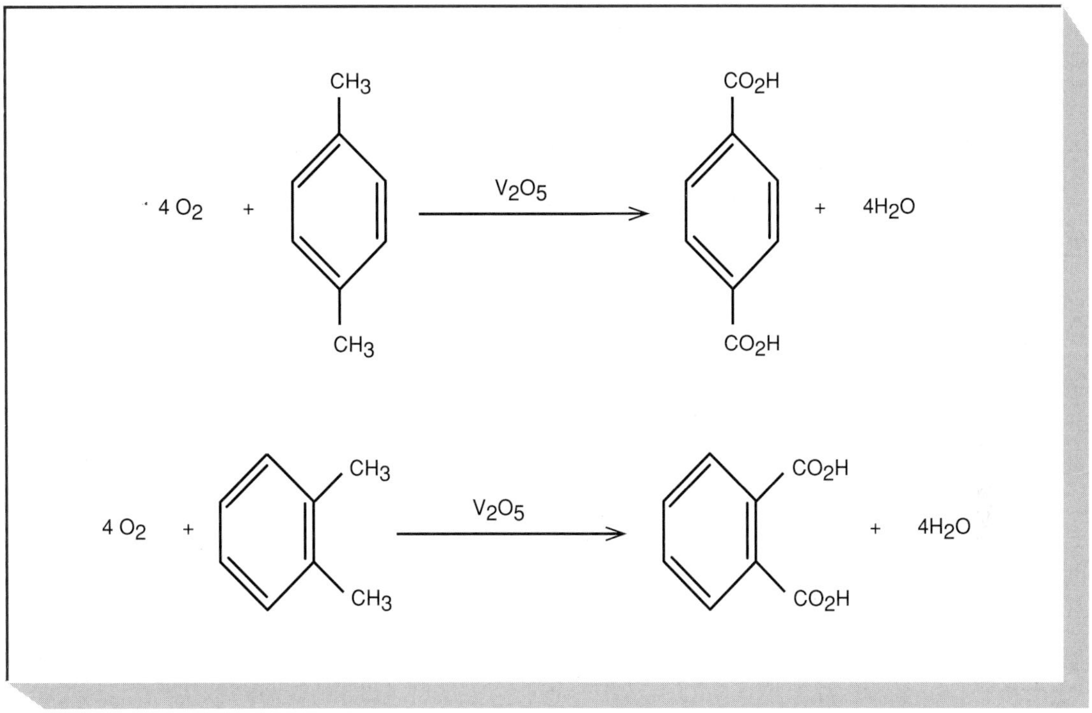

Figure 16.7
The oxidation of xylenes provides compounds used to produce textile fibers and protective coatings for appliances.

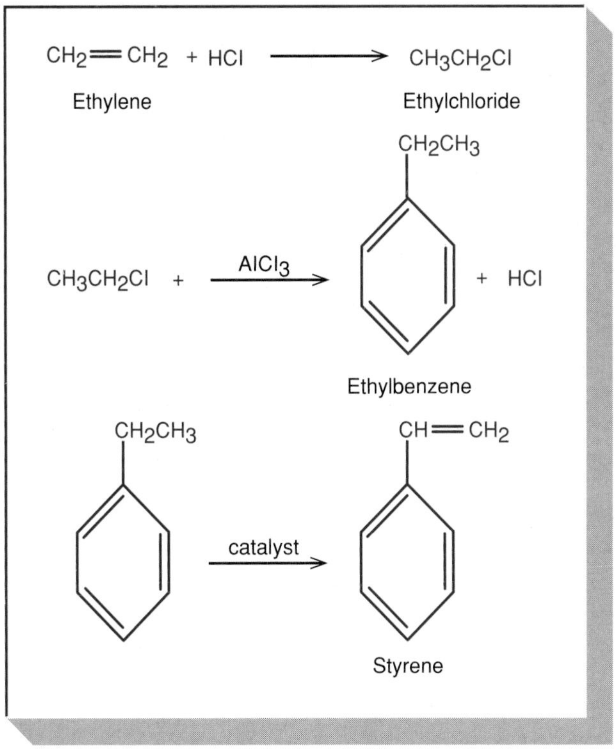

Figure 16.8
The production of styrene utilizes the hydrocarbons ethylene and benzene. Ethylene undergoes an addition reaction, and benzene undergoes a substitution reaction. In a third reaction, hydrogen is catalytically removed from ethyl-benzene to form the final product.

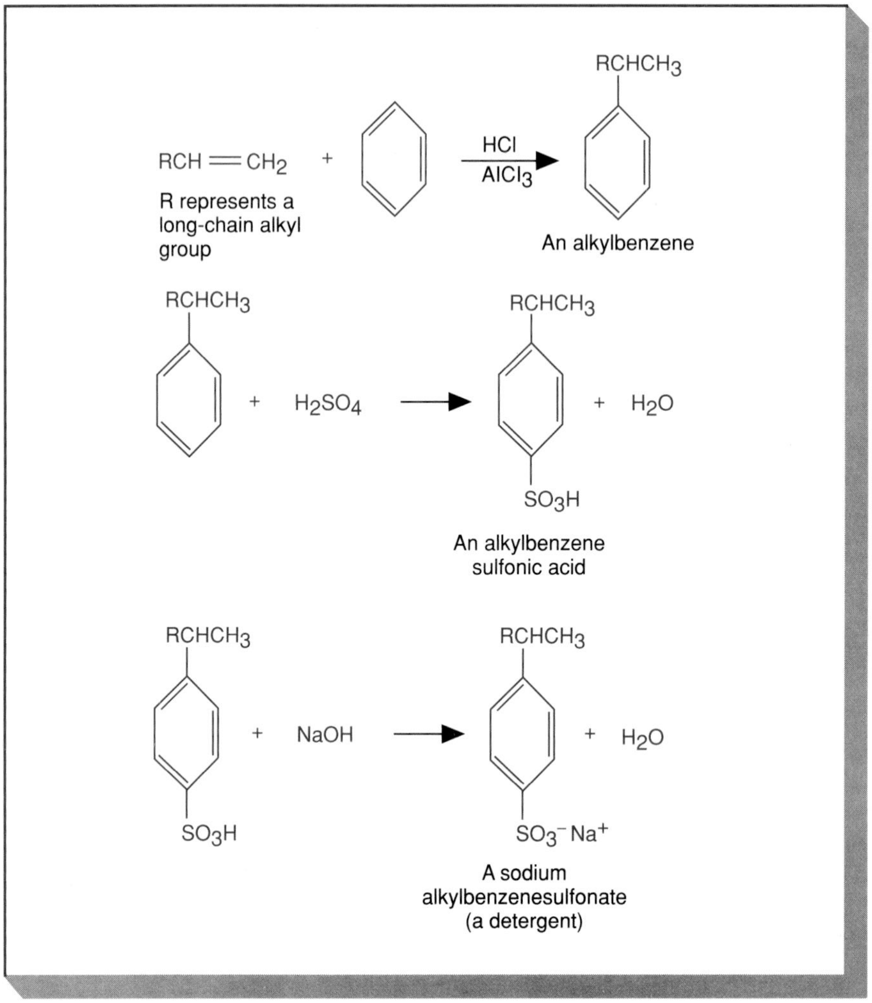

Figure 16.9
The most widely used detergents are made from alkenes, benzene, sulfuric acid, and sodium hydroxide in a series of reactions.

chloride adds to ethylene. In the second, a Lewis acid, $AlCl_3$, catalyzes the reaction of ethyl chloride with benzene. Finally, hydrogen is catalytically removed from ethylbenzene to give styrene. (The structures of polystyrene and other synthetic polymers are discussed in the next chapter.)

Hydrocarbons from the petrochemical industry, together with sulfuric acid and sodium hydroxide, are utilized in the manufacture of sodium alkylbenzene sulfonates, widely used as detergents. The reactions of benzene are used to link first a hydrophobic alkyl group and then a sulfonic acid group onto the ring. In the final reaction, sodium hydroxide converts the acid end of the molecule to its hydrophilic sodium salt (see Fig 16.9).

Synthetic detergents entering streams need to be degraded by microorganisms or the downstream water quality suffers. Microorganisms can oxidize long-chain hydrocarbons, but they cannot readily oxidize highly-branched alkyl chains found in early alkylbenzene sulfonates. During the 1950s, the appearance of detergent foam in streams pointed out that synthetic organic detergents then in use failed to degrade. Now manufacturers produce detergents with more nearly linear alkyl groups that can be degraded by microorganisms (see Fig. 16.10).

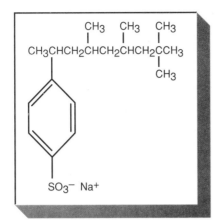

Figure 16.10
Detergents used in the 1950s had highly-branched side chains that were not broken down by microorganisms. The failure to degrade detergents led to foam formation in sewer effluents and in streams. Detergents now in use have more linear side chains and are more readily degraded.

■ Halogenated Hydrocarbons in Agriculture and Industry

The story of the chlorinated insecticide DDT illustrates both the benefits and the costs of the widespread use of halogenated hydrocarbons. The introduction of the insecticide DDT has done much to reduce the incidence of the insect-borne diseases, plague and malaria, and has led to increased yields in western forests and southern cotton fields (see Fig. 16.11).

The 1962 publication of *The Silent Spring* by Rachel Carson, made people aware of the destruction of the environment caused by DDT. DDT was shown to cause environmental damage far from where it is used as an insecticide. People were surprised to learn that DDT is taken up by aquatic plants and concentrated as it moves through the food chain. Aquatic plants low in the food chain have very low concentrations of DDT in their tissues, but concentrations of DDT rise as one ascends the food chain from plants to plant-eating fish and then to fish-eating birds. Predators such as sea gulls are especially at risk since accumulations of DDT interfere with their reproductive cycle.

Like most halogenated hydrocarbons, DDT has little solubility in water because it does not form hydrogen bonds with the water. This lack of solubility is a benefit as an insecticide, for DDT is not readily washed away by rains. However, the very low solubility in water plays an unfortunate role in the biological food chain since DDT is not readily excreted with other wastes. Instead, it becomes concentrated in the fatty tissues where it is more soluble.

The environmental problems associated with DDT apply to many other halogenated hydrocarbons. Some halogenated hydrocarbons react readily with water or with oxygen and are converted into compounds that can be degraded by microorganisms. These compounds pose little danger to the environment. However, many halogenated hydrocarbons are resistant to reaction because most microorganisms that degrade naturally-occurring organic compounds lack enzymes designed to react with these synthetic compounds.

Are there solutions to the problems posed by the wide use of chlorinated hydrocarbons? If there are solutions, will the correct choices be made? These answers are not yet known. Less DDT is used now than in previous decades, but there remain important applications for which it is still the insecticide of

Figure 16.11
DDT was the most widely used chlorinated-hydrocarbon insecticide (a). High volume insecticides are often applied by plane or helicopter (b).
(b) © Steve Kaufman/Peter Arnold, Inc.

(a)

(b)

choice. Compromises between difficult options may be important. If sprayed in the right place at the right time, small amounts of DDT may be highly effective, yet cause minimal environmental damage.

■ Reducing the Release of Toxic Chemicals— Regulations and Goals

Where can chemical wastes from industry be disposed of safely? Too often the approach has been to flush them down the drain and hope for the best. Because natural processes cleanse some wastes, people have hoped that they would cleanse all wastes; this has not proven to be the case. Organic compounds in the wastes buried in the Love Canal region of Buffalo have risen through the soils and volatilized into the basements of homes and schools. An accident at a chemical plant near Seveso, Italy, spread dangerous chemicals over many square miles and caused nearby villages to be abandoned. Wastes from the manufacture of hexa-chlorophene, a compound used to control bacterial growth, were mixed with oil and then used to control dust on dirt roads. The persistence of these wastes has caused the town of Times Beach, Missouri, to be abandoned. Clearly not all chemical wastes are removed by natural processes. Indeed the very processes that may help to clean some of these wastes have made the dimensions of the problem harder to know. Due to their high surface area, soil particles bind some organic and heavy metal ion poisons, thereby slowing their spread so health problems from chemical wastes have been discovered years after a dump site had been used and abandoned.

As we better understand the dangers that chemical wastes pose to our health and environment, policies for the effective disposal or destruction of wastes need to be promulgated and enforced. Those who create hazardous wastes should be required to ensure their safe disposal. Beginning in the late 1970s, chemical companies have contributed to a Super Fund to be used for cleaning up abandoned dump sites. The impact of this Super Fund on cleaning up the environment has been disappointing. Typically there are long delays between the time a site is identified and the cleanup is begun, and the amount of money spent on litigation is close to the amount spent on remediation.

Governmental agencies, environmental and consumer groups, and corporations are attempting to work together in less adversarial ways to achieve significant reductions in the quantities of dangerous chemicals released into the environment. The Environmental Protection Agency (EPA) has asked more than 600 U.S. companies to voluntarily reduce emissions of 17 high-volume chemicals to less than 50% of their 1988 levels by the year 1995. An interim goal of a 33% reduction is to be achieved by the end of 1992.

Process redesign is part of a larger effort to reduce pollution at the source by greatly reducing the release of toxic materials. The petrochemical industry is making major changes in the processing of chemicals to reduce the quantities of waste produced. A large integrated chemical plant can be designed to recycle the chemical by-products from one manufacturing stream and use them as feedstocks in the production of another chemical product.

■ Second and Third Generation Agricultural Chemicals

Increased food production is essential if mass starvation and malnutrition are to be avoided. It is estimated that the world's reserve food supply is now shrinking after holding constant for a number of years. In contrast, the world's population is still growing rapidly. In 1991, the world's population was estimated to be 5.5 billion people, and it is growing at a rate of 100 million people each year. It is estimated that, even now, as much as twenty percent of the world's population is hungry and malnourished.

Most of the available productive land is now being farmed. This land may in fact be decreasing, for it is estimated that more than twenty percent of the world's topsoil has been lost to erosion in the last decade. Much of the new land in undeveloped countries that is being put into food production is marginal land, highly susceptible to erosion.

In developed countries, farmers employ an arsenal of agricultural chemicals in their efforts to increase food production. They use fertilizers to enhance plant growth, insecticides to reduce loss to insects, fungicides to combat plant fungi, other pesticides to combat the nematodes that destroy plant roots, and herbicides to combat weeds that compete for nutrients and sunlight (see Fig. 16.12). Unlike DDT,

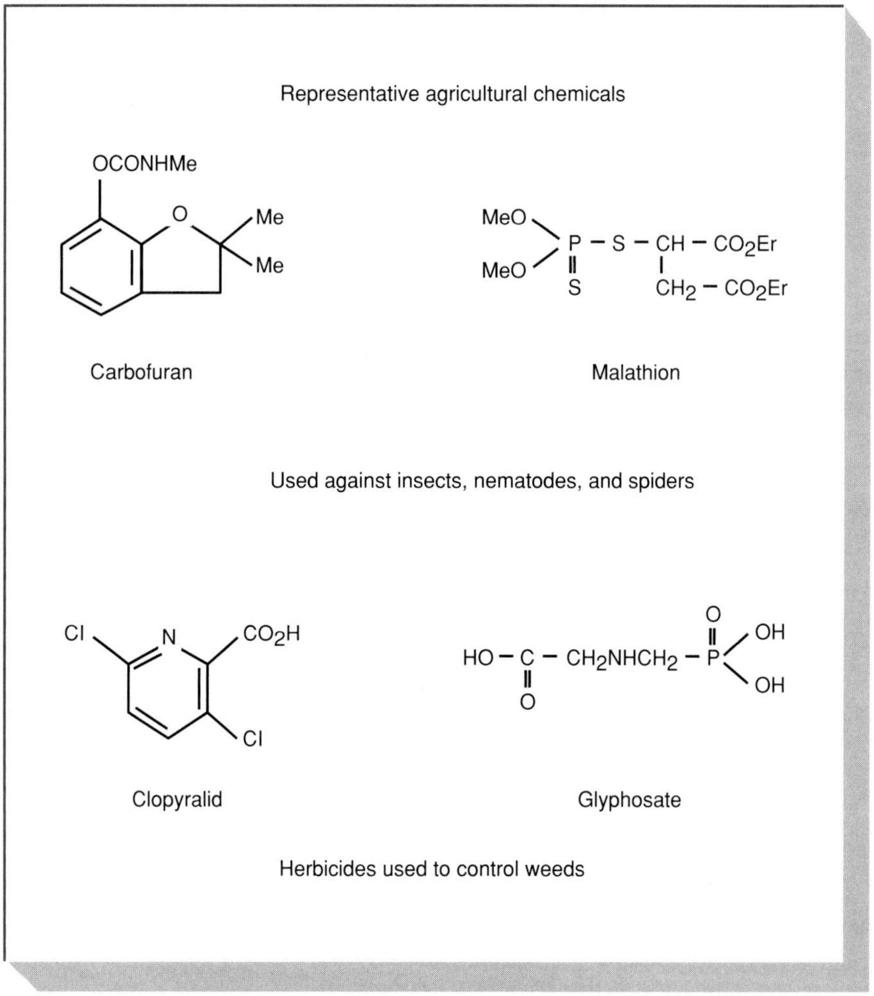

Figure 16.12
Structures of various compounds
used as insecticides and herbicides.

Representative agricultural chemicals

Carbofuran

Malathion

Used against insects, nematodes, and spiders

Clopyralid

Glyphosate

Herbicides used to control weeds

these compounds do not persist in the environment. Some "second generation" agrichemicals are highly-toxic, short-lived chemicals. When applied at the right time and in the right place, they effectively combat infestation and then break down to relatively harmless products. Others resemble the medicines produced by the pharmaceutical industry since they target metabolic reactions specific to the pest rather than to the plant host.

Is it possible to develop chemical compounds that specifically and effectively combat infestation by a destructive insect yet do not adversely affect other forms of life? The effort to find such chemicals is underway. Working together, chemists and biologists have identified some of the compounds called pheromones used by insects to communicate. Researchers monitor insect populations, using traps baited with synthetic pheromones to attract insects. In another approach, scientists seek to find compounds that can act as hormones to interrupt insect growth and reproduction.

■ Antioxidants Are Used to Preserve Food

Oxidation contributes to the spoiling of food. Upon oxidation, unsaturated fats and oils are converted to carboxylic acids that give food an undesirable flavor and odor. To protect foods, antioxidants are sometimes added. Two such compounds, abbreviated BHA and BHT, react with the odd electron species that are intermediates in

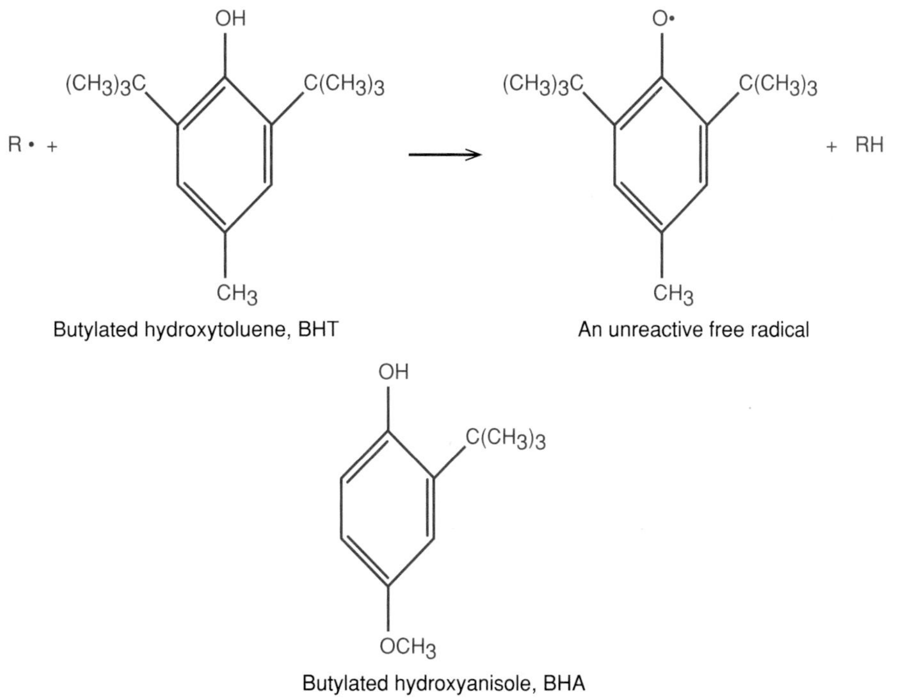

Butylated hydroxytoluene, BHT

An unreactive free radical

Butylated hydroxyanisole, BHA

air oxidation. A BHA or BHT molecule donates a hydrogen atom from its OH group to the reactive intermediate, rendering it unreactive. The BHA or BHT then possesses an odd electron, but the bulkiness of the butyl groups hinder any further reaction. The chain reactions of oxidation leading to spoilage are stopped and the food is preserved (see Fig. 16.13).

Inorganic antioxidants include sodium nitrite and sodium bisulfite. Sodium nitrite is used as a preservative in bacon, and sodium bisulfite is often used in restaurants to help keep freshly cut vegetables and fruit from discoloring. There is a debate whether all food additives should be listed on all product labels since sodium nitrite may act as a carcinogen, some people have intense allergic reactions to sodium bisulfite, and other health concerns are involved.

Figure 16.13
Antioxidants, such as BHT and BHA, are used as food preservatives. These compounds interrupt free-radical chain reactions by donating a hydrogen atom to a reactive radical and forming a relatively unreactive intermediate. The chain reactions leading to food spoilage are interrupted and the food is preserved.

■ **Aside**

Is All Natural Better?

Many people assume that foods grown without the use of pesticides and processed without chemical preservatives are inherently safer to eat, as well as more flavorful, than foods grown and processed with chemicals. Furthermore, these people have been joined by a much larger group of people in efforts to ban the use of compounds such as DDT and ethylene bromide (a compound used to prevent the growth of mold on grain in storage). For these people, the risks of cancer and environmental damages associated with the use of these chemicals far outweigh the benefits of increased food production and decreased storage losses.

Rules governing the use of pesticides widely vary. Different rules apply to pesticides in use before 1978 and those introduced later. Rules for applications to crops and for applications to processed foods also differ. Even traces of carcinogenic substances are banned from processed foods under the terms of the Delaney amendment to the Pure Food and Drug Act. The operational intent, no traces of carcinogenic substances, may be open to question because scientists have increased their ability to detect trace amounts of chemicals more than a thousandfold since the passage of this amendment.

The actual situation is complex, and short-sighted decisions may be made using only a partial understanding of the complexities involved. During the 1980s, Bruce Ames, a scientist at the University of California, and his associates studied the mutagenic properties of large numbers of compounds, both natural and synthetic. Some of the most powerful chemical mutagens are produced by plants themselves and by the molds that grow on plants. Levels of these chemicals may rise when the plant is diseased or attacked by insects. Efforts to develop plants that are more resistant to insects can serve to increase the quantities of some very dangerous, naturally occurring chemicals. Ames argues that we cannot have a risk-free chemical environment, and that the benefits offered by small amounts of synthetic chemicals, even dangerous ones, may far exceed the risks.

■ Questions—*Chapter 16*

1. In most petroleum, the carbon chains are long and predominantly linear. Why is it that the higher the octane number of the gasoline produced by a refinery, fewer gallons are produced per barrel of petroleum?
2. Highly-branched isomers of alkanes have a lower boiling point than linear isomers. Why would it be unsatisfactory to merely vary the degree of branching of the alkanes when shifting from a summer formulation to a winter formulation of gasoline?
3. How would the price of gasoline be different if large-capacity tankers had not been developed for the transport of petroleum? Justify your answer.
4. Why is a refinery built to process sweet, light petroleum unable to process sour, heavy petroleum? (See Chap. 12 for descriptions of "sweet, light" and "heavy, sour" petroleum.)
5. Why are many of the plants for the manufacturing of ammonia located in Texas and Louisiana?
6. Why is it less expensive and less dangerous to transport ammonia for fertilizer than to transport hydrogen gas used in the production of ammonia?
7. Methane is a starting material for the manufacture of methyl alcohol, and ethylene is a starting material for the manufacture of ethyl alcohol. Would natural gas producers or petroleum producers benefit more from the addition of methyl alcohol to gasoline? Who would benefit more from the addition of ethyl alcohol to gasoline?
8. One of the largest plants manufacturing methyl *t*-butyl ether is located in Saudi Arabia. Why would the plant be located there? Where would the methyl *t*-butyl ether manufactured there be used?

9. Why is it necessary to add an alkyl group to benzene in making the detergent sodium alkylbenzene sulfonate?

10. Draw the structure of a detergent micelle.

11. Nitrogen oxides from the engines of supersonic transport planes were predicted to deplete the ozone layer. This concern was a part of the decision not to develop these planes. Why are nitrogen oxides produced by fires and automobiles not as great a threat to the ozone layer?

12. Rachel Carson's book *The Silent Spring* was met with both sympathetic and hostile reviews. Find and read two reviews of this influential book. Do the reviewers agree or disagree?

13. Exposure to light speeds the breakdown of many chlorinated hydrocarbons. Speculate on why the absorption of a photon of light might change the chemistry of a molecule.

14. Vitamin E is a fat-soluble compound that acts to protect tissue from oxidation. How might its chemistry resemble that of BHA and BHT?

15. Read the labels on a jar of peanut butter and on a bottle of wine to find the preservatives in each. What are some other foods that list preservatives on their labels?

17

Structural Polymers

P*olymers* are compounds composed of very large molecules with repeating subunits. The scientific study of these polymers is a development of the twentieth century. The first synthetic polymers, prepared by modifying naturally occurring cellulose fibers, were used as substitutes for ivory and silk. With the advent of automobiles, the petroleum industry began to provide low-cost hydrocarbon starting materials for the synthesis of new polymers. The polymer industry developed as population growth created increased demands for natural products that could not readily be met because of their limited supplies. Shortages during World War II hastened the development of nylon and synthetic rubber. The growth of polymer science has been so rapid that now more than one-half of the chemists in industry work in polymer-related areas. The explosion of knowledge about polymers and the application of that knowledge to technology have occurred in the areas of both synthetic and naturally occurring polymers.

Polymer chemistry is directed toward the production of materials that can meet the needs generated by a wide variety of applications. Whether it be a new textile fiber, a strong, light weight composite for use in transportation, a human growth hormone, or human insulin, the goal is to meet a defined need with a well-designed product. In the next two chapters, we will sample the variety of synthetic and natural polymers. We will also look at the relationships between the properties of these polymers and their chemical structures.

■ Big Molecules Are Different

Before the twentieth century, polymer science was slow to develop. The traditional experimental methods and logic used to study organic compounds were inadequate to deal with polymer problems. Organic chemists had developed experimental methods to analyze compounds composed of small or medium-sized molecules, but polymer molecules, which may consist of hundreds or thousands of atoms, presented new problems. They are hard to dissolve and cannot be purified by distillation or recrystallization. When polymers formed accidently in a reaction mixture, they were often ignored or discarded.

Figure 17.1
Dust in the atmosphere scatters light, causing the daytime sky to look blue and sunsets to look red. Like tiny dust particles, large polymer molecules scatter light. Light-scattering experiments are used to determine the size of polymer molecules.

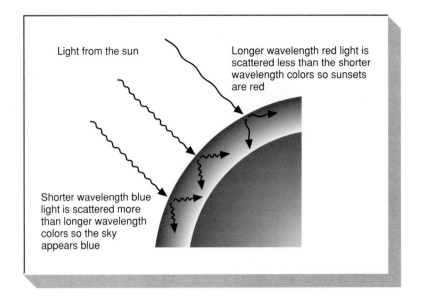

Light from the sun

Longer wavelength red light is scattered less than the shorter wavelength colors so sunsets are red

Shorter wavelength blue light is scattered more than longer wavelength colors so the sky appears blue

To explore the unique properties of polymers, scientists needed to develop new experimental methods and new theories to interpret the experimental results. A polymer is different because it is composed of large molecules. (Indeed, as late as 1920, some scientists were reluctant to believe that polymers were molecules.) To understand polymers, scientists needed to analyze contributions to physical and chemical properties made by size alone as well as those made by functional groups. The methods developed to analyze the size of macromolecules included light scattering, osmotic pressure measurements, and ultracentrifugation.

Light from the sun is scattered by tiny particles such as dust, as shown in figure 17.1. Short-wavelength blue light is scattered more than longer-wavelength red light. Widely scattered blue light from the sun reaches us from many angles so the sky looks blue during the day. At dawn, and at sunset, the red light which has traveled a longer path through the atmosphere colors the sky. Volcano eruptions increase the amount of atmospheric dust so that sunsets are spectacularly colored after large eruptions such as that of Mount Saint Helens. Very large molecules, sometimes called macromolecules, are also large enough to scatter light. Because solutions of polymers contain molecules large enough to scatter light, they appear opalescent blue in color. This can be seen by looking at a dilute solution of powdered milk, which contains protein macromolecules. From measurements of light scattering, scientists can determine the size of polymer molecules.

Osmosis occurs across membranes. Trees draw in water at the roots, and osmotic pressure across cell membranes lifts this water to the leaves. When a polymer solution is separated from an excess solvent by a membrane that is permeable to the solvent, but not the polymer, the solvent will pass into the polymer solution and dilute it. The pressure driving the solvent molecules through the membrane is called osmotic pressure. Scientists use osmotic pressure measurements to analyze the concentrations of very dilute polymer solutions (see Fig. 17.2).

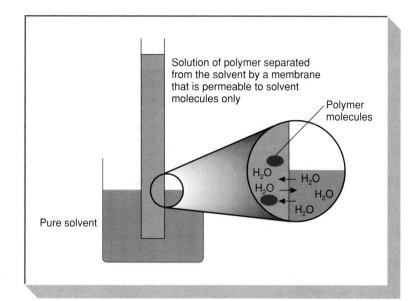

Figure 17.2
Osmotic pressure exists across a semipermeable membrane, as solvent molecules flow from the solvent side of the membrane to dilute the solution of polymer. Osmotic pressure measurements are used to measure the concentration of polymer solutions.

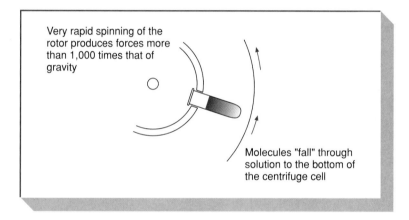

Figure 17.3
An ultracentrifuge can be used to separate molecules according to size and shape. Smaller molecules fall through solution faster than larger molecules. Likewise, spherical molecules encounter less resistance and fall faster than molecules with other shapes.

Scientists also separate polymers using ultracentrifuges. In an ultracentrifuge, samples are spun thousands of times per second and this spinning provides forces thousands of times greater than that of gravity and permits the separation of molecules based on differences in size, shape, and density (see Fig. 17.3).

■ Thermoplastic Polymers

Many, but not all, synthetic polymers are thermoplastic materials. For example, both polyethylene, used in milk containers, and polystyrene, used in disposable coffee cups, are *thermoplastic* polymers. At elevated temperatures, thermoplastics are viscous liquids that can be molded. Upon cooling, they harden and retain the shape of the mold. Note that glass is thermoplastic and clay plastic, according to this definition. Note also that most so-called plastic objects are not plastic at the temperatures at which they are used.

Figure 17.4
The properties of many polymers reflect a structure in which there is a mixture of ordered crystalline regions and more random amorphous regions.

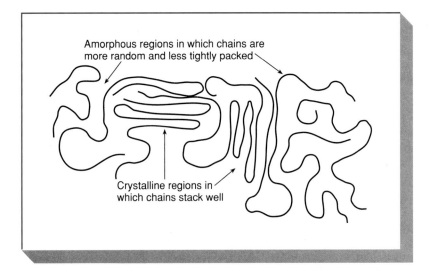

Amorphous regions in which chains are more random and less tightly packed

Crystalline regions in which chains stack well

The properties of many thermoplastic polymers reflect a structure that is a mixture of crystalline and amorphous regions as illustrated in figure 17.4. In the crystalline regions, chains stack in an ordered fashion. These regions contribute greater strength and dimensional stability to a polymer. In the amorphous regions, chains do not stack well and the arrangement of adjacent chains is more random. These regions contribute greater toughness and flexibility to a polymer. Many polymers are translucent or opaque because the crystalline and amorphous regions differ in density, and light is scattered as it passes from region to region.

Because intermolecular forces in polymers are relatively weak, crystalline regions melt at moderate temperatures, allowing the polymers to flow under pressure and to be shaped in molds. Upon cooling, the melted polymer hardens, some crystalline regions reform, but much of the polymer remains amorphous.

Producers control the molecular weights of polymers in order to control their properties. Polymers with very high molecular weights are difficult to fabricate since they are exceedingly viscous even at elevated temperatures. Polymers with too low molecular weights lack strength and toughness.

■ Addition Polymers

Monomers are the small molecules from which polymers are made. The simplest, least expensive, and most widely used synthetic polymer is polyethylene, which is prepared from the low-cost monomer ethylene or ethene. Polyethylene and its derivatives, prepared from these modified ethylene monomers, are called *addition polymers*. The structures of some common addition polymers are shown in table 17.1.

A brief sampling of addition polymers suggests the diversity of their properties and the range of their applications. Polymethyl methacrylate has bulky side groups along the polymer chain. It stacks very poorly and is either entirely glassy or amorphous. It is a tough transparent polymer used in paints and safety glass. Polyvinyl chloride also lacks crystallinity. Due to the large chlorine atoms on every other carbon atom, forces between chains are stronger than in most other addition polymers. Polyvinyl chloride (PVC) is both oil and water resistant and is used widely as pipes, hoses, and phonograph records, where strength, flexibility, and toughness are desired. In contrast, the linear chains in Teflon, the polymer of tetrafluoroethylene, are very straight and stiff. In these extended chains, relatively

• • • • • • • • • • • • • • **Table 17.1** • • • • • • • • • • • • • • •

A Sample of Addition Polymers

Monomer	Polymer	Uses
$CH_2=CH_2$ Ethylene	$(-CH_2CH_2-)_n$ Polyethylene	Low-cost toys, containers for food, films
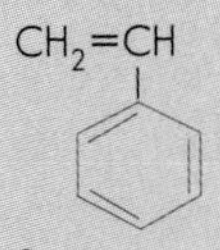 $CH_2=CH$ Styrene	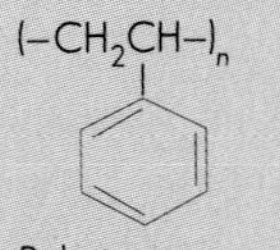 $(-CH_2CH-)_n$ Polystyrene	Packaging, appliance parts, toys
$CH_2=CH$ $\mid$ Cl Vinyl chloride	$(-CH_2CH-)_n$ $\mid$ Cl Polyvinyl chloride (PVC)	Pipes, floor tiles, adhesives, phonograph records
$CH_2=CH$ $\mid$ $C\equiv N$ Acrylonitrile	$(-CH_2CH-)_n$ $\mid$ $C\equiv N$ Polyacrylonitrile	Textile fiber, rugs
$CH_2=CH$ $\mid$ $O-CCH_3$ $\parallel$ O Vinyl acetate	$(-CH_2CH-)_n$ $\mid$ $O-CCH_3$ $\parallel$ O Polyvinyl acetate	Latex paint
$CH_2=CH$ $\mid$ $COCH_3$ $\parallel$ O Methyl methacrylate	$(-CH_2CH-)_n$ $\mid$ $COCH_3$ $\parallel$ O Polymethyl methacrylate	Glass substitute, safety glass
$CF_2=CF_2$ Tetrafluoroethylene	$(-CF_2CF_2-)_n$ Polytetrafluoroethylene	Bearings, nonstick frying pans

negative fluorine atoms are at their maximum distance from one another. The surface of Teflon consists largely of long linear chains so it is remarkably smooth and friction-free (see Fig. 17.5).

■ Formation of Addition Polymers

Addition polymers can be produced by free-radical reactions as shown in figure 17.6. A small amount of an initiator is mixed with a large quantity of monomers. Initiator molecules have weak bonds that break when heated to give radicals, a species with an odd number of electrons. Radicals induce polymerization by adding rapidly to double bonds. The product of each addition is itself a new radical so the additions of monomers continue, generating long chains. Two growing chains may join, or one growing end may remove a hydrogen from another growing chain. Termination reactions end the polymer's growth.

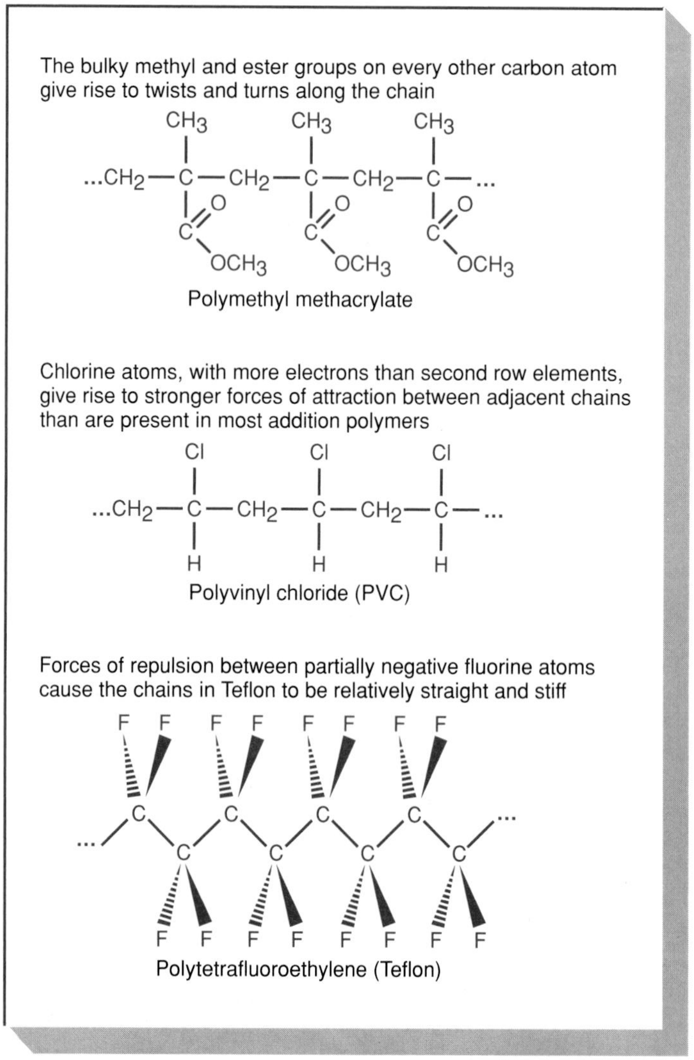

Figure 17.5
The properties of addition polymers depend in part on the substituents along the chain and on how these substituents affect the interaction of neighboring chains.

The bulky methyl and ester groups on every other carbon atom give rise to twists and turns along the chain

Polymethyl methacrylate

Chlorine atoms, with more electrons than second row elements, give rise to stronger forces of attraction between adjacent chains than are present in most addition polymers

Polyvinyl chloride (PVC)

Forces of repulsion between partially negative fluorine atoms cause the chains in Teflon to be relatively straight and stiff

Polytetrafluoroethylene (Teflon)

The ends of a polymer chain may differ from the bulk of the polymer. End-group reactivity can present special problems to the polymer chemist just as end-group twists or catches may present special problems in opening a zipper, or in keeping it closed.

A knowledge of the mechanism of polymer formation helps manufacturers control the properties of the polymers produced. For example, the use of a higher concentration of initiator results in a lower-melting polymer with a lower range of molecular weights, because more and shorter chains are formed. If, instead, a higher-melting polymer is desired, a lower concentration of initiator may be used.

■ Plasticizers

Additives called *plasticizers* are included in many polymer formulations. Plasticizers are added to increase the softness and pliability of some plastics and to increase the useful lifetime of others. For example, an ester of phthalic acid with a branched-chain alcohol, shown in figure 17.7, is used as a plasticizer. Compounds used as plasticizers dissolve in the polymer so their branched structure inhibits the formation

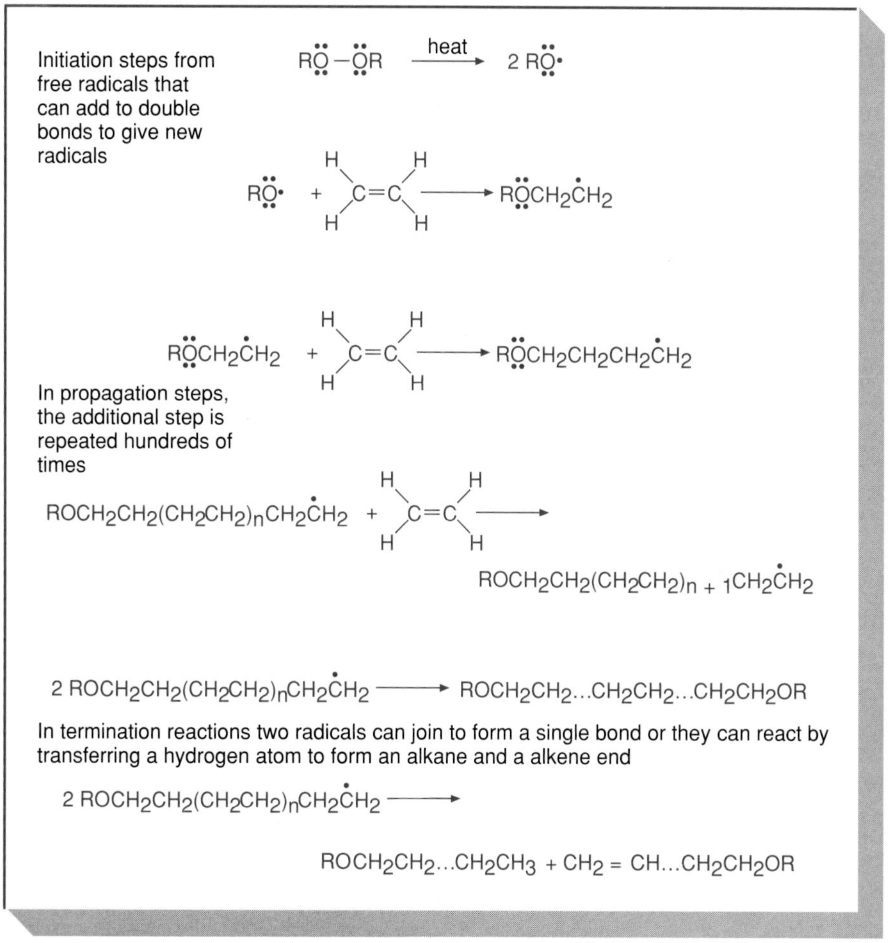

Initiation steps from free radicals that can add to double bonds to give new radicals

$$R\ddot{O}-\ddot{O}R \xrightarrow{\text{heat}} 2\ R\ddot{O}\cdot$$

$$R\ddot{O}\cdot \ +\ \underset{H}{\overset{H}{>}}C=C\underset{H}{\overset{H}{<}} \longrightarrow R\ddot{O}CH_2\dot{C}H_2$$

$$R\ddot{O}CH_2\dot{C}H_2 \ +\ \underset{H}{\overset{H}{>}}C=C\underset{H}{\overset{H}{<}} \longrightarrow R\ddot{O}CH_2CH_2CH_2\dot{C}H_2$$

In propagation steps, the additional step is repeated hundreds of times

$$ROCH_2CH_2(CH_2CH_2)_nCH_2\dot{C}H_2 \ +\ \underset{H}{\overset{H}{>}}C=C\underset{H}{\overset{H}{<}} \longrightarrow$$

$$ROCH_2CH_2(CH_2CH_2)_{n+1}CH_2\dot{C}H_2$$

$$2\ ROCH_2CH_2(CH_2CH_2)_nCH_2\dot{C}H_2 \longrightarrow ROCH_2CH_2...CH_2CH_2...CH_2CH_2OR$$

In termination reactions two radicals can join to form a single bond or they can react by transferring a hydrogen atom to form an alkane and a alkene end

$$2\ ROCH_2CH_2(CH_2CH_2)_nCH_2\dot{C}H_2 \longrightarrow$$

$$ROCH_2CH_2...CH_2CH_3 + CH_2=CH...CH_2CH_2OR$$

Figure 17.6
Addition polymers can be formed by a free-radical mechanism. First, free radicals are generated to initiate chain growth. Then, growing radical chains repeatedly add to double bonds, forming new radicals in each addition step. Finally, two radicals can react to terminate the growth process.

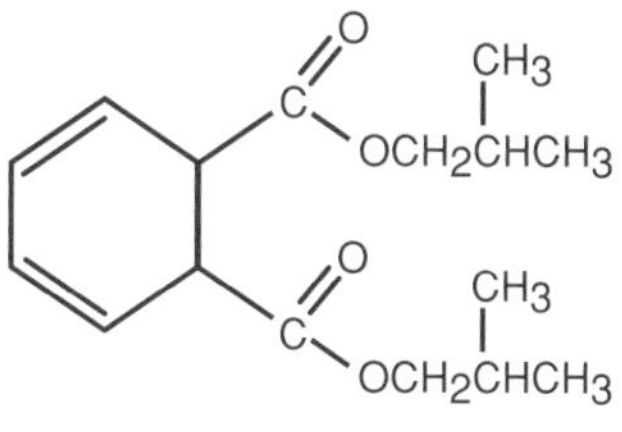

Diisobutyl phthalate

Figure 17.7
Diesters of phthalic acid and branched-chain alcohols are used as plasticizers to increase the softness and flexibility of polymers such as PVC.

of crystalline regions that would cause the polymer to harden and become brittle. Because the plasticizer molecules are not very volatile, they enable the polymer to remain amorphous over long periods of time. The gradual escape of their plasticizers causes some polymers to become more crystalline, often resulting in brittle dashboards and cracked seat covers. The haze on the windshield of a car parked in the sun on a hot summer day may be due to the escape of plasticizers from the plastic dashboard.

Figure 17.8
There are problems with free-radical polymerizations. Polypropylene cannot be prepared since propylene can donate a hydrogen atom to form an unreactive radical, which terminates the chain. Polyethylene prepared by a free-radical process has side chains that cause the polymer to be too low melting for uses that involve sterilization temperatures.

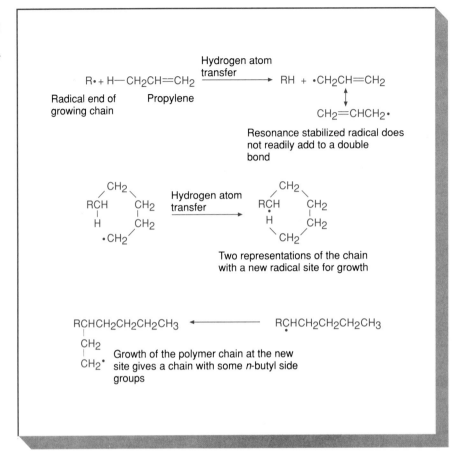

■ Aside

New Catalysts and Improved Properties for Addition Polymers

The free-radical process for forming addition polymers has certain drawbacks. Firstly, monomers such as propylene cannot be used in this process due to a side reaction. A growing radical chain can abstract (remove) a hydrogen atom from a propylene molecule to terminate one chain and form a new radical. The resonance-stabilized radical formed by removing a hydrogen atom from propylene is too unreactive to undergo chain growth. Secondly, a growing chain itself can undergo hydrogen-abstraction reactions. For example, the softening temperature of polyethylene prepared by a free-radical process is lowered by the presence of short side chains resulting from hydrogen-abstraction reactions followed by growth from the resultant new radical sites (see Fig. 17.8).

The use of Lewis acid catalysts led to the development of a new generation of addition polymers designed to meet applications requiring more demanding specifications. Linear polyethylene made using these catalysts is stronger and has a higher density than the branched polymer made by free-radical processes. Unlike the branched polymer, linear

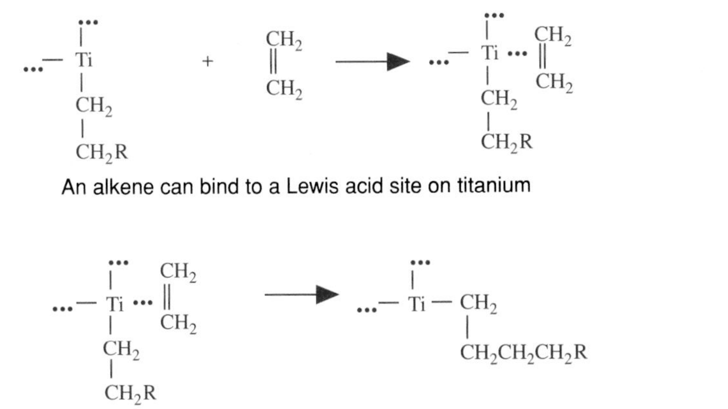
An alkene can bind to a Lewis acid site on titanium
A monomer binds to titanium and then the alkyl group on titanium adds to the bound monomer to give a new longer straight chain alkyl group and free a Lewis acid site for another alkene to bind

Figure 17.9
New Lewis acid catalysts are made to make high-density polyethylene, polypropylene, and other addition polymers with improved properties.

polyethylene can be heated to 100° C without softening, so that it can be sterilized, and then used as a low-cost material for food packaging. Propylene made using these catalysts has found application as outdoor carpeting, and as a mildew- and stain-resistant furniture fabric.

The use of Lewis acid catalysts permits much greater control of polymer geometry since both monomers and the growing chain are bound to the same metal atom. As each monomer is inserted into the chain, a new position on the metal becomes available, first to coordinate, and then to deliver a monomer unit to the growing chain (see Fig. 17.9).

■ Rubber, an Elastic Polymer

Rubber is an unsaturated hydrocarbon polymer, poly-*cis*-isoprene, shown in figure 17.10. When it was first studied, natural rubber was a laboratory curiosity. When the latex from the rubber tree was pressed and dried, the resulting sticky polymer would both bounce and yield. If struck, it behaved elastically and when pressed over a long period of time, it deformed permanently.

The discovery of the vulcanization process of rubber by Charles Goodyear in 1839 led to the many applications we know today. Green rubber is mixed with a small amount of sulfur, shaped into a product and then heated. Vulcanization, which cross-links the polymer chains, occurs during the heating. Sulfur reacts at a few double-bond sites to form cross-links or interchain bonds as shown in figure 17.11. The more highly cross-linked the rubber, the less flexible and harder it is. Cross-links are more numerous and far closer together in rubber used for tire treads than in the rubber used in household gloves.

The polymer chains between the cross-links in rubber are extremely flexible. Weak forces between chain segments and the *cis* geometry of the double bond result in chains that turn and twist but do not form ordered crystalline regions.

Figure 17.10
Rubber is a hydrocarbon polymer with a 5-carbon alkene as the repeating unit.

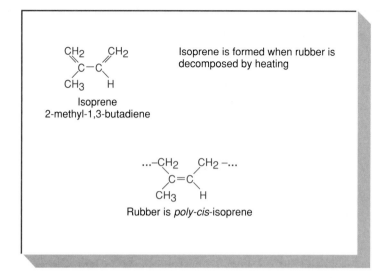

Isoprene is formed when rubber is decomposed by heating

Isoprene
2-methyl-1,3-butadiene

Rubber is *poly-cis*-isoprene

Figure 17.11
In the vulcanization process, green rubber is heated with sulfur. In reactions that involve the double bonds of rubber, some −SS− bridges form to cross-link the chains.

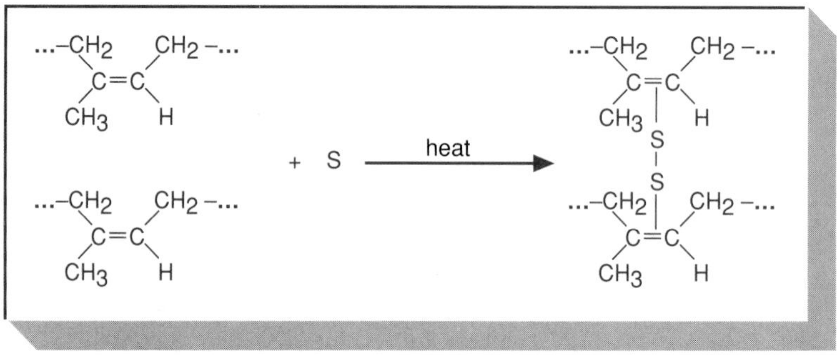

Figure 17.12
When rubber is stretched, the chain segments between cross-links become straighter and more ordered. When the tension is removed, the chains return to a more random orientation.

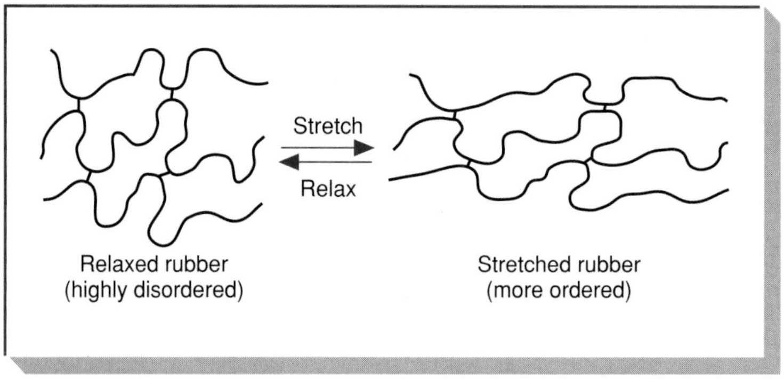

Stretch

Relax

Relaxed rubber
(highly disordered)

Stretched rubber
(more ordered)

When rubber is stretched, its molecular structure becomes more ordered. Randomly coiled chains are extended in the direction of stretching, but because of the cross-links, they cannot slide past one another. When the stretched rubber is released, random motions along the polymer backbone cause the chains to shorten. This motion draws cross-links closer together and causes the stretched rubber to retract (see Fig. 17.12).

Since increased randomness (higher entropy) is favored more at higher temperatures, the restoring forces in rubber increase with increasing temperature. A hot golf ball can be driven farther than a cold golf ball. If rubber objects are cooled sufficiently, they become leathery. The rate of relaxation of the extended chains slows down and the rubber loses its elasticity.

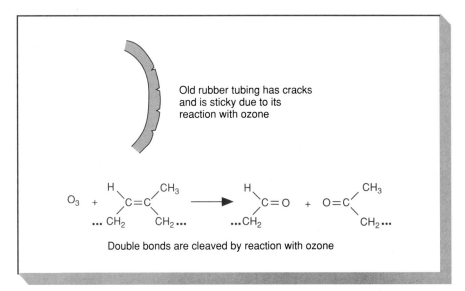

Figure 17.13
Ozone, a component of air pollution, shortens the useful lifetime of rubber.

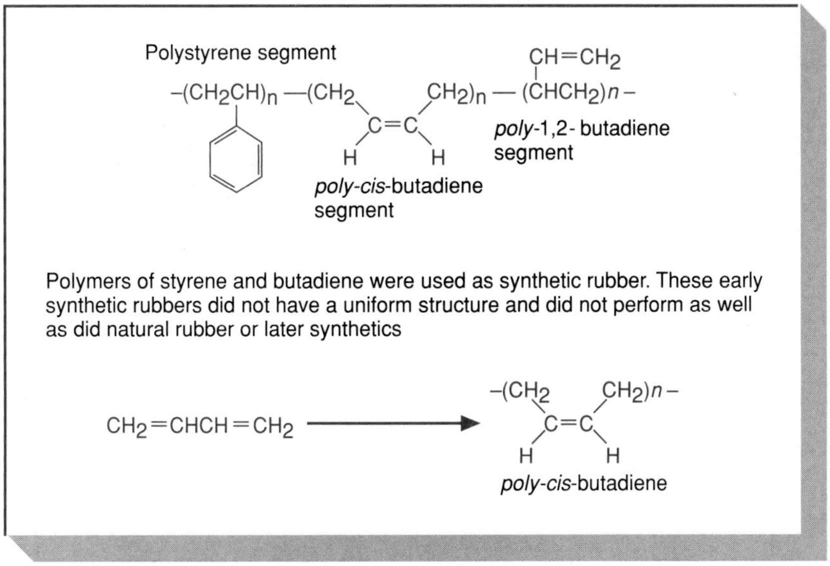

Old rubber tubing has cracks and is sticky due to its reaction with ozone

Double bonds are cleaved by reaction with ozone

Figure 17.14
Synthetic rubbers are made from 1,3-butadiene and styrene and from 1,3-butadiene alone. By using Lewis acid catalysts, a high-quality synthetic polymer with the same *cis* geometry as natural rubber can be prepared.

Polystyrene segment

poly-cis-butadiene segment

poly-1,2- butadiene segment

Polymers of styrene and butadiene were used as synthetic rubber. These early synthetic rubbers did not have a uniform structure and did not perform as well as did natural rubber or later synthetics

poly-cis-butadiene

Ozone, a component of air pollution, attacks rubber by adding to the double bonds of rubber, resulting in their cleavage. Increased ozone concentrations are associated with photochemical smog. An indirect economic cost of air pollution is a decreased useful life for rubber hoses, windshield wipers, and tires (see Fig. 17.13).

Today, synthetic rubbers are used more widely than natural rubber. The shift has been dramatic. Before World War II, natural rubber accounted for more than 95% of all use. Today, synthetics comprise 80% of the rubber in use. An early synthetic rubber was a co-polymer of styrene and butadiene. However, early synthetic rubbers tended to build up more heat when flexing than natural rubbers, and excessive heat destroys the cross-links and weakens the rubber. With the introduction of Lewis acid catalysts in 1955, improved stereochemical control of polymerization became possible. Since then, poly-*cis*-butadiene and poly-*cis*-isoprene have been synthesized. These polymers are superior to earlier synthetic rubbers and are widely used. However, vulcanized natural rubber remains an attractive product for demanding uses such as tire casings, while synthetic rubbers are used in treads (see Fig. 17.14).

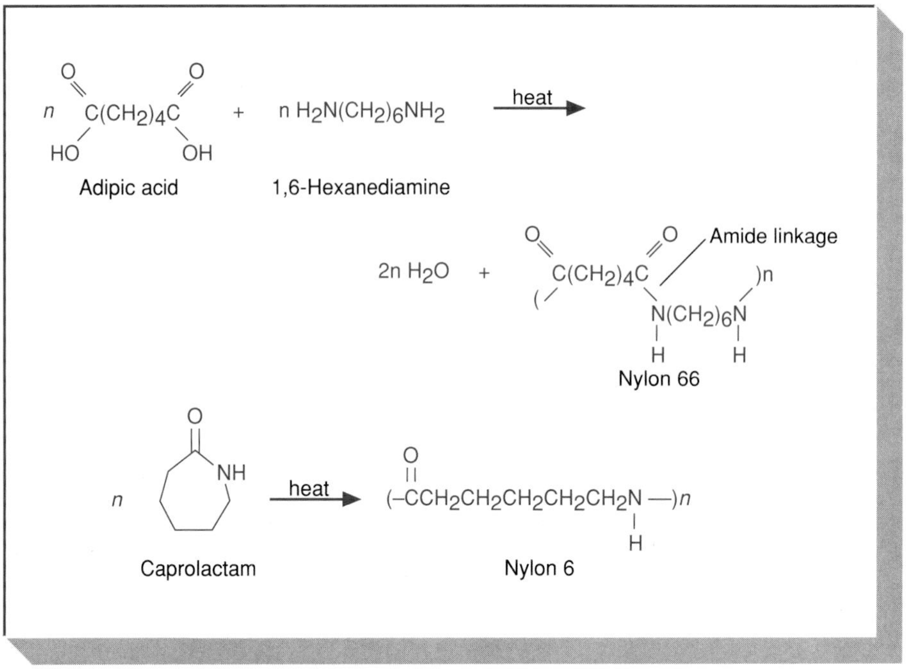

Figure 17.15
Synthetic polyamide polymers are called nylons. Nylons may be prepared from a diacid and a diamine as shown for nylon 66, or they may be prepared from cyclic amides as shown for nylon 6.

■ Condensation Polymers

Polyamide and polyester polymers (discussed below) are sometimes referred to as *condensation polymers*. Recall that an amide can be prepared by a condensation reaction in which a carboxylic acid combines with an amine to split out a small molecule (H_2O), and that an ester can be prepared by the condensation reaction of a carboxylic acid with an alcohol. By combining monomers with two different functional groups, one at each end of the molecule, condensation reactions can be used to prepare long, chainlike molecules having many functional groups along the polymer backbone.

Nylons constitute a family of synthetic polyamides. (Proteins, naturally occurring polyamides, are discussed in Chap. 18.) Nylons may be prepared by condensation reactions of diamines with dicarboxylic acids. The formation and structure of nylon 6 and nylon 66 are shown in figure 17.15.

One important use of nylon polymers is in the production of textiles. Synthetic textile fibers are oriented, highly crystalline polymers with molecular weights near 10,000. The melted polymer is extruded through small holes and immediately stretched. The polymer chains become oriented along the axis of the fiber. The intermolecular forces acting along the length of the fiber and those acting across the cross section of the fiber differ since cross-sectional strength arises from the covalent bonds of the many parallel chains. Longitudinal strength, or the fiber's resistance to rupture upon stretching, is dependent on strong interchain forces (see Fig. 17.16).

The amide groups provide strong interchain hydrogen bonds as shown in figure 17.17. In contrast, the hydrocarbon portions of the nylon chain contribute flexibility to the polymer. By varying the length of the hydrocarbon segments between the amide groups, polymer scientists prepare nylons with different properties.

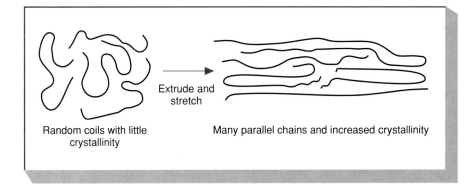

Figure 17.16
Textile fibers are prepared by squeezing the molten polymer through small dies and then quickly stretching and cooling the emerging thread. Because of the stretching, polymer strands are aligned parallel to the length of the fiber. A textile fiber has greater crystallinity and strength than the bulk polymer with its more random orientation of chains.

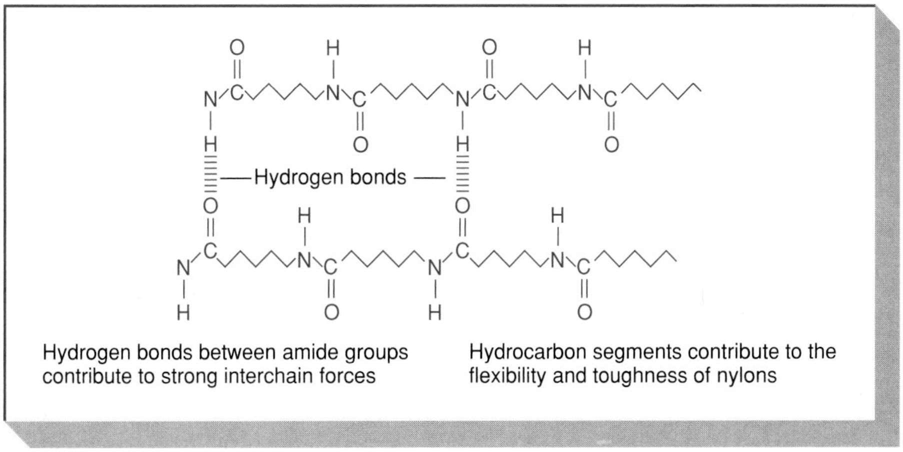

Random coils with little crystallinity

Extrude and stretch

Many parallel chains and increased crystallinity

Figure 17.17
The combination of strong interchain hydrogen bonds and flexible hydrocarbon segments gives nylon 6 strength and flexibility for use in textile fibers.

—Hydrogen bonds—

Hydrogen bonds between amide groups contribute to strong interchain forces

Hydrocarbon segments contribute to the flexibility and toughness of nylons

Shorter chains result in a harder, more crystalline, rigid product. For example, the nylon used in a sweater may be 15–35% crystalline, but that used in a fishing line is more than 90% crystalline.

Polyester fibers are made by the condensation polymerization of a diester (dimethyl terephthalate) and a diol (ethylene glycol). The interchain forces in polyesters are potentially weaker than those in nylons. Although polyester's ester groups are polar, there are no hydrogen bonds present. To provide satisfactory strength in a fiber, the hydrocarbon portions of the chains between the functional groups are shorter and more rigid than those in nylons. A two-carbon diol is used and the diacid contains a planar benzene ring. Chains stack better in a polyester than in a polyamide (see Fig. 17.18).

In the early 1980s, aromatic amide polymers were introduced in tire reinforcing cords and in other applications requiring high-strength, lightweight polymers. Both the rigidity of aromatic rings and the attraction of hydrogen bonds contribute to the high strength of aromatic amide polymers.

■ Thermosetting Polymers

Most polymers used for high temperature applications or for applications requiring high strength and durability are *thermosetting plastic.* Thermosetting plastics are rigid, highly cross-linked three-dimensional arrays that cannot be remelted or reshaped. The final cross-linking steps occur in the molding or in a final heat treatment.

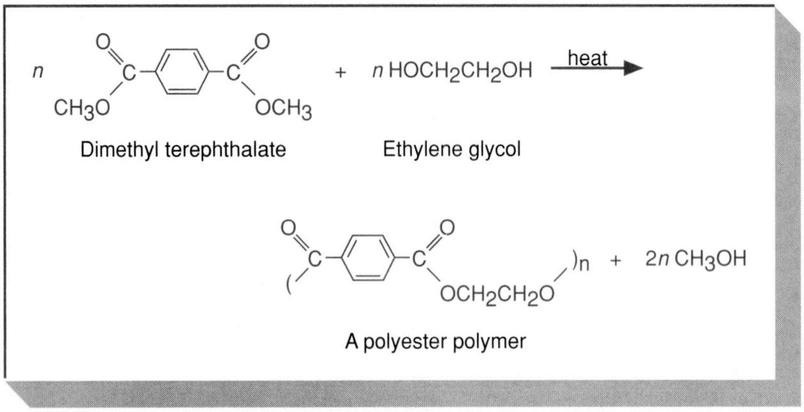

Figure 17.18
Polyester textile fibers are prepared by reacting the diester of an aromatic dicarboxylic acid with a short-chain diol.

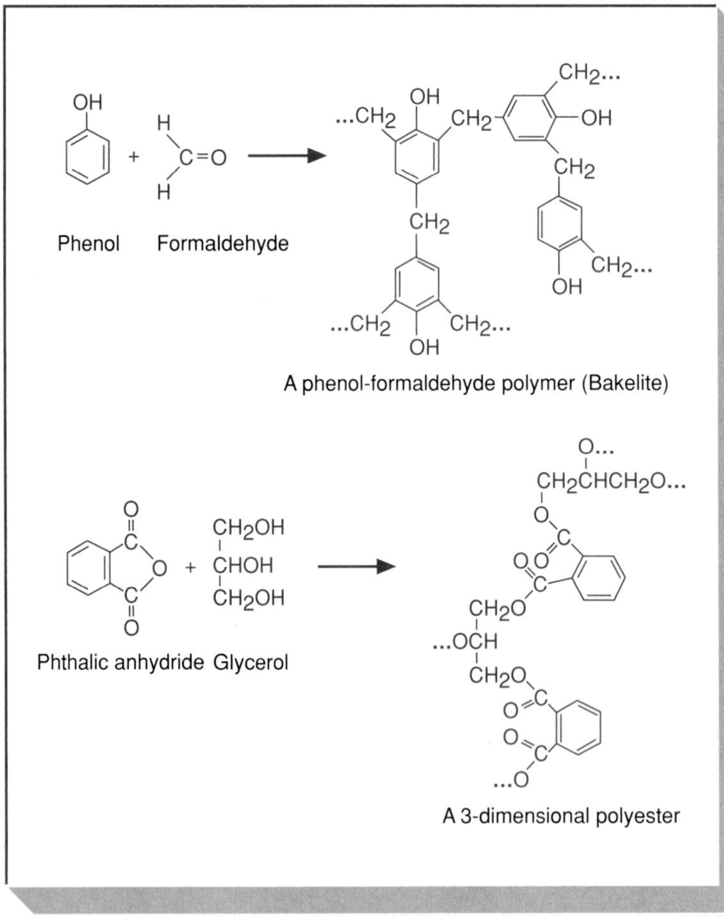

Dimethyl terephthalate   Ethylene glycol

A polyester polymer

Figure 17.19
Thermosetting polymers have three-dimensional structures. Once a thermosetting resin is fully formed, it cannot be melted and reshaped. Phenol formaldehyde polymers and polyesters made using the triol glycerol are among the most widely used thermosetting polymers.

Phenol   Formaldehyde

A phenol-formaldehyde polymer (Bakelite)

Phthalic anhydride Glycerol

A 3-dimensional polyester

The structures of two widely used thermosetting resins are illustrated in figure 17.19. Phenol-formaldehyde resins, known as Bakelite resins, were among the most widely used of thermosetting resins. A polyester prepared from glycerol and phthalic acid is used in the baked-enamel finishes of appliances. This three-dimensional polyester is prepared by substituting the triol, glycerol, for the diol, ethylene glycol, used in textile fibers and by substituting phthalic acid for terephthalic acid.

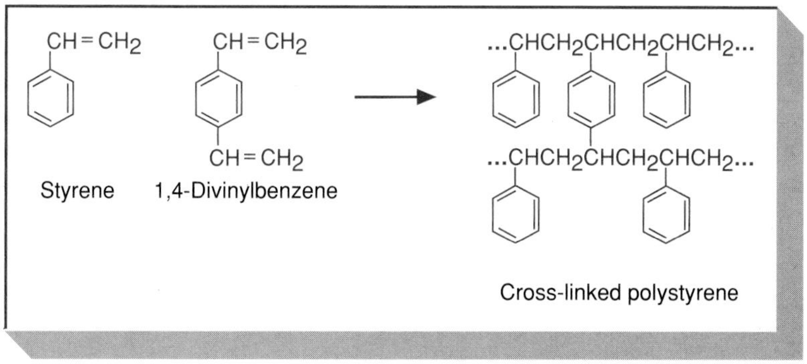

Figure 17.20
Three-dimensional, cross-linked polystyrene can be prepared by the polymerization of styrene to which a small amount of divinyl benzene has been added. Each of the alkene groups of divinylbenzene becomes part of a different polystyrene chain. To produce a polymer with closer cross-links, more divinyl benzene is added.

■ Aside

New Composite Materials Meet Demanding Uses

For uses ranging from fishing rods to sports cars, the use of polymeric materials is growing rapidly. Materials are being combined to give a *composite* that has properties superior to either material alone. For example, new fishing rods are light, strong, and supple because fibers of a borosilicate glass (low thermal expansion) or graphite run the length of each rod segment, and because of the three-dimensional polymer matrix that binds them together.

Because composite materials can be molded into complex shapes and are highly resistant to corrosion, they are finding new uses in automotive manufacturing. The stylish front end of a sports car may be a composite while straighter side panels are sheet metal. In contrast to the example of the fishing rod, the reinforcing fibers of the sports car may be woven to give strength in two dimensions and laminated layers also give increased strength. Thermosetting polyesters are the most widely used of the matrix materials.

■ Polymers for Ion Exchange

Synthetic ion exchange resins are used in water softening and purification systems as well as in chemical separation applications. Ion-exchange resins are chemically-modified polystyrene resins. By including a very small amount of divinylbenzene with the styrene monomer, a cross-linked polymer is formed as shown in figure 17.20. The benzene rings in the polymer are then chemically modified to introduce charged substituents, and depending on the reactions used, to convert the polymer into either a giant anion or a giant cation. With charged groups present on the polymer, the beads of ion-exchange polymers swell with water. If the polymer network is anionic, then the surrounding water contains dissolved cations attracted to the negative charges on the polymer. If the polymer is a giant cation, there are bound anions in solution. Due to the strong attraction of opposite charges, these dissolved ions cannot wash out, but they can be exchanged for other ions of like charge (see Fig. 17.21).

Hard water contains dissolved calcium or magnesium compounds that are less soluble in hot water than in cold. When hard water is heated, it leaves deposits that can narrow or close pipes and cause steam-producing boilers to lose efficiency. However, when the hard water containing dissolved calcium or magnesium salts passes through a cation-exchange resin loaded with sodium ions, calcium and

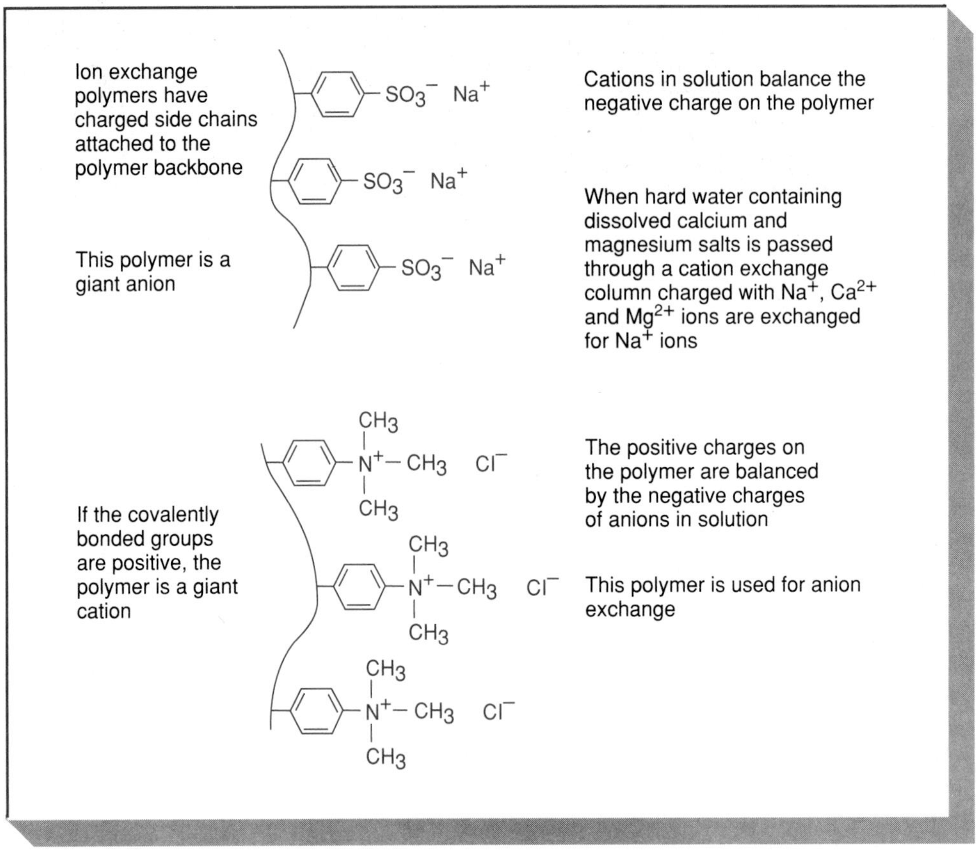

Figure 17.21
Chemists chemically modify polymers to produce ion-exchange resins for use in chemical separations.

magnesium ions are retained on the column and sodium ions replace them in the solution eluted from the column. The resulting soft water is suitable for use in boilers or for washing, although, because of its high sodium content, it is not desirable for drinking where dietary restrictions on salt intake are prescribed.

Ion-free, or deionized water, can be prepared by using first a cation-exchange resin and then an anion-exchange resin. Mineral cations such as Na^+ and Ca^{2+} are exchanged for hydronium ions, and mineral anions such as Cl^- and SO_4^{2-} are exchanged for hydroxide ions. The hydronium ions and hydroxide ions combine to form ion-free water.

Reflection

Is Recycling Plastics the Answer to Disposal Problems?

There is growing public concern about the accumulation of plastic wastes. What is to be done with discarded packaging materials? Many people view the current solution—burial at a sanitary landfill—as increasingly unsatisfactory for the long term, because most plastics do not degrade. The problem is not new nor is it growing. The fraction of the waste buried in urban landfills each year that is plastic remains nearly constant. Even though plastic is being put to more and more uses, many current applications use smaller quantities of stronger plastics than were available in the past.

Efforts are underway to develop recycling programs for plastics. These recycling programs face formidable obstacles because polymers differ widely in properties. When different polymers are combined, the resultant mixture is of limited utility. If, instead, the recycled plastics are first sorted according to type of polymer, then the recycled material has greater value and can be combined with virgin polymer and reused in higher quality applications. The success of plastics recycling programs may well depend on the cooperation of individuals in sorting and recycling their discarded plastic articles.

If people could recycle all their plastics, should they do so? To investigate the complexity of this issue, it is useful to consider both the environmental benefits and the environmental costs of such a change. Note that many plastic packaging materials were introduced to lower transportation costs by reducing the volume and the weight of container materials. The transportation of all materials has adverse environmental impact, so that the greater the distances involved in the collection and return of recycled material, the greater the adverse environmental impact. Thus it may be both more economically advantageous and environmentally desirable to recycle plastics in urban areas than in small towns or rural areas.

If not recycled, should the use of plastics be limited? In the 1990s, fast food restaurants are returning to the use of paper containers rather than ones made of styrofoam. Here, too, the gain to the environment made by reducing the quantity of plastics used is partially, or fully, offset by the costs to the environment made in producing their paper replacement.

There are no pollution-free ways to package and distribute essential goods such as food, nor are there quick technological cures for the problems of waste disposal. The best way to reduce the volume of plastic wastes may be to reduce the excess packaging of materials in marketing and in distribution.

■ Questions—*Chapter 17*

1. In an early approach to determine the molecular weight of hemoglobin, a weighed sample of the protein was burned and the ash was analyzed for iron, showing that hemoglobin contains 0.34% iron by weight. If there is one iron per hemoglobin molecule, what would be the molecular weight of the hemoglobin? The actual molecular weight of hemoglobin is 64,500. How many iron atoms are present in each hemoglobin molecule?

2. The monomer $CH_2=CHCN$ is used to make Orlon, a polymer used in textile fibers. Draw the structure of a segment of this polymer.

3. The polymer—$(CH_2CCl_2)_n$—is used to manufacture films for wrapping food. Draw a Lewis structure of the monomer used to make this addition polymer.

4. Polyethylene glycol—$(CH_2CH_2O)_n$—is soluble in water. Why is this polymer water-soluble while polyethylene is not?

5. Which of the following properties of polymers is most dependent on the presence of both crystalline and amorphous regions in a polymer?
   (a) Dimensional stability
   (b) Flexibility
   (c) Toughness

6. Common foam plastic coffee cups are made of polystyrene. How do the benzene rings contribute to the high-temperature strength of polystyrene compared to that of polyethylene?

7. Why is linear, or high-density, polyethylene stronger than low-density polyethylene that has alkyl side groups?

8. In the free-radical polymerization of ethylene, would an increase in the concentration of initiator used result in a lower-molecular-weight or a higher-molecular-weight polymer? Explain briefly.

9. A compound with the structure $C_4H_9N=NC_4H_9$ is used to initiate the formation of some addition polymers. When heated, this compound decomposes to give N_2 and other products. How does it act to initiate the polymerization of ethylene?

10. The number of polymer chains formed using the initiator described in question nine is less than predicted from the consumption of initiator. In addition to polymers, some product having the formula C_8H_{18} is also formed. What is the mechanism of formation of this by-product?

11. How would the addition of the compound di-*n*-octyl phthalate to polyvinyl chloride affect the properties of the polymer?

12. Rubber swells when in contact with hydrocarbon solvents. Why? Why does it not swell in contact with water?

13. A highly cross-linked rubber is (less, more) rigid and is (easier, harder) to stretch than a moderately cross-linked rubber.

14. Is a synthetic rubber made from butadiene and styrene subject to attack by ozone? Why or why not?

15. How does the arrangement of polymer chains of nylon differ in a textile fiber and in a molded object?

16. Nylons made using diacids with longer hydrocarbon sequences between acid groups are (less, more) water resistant and (less, more) flexible than those made using shorter chain diacids. Which nylon formulation would be more suitable for the bristles of a toothbrush? For the handle?

17. Draw the structure of a polymer prepared by reacting $NH_2CH_2CH_2NH_2$ with a four-carbon diacid.

18. Chemists developed nylon as a substitute for silk and wool. What structure do you expect to be present in these protein fibers based on analogy to the structure of nylon?

19. Why is each of the following compounds unsuitable for use in making polyester for use as a textile fiber?
    (a) $HOCH_2(CH_2)_8CH_2OH$
    (b) Glycerol
    (c) Phthalic acid

20. When the triol, glycerol, replaces the diol, ethylene glycol, in making a polyester, the resulting polymer can no longer be melted. Why?

21. A cross-linked polystyrene polymer having $-CH_2N(CH_3)_3^+$ side groups can be used to separate (anions, cations) by ion exchange.

22. Why are Ca^{2+} and Mg^{2+} more tightly bound to cation-exchange resins than is Na^+?

18

Carbohydrates, Lipids, and Proteins

We are what we eat. This saying serves to introduce the presentation and discussion of three important classes of biochemical compounds: carbohydrates, lipids (fats and oils), and proteins. We often consider these compounds in a nutrition context when assessing our diets and in setting nutritional standards. The saying then serves to connect our food intake to our biochemical composition, for compounds classified as carbohydrates, lipids, and proteins comprise much of our bodies.

This transformation of the complex molecules in food into the complex molecules in our bodies does not take place directly. In the process of digestion, the complex molecules of carbohydrates, lipids, and proteins in food are broken down into smaller, simpler molecules. These smaller molecules enter the bloodstream to be burned as a source of energy or to be reassembled into new complex carbohydrates, lipids or proteins. Like all other living organisms, our bodies create complex chemical structures using a relatively small number of simple chemical building blocks.

This chapter presents the structures of representative carbohydrates, lipids, and proteins. It emphasizes the relationship between the structures of biochemical monomers and biochemical polymers. It also relates the physical properties of these compounds to their chemical structures.

■ Carbohydrates

Before the structures of sugars were known, their formulas (for example, $C_{12}H_{22}O_{11}$ or $C_{12}(H_2O)_{11}$) suggested the name "carbohydrate" or "hydrate of carbon." *Carbohydrates* are sugars and their derivatives. So to understand the structures of the more complex carbohydrates cellulose and starch, we must first consider the structures of simple sugars. *Sugars* are polyhydroxy aldehydes and ketones (see Fig. 18.1).

Simple sugars are extremely soluble in water because they have many —OH groups that form strong hydrogen bonds to water molecules. The structure of sugars in solution can be complicated by the fact that an —OH group can undergo an intramolecular reaction with the carbon–oxygen double bond to form a five- or six-membered ring. A mixture of cyclic forms, together with their high solubility, make simple sugars difficult to recrystallize from their syrups (a concentrated solution of sugar in water is called a syrup).

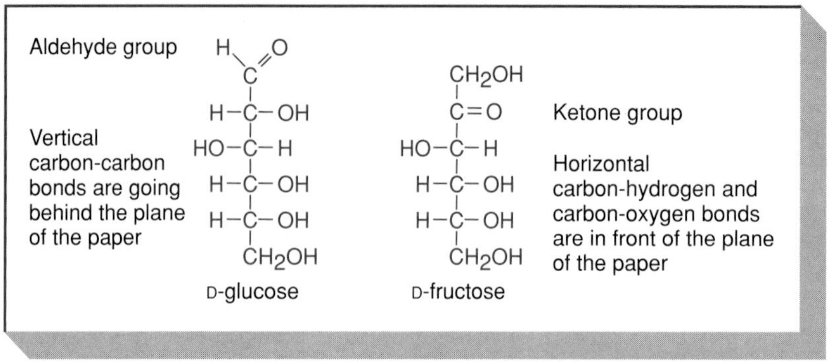

Figure 18.1

Sugars are polyhydroxy aldehydes and ketones. The structures of two important sugars, D-glucose and D-fructose, are shown. The carbons shown to be bonded to four different groups are chiral. The geometry at these centers is fixed in each sugar, and sugars differing in geometry are isomers.

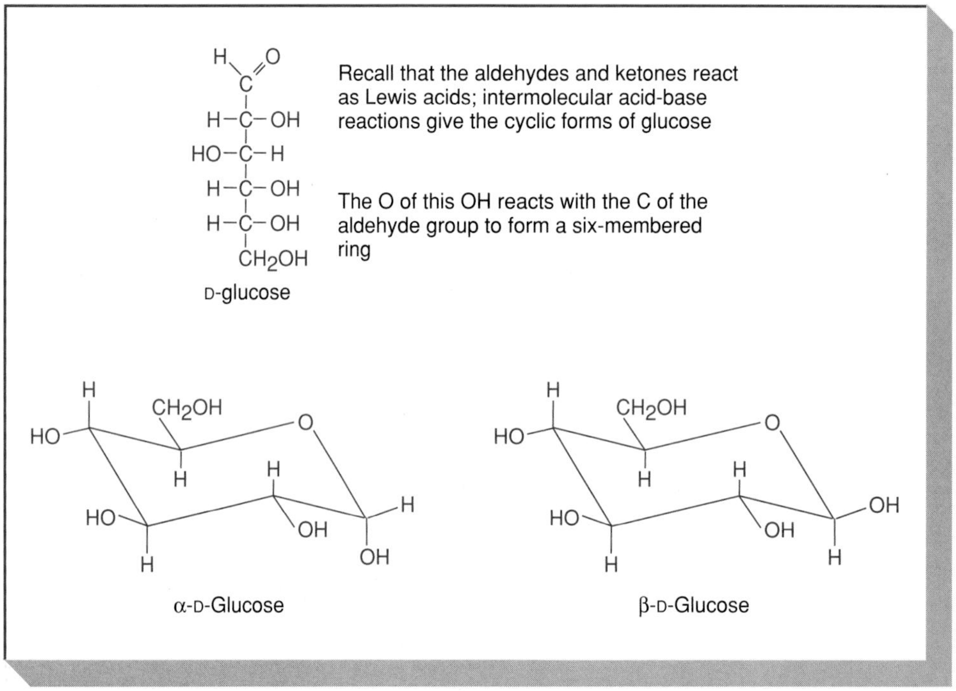

Figure 18.2

Glucose can exist as one of two cyclic isomers that can interconvert through an acyclic form. The cyclic isomers differ in their geometry at C-1, the former aldehyde carbon atom.

Glucose, shown in figure 18.2, is the most widely occurring simple sugar. It plays a central role in energy metabolism; other sugars are converted to glucose to be oxidized by a common route. Glucose is a six-carbon sugar that contains an aldehyde group. In solution, it is a mixture of two cyclic structures, α- and β-glucose, that interconvert through an acyclic form.

The metabolism of glucose differs in aerobic and anaerobic organisms. In both cases, glucose is first converted to three-carbon sugars by a rapid anaerobic pathway called glycolysis. The products of this oxidation are then further oxidized to carbon dioxide and water by a second, slower pathway in aerobic organisms. (In vigorous exercise, lactic acid may accumulate when the aerobic oxidation fails to keep pace with the glycolysis.) In contrast, yeast converts these three-carbon sugars to ethyl alcohol and carbon dioxide in the absence of oxygen.

Simple sugars can combine to form dimers, trimers, and polymers by further reaction at the oxygen atom of the aldehyde or ketone group. For example, sucrose, or cane sugar, is a dimer of glucose and fructose. Fructose (fruit sugar) is a six-carbon sugar isomeric to glucose that has a ketone group. The structures of two other dimeric sugars, lactose (the sugar in milk) and maltose (obtained from starch), are also shown in figure 18.3.

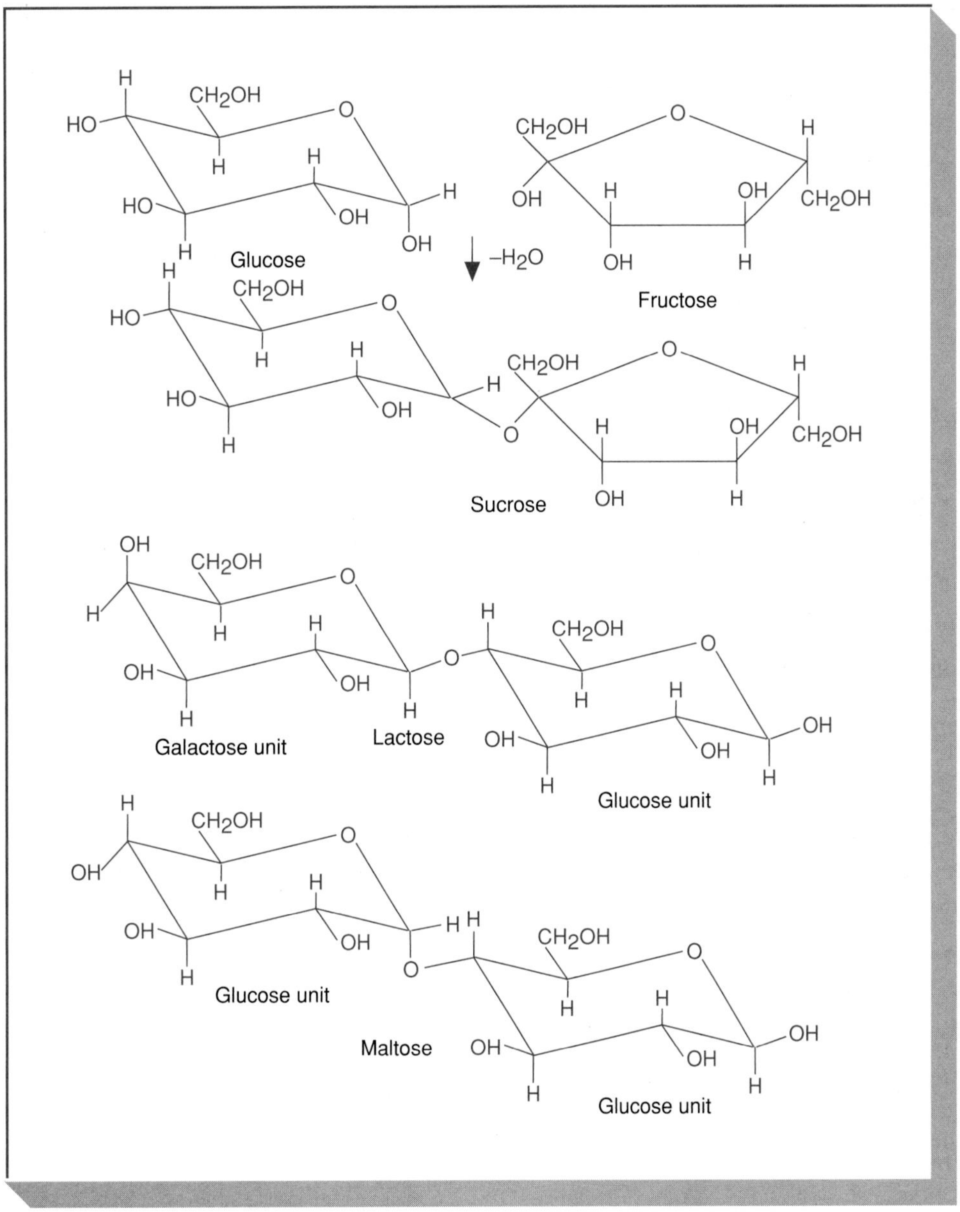

Figure 18.3
More complex sugars are formed from simple sugars by further reaction at the oxygen atom of an —OH group, bonded to the carbon of an aldehyde or ketone in the simple sugar.

■ Cellulose and Starch

Cellulose and starch are similar in that they are both polymers of the sugar glucose, but they serve very different biochemical functions. Cellulose is a structural polymer found in wood and cotton that gives strength and rigidity to plant cell walls. Starch is an energy storage polymer from plants. When sugar is needed to fuel metabolism, it is released from starch; when excess sugar is present, it is stored in the form of starch. Glycogen, a polymer with a structure closely related to that of starch, serves a similar energy-storage function in animals.

Cellulose and starch differ in their properties because they differ in the geometry of the glucose monomer units joined in the polymers. Cellulose is a polymer of β-glucose, and starch is a polymer of α-glucose. The different geometry of the α and β linkages and the different arrangements of their polymer chains result in the very different properties of cellulose and starch.

Figure 18.4

Cellulose is a polymer of glucose that features long parallel chains and strong interchain hydrogen bonds.

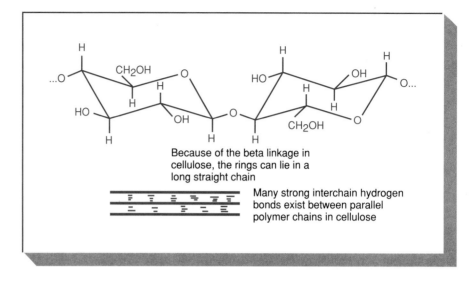

Because of the beta linkage in cellulose, the rings can lie in a long straight chain

Many strong interchain hydrogen bonds exist between parallel polymer chains in cellulose

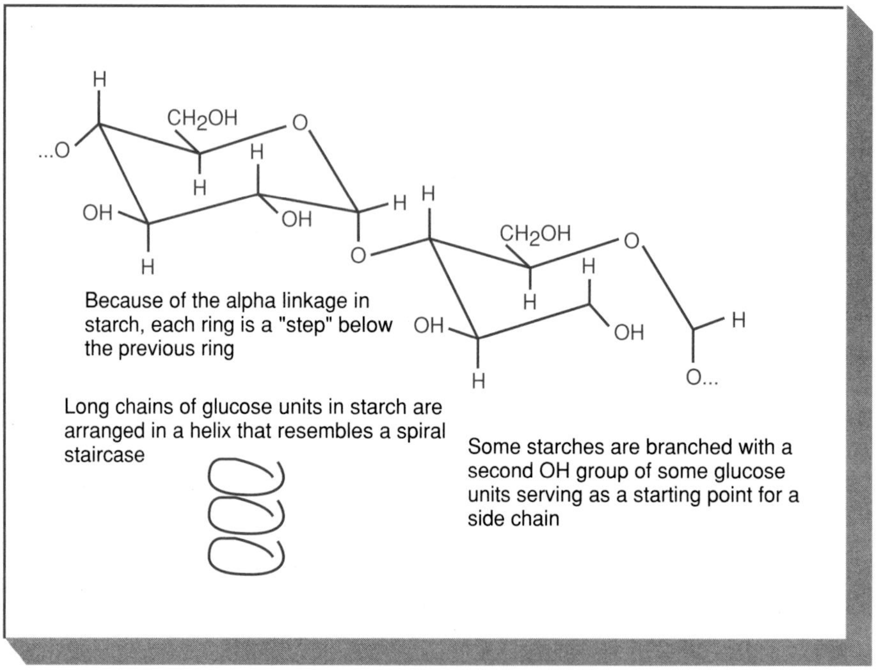

Figure 18.5

Starch is a polymer of glucose and has a helical structure. These chains can unwind so that the —OH groups of starch can hydrogen bond to water.

Because of the alpha linkage in starch, each ring is a "step" below the previous ring

Long chains of glucose units in starch are arranged in a helix that resembles a spiral staircase

Some starches are branched with a second OH group of some glucose units serving as a starting point for a side chain

Cellulose is a highly crystalline polymer. Cellulose chains are linear, stack well, and form strong interchain hydrogen bonds. Cellulose is insoluble in water even though it has many —OH groups. Because polymer chains are close and parallel, many hydrogen bonds would need to be broken at the same time in order to separate the polymer chains (see Fig. 18.4).

In contrast, starch is far less crystalline. Its polymer chains are wound in helices as shown in figure 18.5. Starch is more soluble in water than cellulose because its —OH groups are better exposed for interaction with water than those in cellulose. Some starches are linear and others are highly branched. Linear starches may dissolve completely; other, more branched starches swell with water. These starches may be used as thickening agents, for example, in the making of gravy.

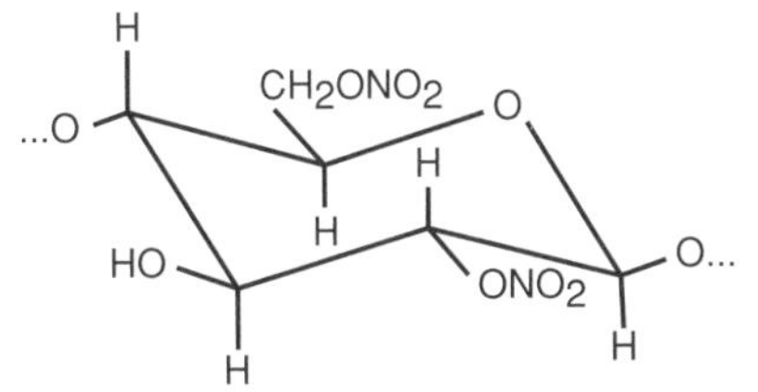

Cellulose nitrate is prepared by modifying some of the OH groups of cellulose by reaction with nitric acid. Cellulose nitrate has been used for billiard balls, as backing for photographic film, and as smokeless gunpowder

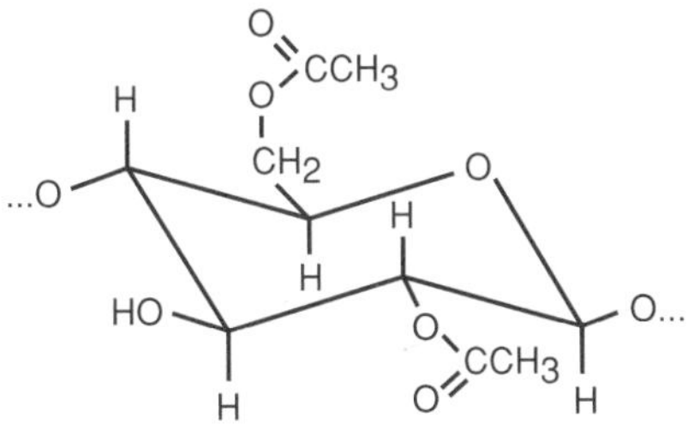

Cellulose nitrate chain

Cellulose acetate has fewer interchain hydrogen bonds than cellulose, so it can be dissolved and then spun to form monofilaments

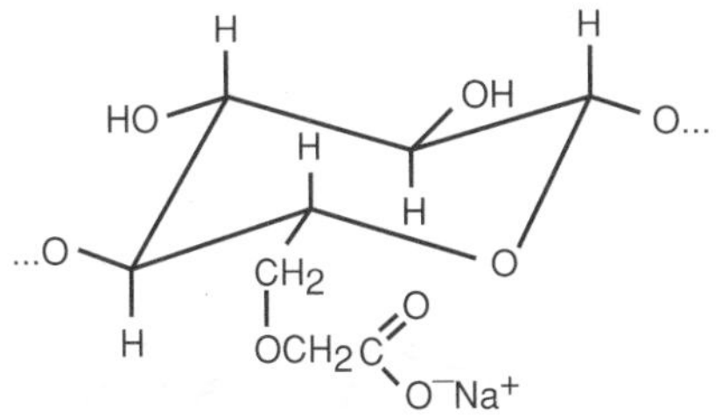

Cellulose acetate chain

Sodium carboxymethylcellulose has the beta linkage of the cellulose chain so it is not digested. It is used as a food additive in some diet foods

Sodium carboxymethylcellulose chain

Figure 18.6
A wide variety of polymers have been prepared by modifying some of the —OH groups of cellulose.

Why is starch available to us as a food, and cellulose is not? The difference is not in the energy content of the compounds, for the number of calories in a gram of starch and in a gram of cellulose are almost identical. The difference is that we have enzymes that cleave the α-glucose linkages in starch, but we lack enzymes that cleave the β-linkage in cellulose. Even termites can utilize the sugar content of cellulose for food only because they harbor microorganisms that have the enzymes needed to break down cellulose.

There is growing interest in the chemical modification of starches for use in polymer applications such as packing materials and garbage bags. Starch-based polymers would have an advantage in that they are biodegradable, because the α-linkages of the starch backbone are readily cleaved by enzymes.

■ Cellulose Is Modified for a Variety of Uses

Although cellulose is a strong, abundant polymer, it cannot be readily fabricated because the molecular weight is too high and the interchain forces are too great. Chemists modify cellulose to prepare polymers with widely different properties. They lower the molecular weight of cellulose by treatment with sodium hydroxide. They chemically modify some of the —OH groups in cellulose to produce new polymers as illustrated in figure 18.6.

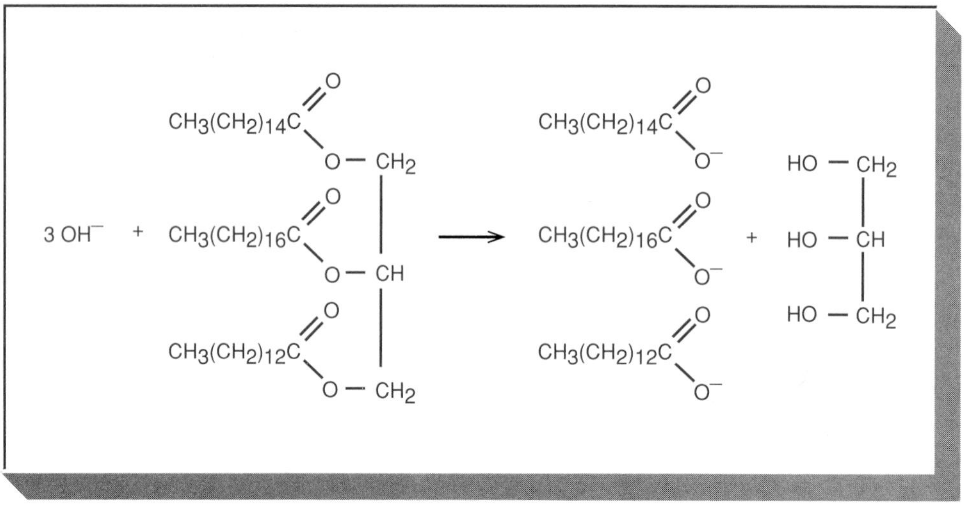

Figure 18.7
Fats are esters of long-chain carboxylic acids and glycerol. They are cleaved by reaction with hydroxide ion to form soaps and glycerol.

Cellulose nitrate, the product of the reaction of nitric acid and cellulose, was one of the first commercial synthetic polymers. Nitrate esters of cellulose known as gun cotton, or cordite, are used to make smokeless gunpowder for artillery shells. In this polymer, the oxidizing agent (nitrate groups) and the reducing agent (sugar units) are joined chemically. In the early days of the film industry, cellulose nitrate was used as the backing for camera films. The slow self-oxidation of this polymer has damaged or destroyed prints of many early motion pictures. Cellulose nitrate film is also highly flammable and has resulted in disastrous fires. Camera film is now made of cellulose acetate, which is much less flammable.

Rayon and acetate textile fibers are prepared by converting some —OH groups of cellulose to esters. Because esters do not hydrogen bond to one another, they have weaker interchain forces and increased solubility. The resulting polymers can be spun for monofilament applications. For example, the strength of a cotton thread is much less than that of its cellulose fibers; its strength depends instead on the frictional forces holding these fibers together.

Carboxymethylcellulose is a major food additive used to add bulk to diet foods. Because it has the β-linkage of the cellulose chain, it is not attacked by the enzymes that break down starches. Although it is made up of sugar units, carboxymethylcellulose has no nutritional value.

■ Lipids

The simplest lipids are fats and oils, compounds that are esters. Fats are esters of carboxylic acids and the triol, glycerol. The acids commonly found in fats have long, unbranched chains consisting of an even number of carbon atoms (see Fig. 18.7).

Fats and vegetable oil serve as long-term, high-energy storage depots. Fatty acids have an even number of carbon atoms because they are built up, and broken down, two carbons at a time. These two-carbon fragments can be used for energy production by some of the same enzymes that function in the aerobic oxidation of sugars. Fats are reduced more than carbohydrates. More than twice as much oxygen is required to burn a gram of fat than is needed to burn a gram of carbohydrate, and a gram of fat furnishes more than twice the calories of a gram of carbohydrate.

When the pioneers made soap using wood ashes (or lye) and animal fats, they were unaware that they were doing chemistry with esters. The wood ashes were a source of the base potassium carbonate, and the animal fat consists of triglyceride esters. The hydrolysis (reaction with water) of an ester in base, the key reaction in making soap, is called *saponification.*

Synthetic detergents can be made from animal or vegetable lipids. Fatty acids can be reduced to produce long-chain alcohols (these alcohols are also obtained by the saponification of some vegetable oils). The alcohols are converted to detergents first by reaction with sulfuric acid, and then with sodium hydroxide:

$$CH_3(CH_2)_{10}CH_2OH + H_2SO_4 \rightarrow CH_3(CH_2)_{10}CH_2OSO_3H + H_2O$$
$$\text{lauryl alcohol} \qquad\qquad\qquad \text{lauryl sulfate}$$

$$CH_3(CH_2)_{10}CH_2OSO_3H + NaOH \rightarrow CH_3(CH_2)_{10}CH_2OSO_3^-Na^+$$
$$\text{sodium lauryl sulfate}$$

Lauryl sulfate is an ester of sulfuric acid prepared by the condensation reaction of lauryl alcohol with sulfuric acid. Sodium, potassium, and ammonium salts of lauryl sulfate are mild detergents frequently used in shampoos.

■ Vegetable Oils Are Unsaturated

The fatty acids found in plants differ from those found in animals because lipids need to be flexible, or liquid, under different physiological conditions. Vegetable oils remain liquid at temperatures at which fats from warm-blooded animals are solids. Most fatty acid residues in animal fats are saturated and those obtained from vegetables contain double bonds, or unsaturation.

Stearic acid, a C-18 straight-chain acid, melts at 70° C. In contrast, oleic acid with one double bond melts at 14° C. The double bonds in naturally occurring unsaturated fatty acids have the *cis* configuration. The two hydrogen atoms are on one side of the rigid double bond and the two alkyl groups are on the other. The *cis* double bond imparts a bend to the tail of the acid. As a result of this bending, they do not stack well in lattices, so unsaturated lipids remain liquid at temperatures where their saturated counterparts are solids (see Fig. 18.8).

Margarine, a butter substitute, is made from vegetable oils by hydrogenating the double bonds of some of the unsaturated fatty acid esters. Adding hydrogen to the double bonds gives higher-melting compounds. Since we have a dietary requirement for some polyunsaturated fatty acids, the oils used to make many softer margarines are only partially reduced. Some nutritionists express concern that these margarines contain unsaturated lipids having *trans* double bonds. *Trans* isomers, not commonly found in nature, are a by-product of the hydrogenation produced by a rearrangement occurring on the catalyst.

■ Proteins

Proteins are nitrogen-containing biopolymers that play very diverse roles in living organisms. Some proteins play structural roles such as that of collagen in fingernails and hair. Other proteins are dynamic, changing shape to provide the pull of muscle. Still other proteins, called enzymes, catalyze biochemical reactions.

When proteins are heated with aqueous solutions of strong acids, they cleave to give a mixture of *amino acids.* Each of these amino acids has an amine group and a carboxylic acid group attached to a central carbon atom called the *alpha* carbon. The twenty amino acids obtained from proteins differ in the substituents attached at the *alpha* carbon atom. Substituent groups vary widely. Some are polar, and others are not. Most substituent groups are neutral, a few are acidic, and a few others are basic.

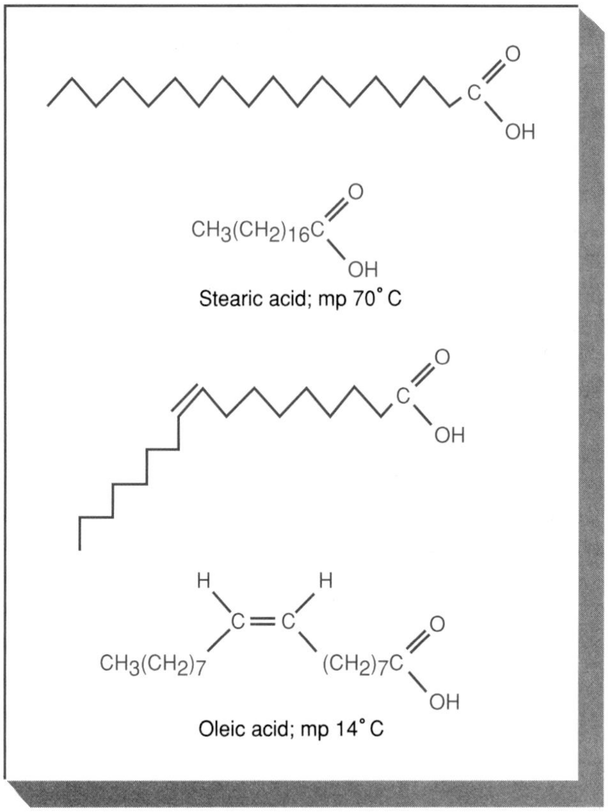

Figure 18.8
Stearic acid, with a saturated straight chain, melts at 70°C. In contrast, oleic acid, with one double bond, melts at 14°C. Unsaturated lipids do not stack well in lattices and remain liquid at temperatures where their saturated counterparts are solids.

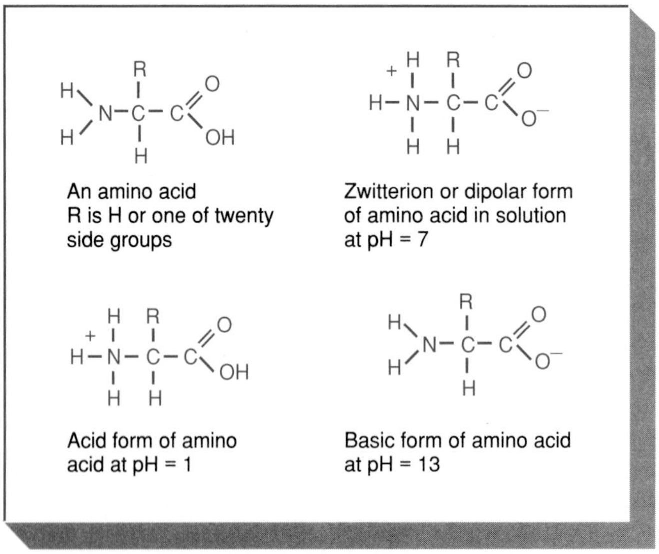

Stearic acid; mp 70° C

Oleic acid; mp 14° C

Figure 18.9
Amino acids are the building blocks of proteins. Because the carboxylic acid group is acidic and the amine group is basic, the form of an amino acid solution depends on pH.

An amino acid
R is H or one of twenty side groups

Zwitterion or dipolar form of amino acid in solution at pH = 7

Acid form of amino acid at pH = 1

Basic form of amino acid at pH = 13

Most microorganisms synthesize all their own amino acids, but many animals (including humans) have lost the ability to synthesize a number of them. These animals depend on their diet to provide these *essential amino acids*.

Amino acids are high-melting solids, and they are more soluble in water than in nonpolar organic solvents. They occur as dipolar ions, sometimes called zwitterions, rather than as neutral molecules (see Fig. 18.9). These positive and negative charges cause forces in the crystal lattice and forces in the aqueous solution to be

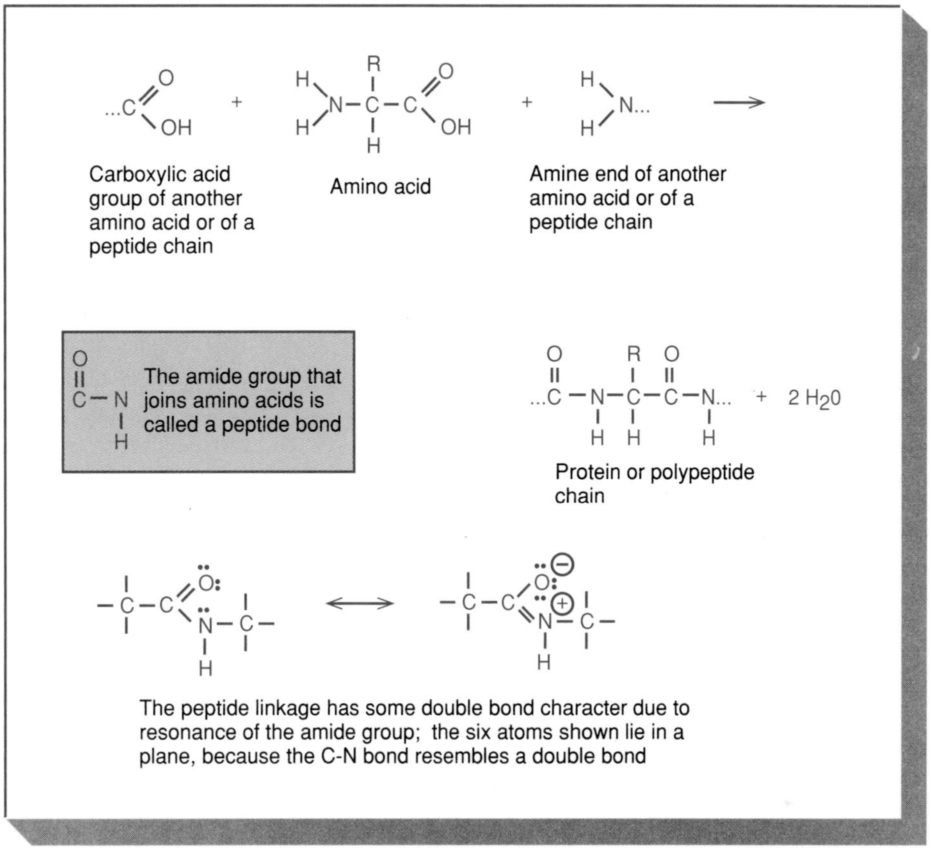

Figure 18.10
Amino acids are joined together in proteins by amide linkages, also known as peptide bonds.

strong. An amino acid can react either as an acid or as a base. The dipolar form of amino acids predominate in solution at a pH near 7. At pH 1, amino acids have a net positive charge; at pH 13, they carry a net negative charge.

Amino acids (with the exception of glycine) obtained from proteins are chiral. Only L-amino acids occur in proteins; their nonsuperposable mirror image isomers are absent or play specialized roles in living systems. The chirality of amino acid building blocks is important for assembling efficient enzymes. If both right- and left-handed amino acids were joined in proteins, a much larger number of proteins could be formed. Proteins with the incorrect stereochemistry would be molecular debris. By using only L isomers of amino acids, living organisms are more efficient in assembling large protein molecules.

■ Secondary Structures of Proteins

Amino acids are joined together in proteins by amide bonds as shown in figure 18.10. These amide linkages are also called *peptide bonds;* consequently, proteins are called *polypeptides*. The peptide linkage of a protein is strong. The atoms bonded to the carbon and nitrogen of the peptide bond lie in a plane, for the C–N bond has some double-bond character. The chain of a protein is stiff and bends only at the two bonds to the α carbon of each amino acid. Consequently, only a limited number of shapes can be taken by the protein chain. Even though the amino acids along the protein chain may vary, the arrangements of the polymer chain fall into a small number of patterns, referred to as the *secondary structures* in proteins. (The sequence of amino acids along the polymer backbone is referred to as the primary structure of proteins.)

The β-pleated sheets, like those found in silk, and the α-helix, like those found in wool, illustrate the two major types of secondary structures found in protein molecules (see Fig. 18.11). In silk, the secondary structure is described as an antiparallel β-pleated sheet. Segments of chains running in opposite directions are held together by strong interchain hydrogen bonds. Small side chain groups of amino acids extend above and below the folded sheet. In contrast, the polypeptide chain of the major protein in wool is arranged in an α-helix. In an α-helix, hydrogen bonding occurs between the C=O and N–H that are located one turn, or 3.6 residues, apart. The side chain groups of amino acids extend out from the helix. The hydrogen bonding of peptide groups favors the formation of both α-helices and β-pleated sheets in proteins.

■ Tertiary and Quaternary Structures of Proteins

The term, tertiary structure, is used to describe the three-dimensional shape of proteins resulting from the folds and bends. The shapes of a number of the proteins that play important roles in metabolism are known from X-ray diffraction studies. Heme, the protein in hemoglobin, has several segments of α-helices; other proteins have few. Lysozyme is an example of an enzyme with segments of β-pleated sheets. Enzymes are folded in such a way that amino acids which are far apart on the polypeptide chain may be close together in space (see Fig. 18.12).

Proteins assume their shapes as they are being synthesized. The folding of proteins is, in part, a consequence of the solubility of amino acid side chains. For highly folded, globular proteins, the groups that interact strongly with water (acidic and basic groups as well as polar amide and alcohol groups, which favor strong hydrogen bonding) are usually at the surface, and the nonpolar, hydrophobic side chains are in the interior (see Fig. 18.13).

Many proteins found outside cells are cross-linked. The oxidation of two nearby thiol (—SH) groups in cysteine amino acids to give a disulfide (—SS—) bond forms a cross-link. Cross-links hold separate peptide chains together in some enzymes and hormones. For example, enzymes that digest proteins are synthesized as *zymogens* (inactive precursors larger than the active enzymes). After synthesis, zymogens are cross-linked by oxidation, and then converted to active enzymes by the cleavage of ends or small loops. The resulting small changes in shape cause large increases in enzymatic activity.

Permanents, used to modify the curl of an individual's hair, change the spacing of cross-links in hair proteins. First, a reducing solution cleaves the sulfur–sulfur bonds of the cross-links already present in the hair. The hair can then be curled or straightened to attain a new arrangement of neighboring —SH groups. Finally, an oxidizing agent forms new cross-links making the modified shape "permanent" (see Fig. 18.14).

■ Cell Membranes Combine Lipids and Proteins (and Sometimes Carbohydrates)

The structure of cell membranes is described by a fluid mosaic model. Membranes consist of lipid bilayers in which various proteins are embedded. The lipids found in cell membranes have hydrophilic heads and hydrophobic tails. The hydrophilic heads of the lipids extend into the aqueous layer on either side of the membrane barrier. The hydrophobic tails form a barrier to the passage of ions and polar molecules.

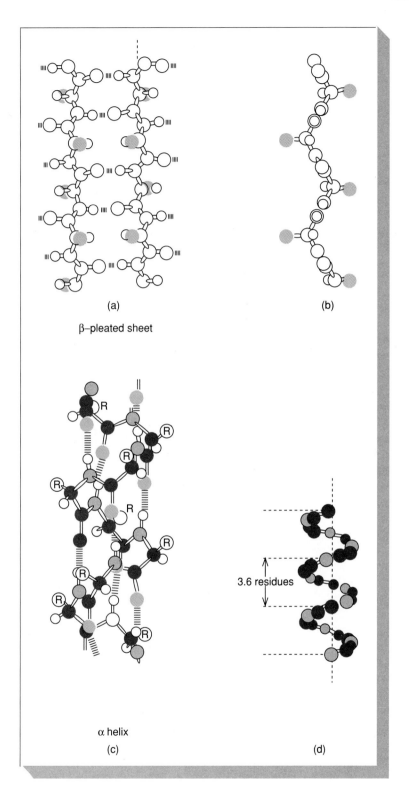

(a)

β–pleated sheet

(b)

α helix

(c)

3.6 residues

(d)

Figure 18.11
Two types of secondary structures found in proteins. (a) Top view of two chains arranged in a pleated sheet, showing the hydrogen-bond cross links between adjacent chains. The R groups are indicated in color. (b) An edge view showing the R groups extending out from the pleated sheet. (c) Ball-and-stick model of an α-helix showing the intrachain hydrogen bonds. (d) The repeat unit is a single turn of the helix.

Figure 18.12
This diagram depicts the folding of the protein chain of myoglobin, an oxygen-binding protein closely related to hemoglobin. The planar heme ring of myoglobin appears in the upper middle of the drawing.

From Richard E. Dickerson, in *The Proteins*, Second Edition, Vol. II, H. Neurath, Ed. Copyright © 1964 Academic Press, Inc., Orlando, Florida. Reprinted by permission.

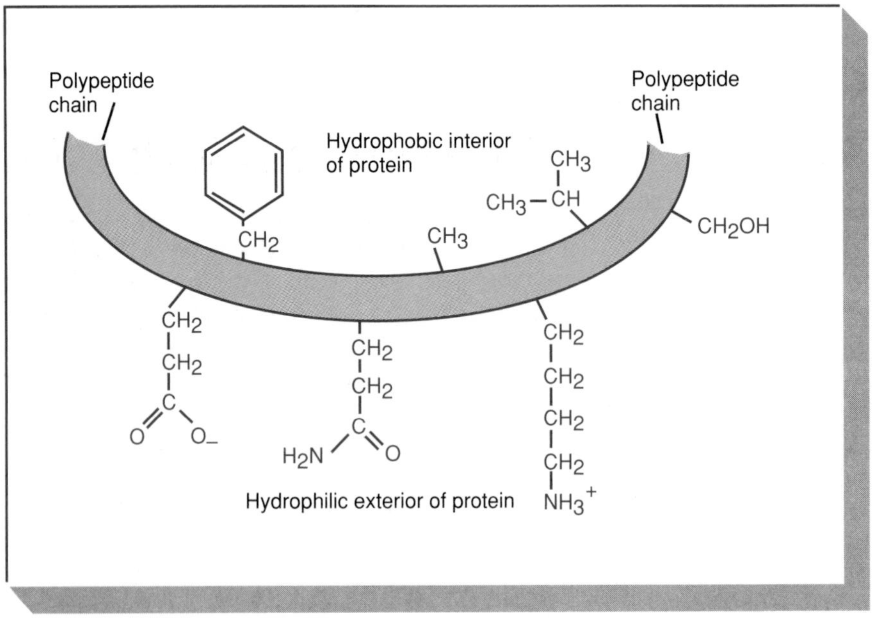

Figure 18.13
Proteins fold in such a way that the side chains of hydrophobic amino acids are in the interior and the side chains of charged or polar amino acids are on the outer surface.

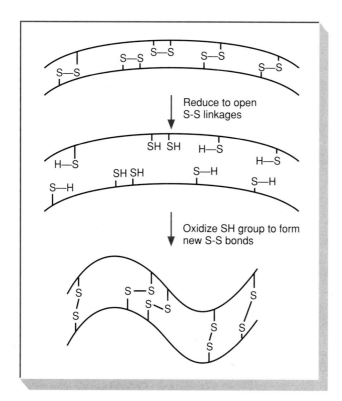

Figure 18.14
In proteins found outside of cells, the —SH group of cysteine amino acids are often oxidized to disulfide linkages. In a permanent, the disulfide groups bonding protein chains in hair are opened by reduction, the hair is set to change the arrangement of —SH neighbors, and new disulfide linkages are then formed by oxidation.

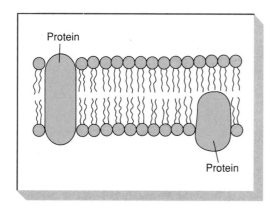

Figure 18.15
According to the fluid mosaic model, cell membranes are composed of lipid bilayers and membrane-bound proteins. Hydrophilic heads of the lipids extend into the aqueous layer, and hydrophobic tails form a barrier to the passage of ions and polar molecules. Proteins in the membrane serve as channels and pumps to pass polar species across the membrane.

Proteins in the membranes serve as channels and pumps to pass polar species across the membrane (see Fig. 18.15). (Among the chemical tools used for the investigation of membrane lipids are enzymes isolated from snake venom. These enzymes catalyze the cleavage of membrane lipids at selected sites and assist the spread of venom from a snakebite by breaking down membrane lipids.)

Many membrane molecules are esters of phosphotidic acid. In this compound, as in fats, two hydrophobic fatty acids are esterified with glycerol. However, the third —OH group of glycerol is esterified to phosphoric acid. The phosphate portion of the ester bears a negative charge and is linked to an ionic or polar residue that serves to increase the size of the hydrophilic head (see Fig. 18.16).

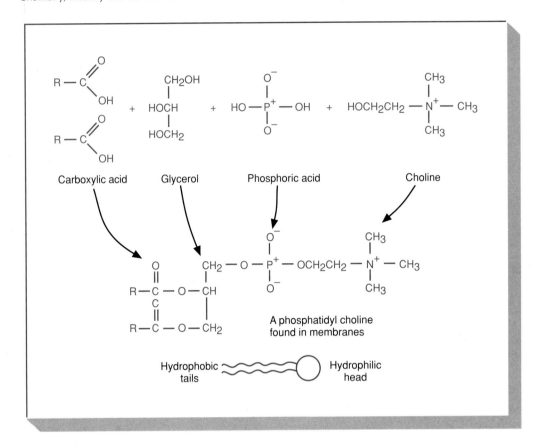

Figure 18.16

Esters of phosphotidic acid are found in membranes. These esters have nonpolar hydrophobic tails and ionic hydrophilic heads. In water, these compounds readily form bilayers.

Membrane proteins serve two very different roles. Some proteins span the lipid bilayer, serving as channels so that ions and compounds can pass across the membrane bilayer. Many of these protein channels have gates to control the cross-membrane flow. Other membrane proteins serve as molecular communication devices to link the interior of a cell with its outside environment. For example, an outer membrane protein may have a receptor site that selectively binds a hormone, much as an enzyme binds its substrate. When a signal molecule binds to the receptor site, the protein changes shape, which in turn causes a protein on the inner surface to change shape. If this inner protein is an enzyme, it may begin to catalyze a reaction that produces a second messenger within the cell.

Carbohydrates bind to the outer surface of some cells. The identity and linkages of sugar molecules on the cell's surfaces can serve as identification tags. For example, the blood groups A, B, AB, and O are based on the presence or absence of certain carbohydrates on the surfaces of red blood cells.

The study of carbohydrates, proteins, and lipids continues to play an important role in biochemistry, but the emphasis is shifting. As our knowledge of chemical structures has grown, attention has moved from the study of the individual classes of compounds to the study of the interactions of these compounds, both where they join together and where they interact at surfaces.

1. Draw the structure of a four-carbon sugar having a ketone functional group.
2. Which of the 2 three-carbon sugars is chiral?
3. The formula of the open-chain form of glucose is $C_6H_{12}O_6$. What are the formulas of α- and β-glucose?
4. Maltose is a disaccharide obtained from starch by the enzymatic cleavage of every second α-glycoside linkage. Draw the structure of maltose.
5. What enzymatic activity (of symbiotic microorganisms) enables termites to digest wood?
6. Why does the esterification of OH groups in cellulose reduce the attraction between polymer chains? Why would polymers that differ in their extent of esterification differ in their properties?
7. Why should the oil from a cod or tuna contain a higher portion of unsaturated fatty acids than the fat from a cow?
8. Read the labels on a soft and a hard margarine. Why do the margarines differ in softening temperature? What chemical reaction has been used to modify vegetable oils for use in margarines?
9. Plant proteins differ widely in their amino acid content. Suggest a reason that a vegetarian diet that features a combination of foods such as corn and beans might be more nutritional than a diet featuring a single source of protein.
10. Glutamic acid is an amino acid with side chain ($-CH_2CH_2CO_2H$) ending in a carboxylic acid group. Draw the structure of glutamic acid in solution at pH 7. What is the net charge on the ion?
11. The amino acid lysine has a $-CH_2CH_2CH_2CH_2NH_2$ side chain. Draw the structure of lysine, present at low pH, that has a +2 charge. Draw the structure of lysine, present at high pH, that has a −1 charge.
12. The isoelectric pH of an amino acid is the pH at which its average net charge equals zero. (Amino acids with neutral side chains have isoelectric pHs near seven.) Rank the following amino acids in order of increasing isoelectric pH: glutamic acid, glycine, and lysine. Justify your order.
13. Draw a partial structure of a peptide showing only the six atoms that are coplanar.
14. Describe the importance of hydrogen bonding in cotton and silk textile fibers. In what ways is the hydrogen bonding similar? How is it different?
15. Why is it essential that a digestive enzyme and its zymogen precursor have similar shapes? Why is it important they differ in shape?
16. Why is the lipid bilayer found in membranes impermeable to the passage of ions?
17. Explain the observation that membrane-bound proteins contain a higher fraction of amino acids having nonpolar side chains than water-soluble proteins do.

19

Informational Biopolymers, Pharmaceuticals, and Biotechnology

The enzymes that catalyze metabolic reactions, and the genetic material that governs inheritance are polymers formed from a small number of monomer units. The sequence of monomer units along the polymer chain is critical to the function of these biopolymers. Scientists have learned to read these sequences, both in proteins and in the nucleic acids that carry genetic blueprints for proteins. By studying these sequences, scientists are learning the molecular basis for genetic inheritance. This knowledge is leading to improvements in diagnosis, and in some cases, to new or improved therapies.

■ Drug Design in the Pharmaceutical Industry

Recall the general features of enzymatic catalysis that were introduced in chapter 10. Enzymes first bind substrates to a region of the enzyme surface known as the active site. They then catalyze the reaction of the substrate and release products into solution. Remember also that the drugs sulfanilamide and penicillin are effective because they bind to the active sites of enzymes to block catalysis.

Pharmaceutical companies engage in an ongoing search for new drugs to combat disease. As part of this effort, they screen extracts from plants used as traditional medicines as well as molds and fungi from a wide variety of sources. They screen these crude extracts for biological activity. When activity is found, they set out to isolate the active components of the extract and to determine their molecular structure. They also seek to determine the mechanism of the activity by finding the enzymes affected by the compound.

Often a compound that combats a disease-causing microorganism is toxic. Chemists then construct and test molecular compounds analogous to the original. They synthesize compounds with similar, but different, structures from the original compound. Then they screen these new compounds to find a structural modification that retains the activity against microorganisms but is less toxic to the host.

Let us take a closer look at how enzymes catalyze reactions. The pathway depends upon the particular reaction and enzyme, but, in general, it occurs in a series of steps. In all cases, a reaction becomes more favorable if binding to the enzyme first increases, then decreases, along the pathway from reactant to product. The structure occurring at the midpoint of the reaction pathway is called the transition state.

A recent approach to drug design is based on the insight that enzymes bind transition states more tightly than they bind reactants or products. Many compounds that resemble the transition states of enzymatic reactions are potent enzyme inhibitors. When the structure of an enzyme is known, and the reaction it catalyses has been determined, chemists may be able to construct models of the enzyme's active site. Then, in the search for new drugs, a chemist may use a high-speed computer to search a database of organic structures to generate a list of possible compounds that might fit the active site closely and inhibit the enzyme activity. Candidates for such transition-state analogs should have the right size and shape, the right combination of polar and non-polar regions, and the right ability to form hydrogen bonds to fit the active site. Chemists then prepare these compounds and test them for biological activity. If they find activity, they synthesize similar compounds in a search to find a still better drug.

■ Aside

Morphine Imitates Natural Pain Killing Peptides

Alkaloids are amines isolated from plants; many have strong physiological activity. The presence of alkaloids in plants and the molecular basis of their physiological activity are puzzling. Scientists do not yet know why plants should synthesize alkaloids, but they are beginning to understand why alkaloids produce such dramatic responses—alkaloids bind to receptor sites present in the brain.

For example, morphine obtained from opium poppies dulls pain. In fact, before the addictive nature of morphine had been discovered, morphine was widely used in medicine as a pain killer. Morphine binds to the same pain repressing site to which an endorphin binds. *Endorphins* are small peptides produced by the body to suppress pain. The body produces endorphins in response to injury, in response to the twirling needles used in acupuncture, and, perhaps, during the course of long, strenuous exercise. Morphine mimics the shape taken by the endorphin. When the rigid morphine molecule is compared to the endorphin peptide, there are similarities. There are corresponding hydrophobic regions, and corresponding groups with the potential to form hydrogen bonds. Morphine is shaped to bind to an endorphin receptor site. Unfortunately, the euphoria produced by morphine led to drug addiction and its medical use was curtailed.

■ Determining the Sequence of Amino Acids in Proteins

Enzymes (as well as polypeptide hormones) can be described as *informational macromolecules*. The sequence of amino acids is unique in each enzyme, and the shape and function of each enzyme is a consequence of its amino acid sequence. The sequence along the polypeptide chain is called the *primary structure* of the protein.

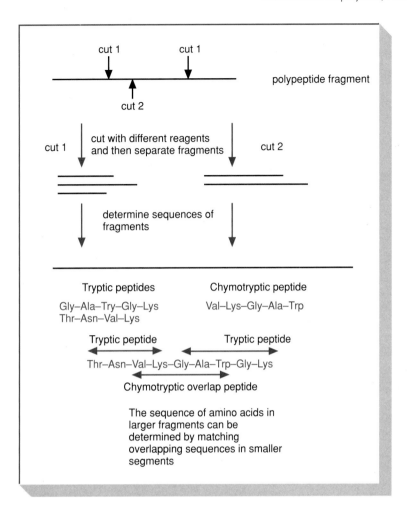

Figure 19.1
The general strategy for determining the sequence of amino acids in a protein.

In 1953, Frederick Sanger (b. 1918) completed the first determination of the sequence of amino acids in a protein, the hormone insulin. Since that time, methods for determining the sequence of amino acids in proteins have become faster and more sensitive, but important elements of Sanger's strategy remain the same. A protein or a large polypeptide is cut into smaller segments using enzymes or chemical reagents. Small peptides are then isolated, and their amino acid content and sequence is determined (see Fig. 19.1).

Chemical and enzymatic reactions are used to determine the sequence of amino acids in these fragments. For example, the free amine group at one end of a peptide may be labeled by a chemical reaction that attaches a marker group. After the peptide bonds are cleaved, the marker group serves to identify the terminal amino acid. In another approach, chemical reactions are used to label and cleave amino acids one at a time. By using a variety of different methods to cut the polypeptide chain, overlapping sequences can be determined. This process is continued until one sequence best fits all the data.

The determination of amino acid sequences of a number of mutated hemoglobins, the oxygen-carrying protein in red blood cells, has led to important insights

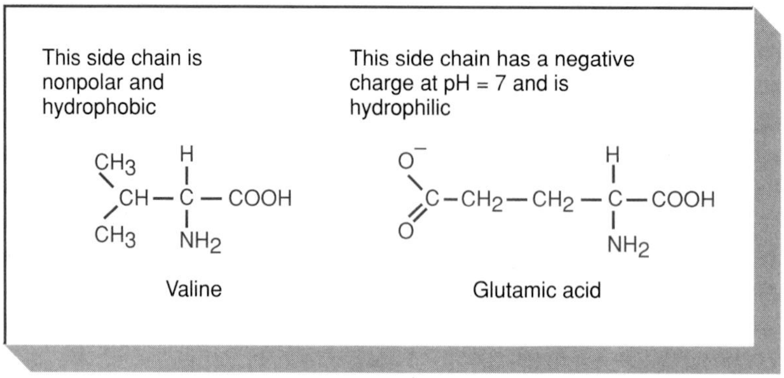

Figure 19.2

In the mutation responsible for sickle-cell anemia, valine is substituted for glutamic acid in position 6 of a hemoglobin chain. The mutated hemoglobin is less soluble in water and can precipitate, causing round, red blood cells to change to sickle-shaped cells that can aggregate and block capillaries.

This side chain is nonpolar and hydrophobic

This side chain has a negative charge at pH = 7 and is hydrophilic

Valine

Glutamic acid

into genetic diseases. (Genetic mutations often result in the synthesis of altered proteins and, since the enzymes that mediate metabolism are proteins, they also result in metabolic deficiencies.) Most of these altered hemoglobins differ from normal hemoglobin by the substitution of a single amino acid. For example, the substitution of a non-polar amino acid for a polar amino acid at a single position in the polypeptide chain converts normal hemoglobin to hemoglobin S, the protein found in persons with the genetic disease sickle cell anemia. Hemoglobin S is less soluble than normal hemoglobin so if it crystallizes in red blood cells, they become distorted and clog capillaries, blocking blood flow (see Fig. 19.2).

■ Nucleic Acids and Genetic Transmission

Nucleic acids are polymers involved in the transmission and expression of genetic information. Nucleic acids can be cleaved to give phosphoric acid, a simple sugar, and a small number of organic bases. In deoxyribonucleic acid (DNA), ester linkages join phosphoric acid and deoxyribose, a 5-carbon sugar lacking the O in one —OH group. One of four organic bases is attached to each sugar unit. The four bases are represented by their initials: A, T, C, and G, which stand for adenine, thymine, cytosine, and guanine. Like DNA, RNA is a nucleic acid polymer. Ribonucleic acid (RNA) contains the sugar ribose rather than the deoxyribose in DNA, and it contains the base uracil, U, rather than thymine as one of its four bases (see Fig. 19.3).

X-ray diffraction patterns of DNA fibers show evidence of α-helical structure. Working from scale molecular models, James Watson (b. 1928) and Francis Crick (b. 1916) formulated a double helix structure for deoxyribonucleic acid in 1953. The bases are in the center of the double helix, and two polymer strands formed from the sugars and phosphates wind in opposite directions around the outside (see Fig. 19.4).

Base pairing provides the mechanism for recognition and copying. The pairing involves both base size and hydrogen bonds. Both AT and GC pairs span the distance between the backbone chains. A pair consisting of the bases A and G would be too large to fit in the double helix, and a pair between C and T would fail to bridge the gap. Hydrogen bonds favor AT and GC pairs. These bases pair with great fidelity. Adenine, A, on one strand always pairs to a thymine, T, on the second strand; guanine, G, always pairs with cytosine, C.

The Watson and Crick structure stimulated a rapid growth in the understanding of the molecular basis of genetics. Information is conveyed by the sequence of bases on a DNA polymer chain. Each strand carries the information needed to construct its complement. Strand separation and duplication provide for the transmission of genetic information from one generation to the next.

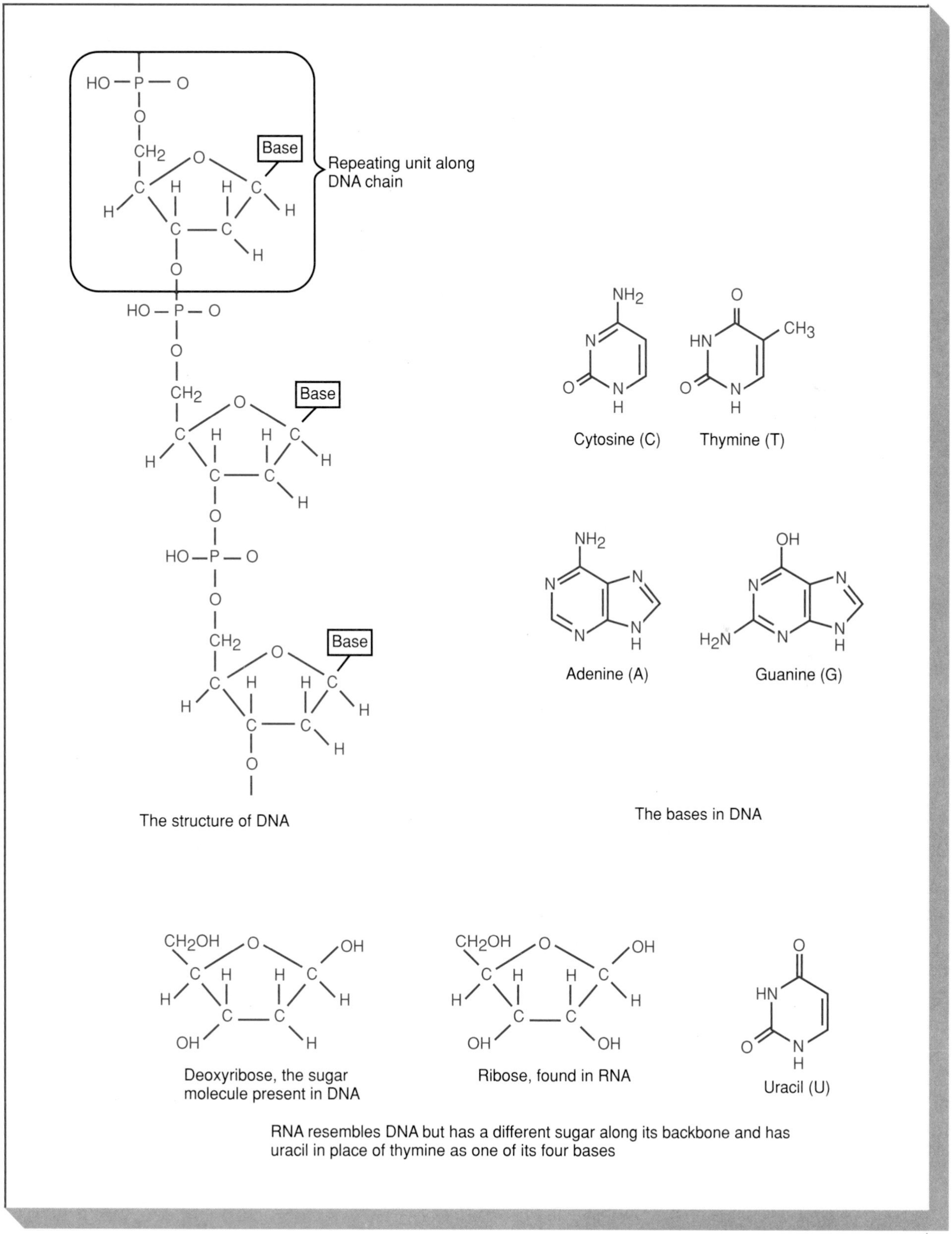
Repeating unit along DNA chain
The structure of DNA
Cytosine (C)
Thymine (T)
Adenine (A)
Guanine (G)
The bases in DNA
Deoxyribose, the sugar molecule present in DNA
Ribose, found in RNA
Uracil (U)
RNA resembles DNA but has a different sugar along its backbone and has uracil in place of thymine as one of its four bases

Figure 19.3
The nucleic acids DNA and RNA are involved in the storage and transcription of genetic information. Each of these molecules has a backbone built up from phosphoric acid and a 5-carbon sugar, and each has one of four bases attached to each sugar along the polymer backbone.

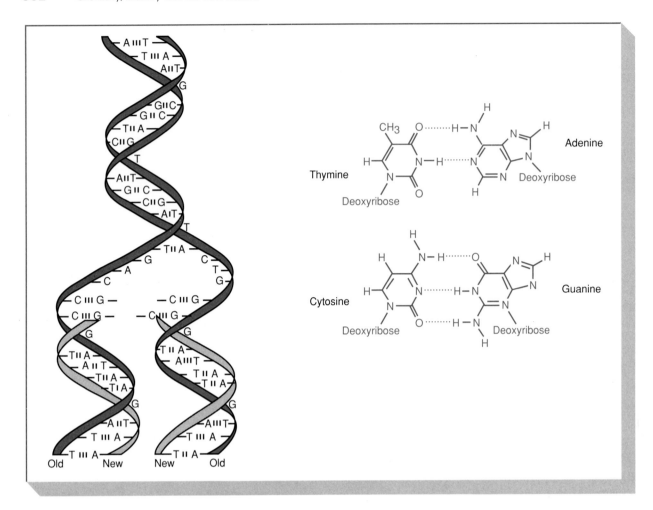

Figure 19.4
Base pairing provides a method for copying the information of DNA onto new DNA.

Genes are expressed in proteins. The phrase, "one gene, one protein" once expressed the hypothesized correspondence. Genetic information flows from the linear DNA to the three-dimensional protein. The information encoded in the sequence of letters (bases) along a strand of DNA prescribes which one of the twenty amino acids is to be used at each stage in the construction of a protein.

Scientists have learned to read the genetic code. They synthesized short synthetic DNA segments with known base sequences, and then they used these segments as templates to make peptides. They then analyzed the peptides to determine which amino acids were incorporated. Information is encoded in base triplets such as GCA, CCG, etc. The number of different three-letter words that can be encoded using four letters is $4 \times 4 \times 4 = 64$. Start and stop signals, together with a redundant code specifying amino acids, have been determined (redundancy helps to protect messages from being mistranslated).

Since protein synthesis occurs in the cytoplasm of cells and since DNA is located in the nucleus, a messenger molecule is required. Scientists looked for, and found messenger RNA. Cells synthesize single-stranded mRNA in the nucleus and the mRNA then moves to the cytoplasm where it serves as a template for protein syntheses.

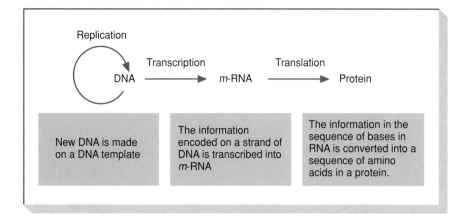

Figure 19.5
The *central dogma* describes the flow of genetic information from generation to generation through replication and the expression of that information in proteins by transcription and translation.

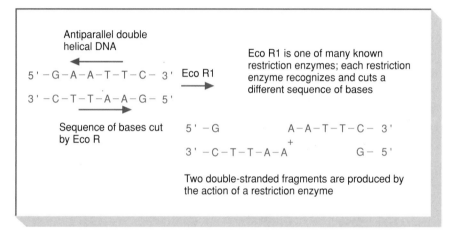

Figure 19.6
Double-stranded DNA is cut by restriction enzymes. Molecular biologists use these enzymes to make smaller fragments of DNA as an aid in determining the sequence of bases in DNA.

The discovery of mRNA was the final step in the development of the model for genetic expression named *the central dogma*. According to this model, information encoded in DNA is first transcribed into mRNA and then translated into protein (see Fig. 19.5).

■ Reading the Sequences of DNA and RNA

A major revolution in the way we understand the mechanism of gene expression is under way. New experimental techniques have enabled scientists to read the base sequences of DNA and RNA, in addition to determining the amino acid sequence of an encoded protein. The simultaneous translation of DNA, messenger RNA, and the protein for a gene, has given surprising results and has led to fundamental changes in the way we understand genetic expression and change.

Scientists discovered and studied restriction enzymes that shred the DNA of invading organisms while the host DNA itself is protected. Restriction enzymes cut double helical DNA at specific short sequences. By changing conditions, either double-stranded fragments or single-stranded fragments can be recovered from the reaction (see Fig. 19.6).

Figure 19.7
Restriction enzymes can recognize and destroy foreign DNA. Some restriction enzymes recognize their "own" DNA by the presence of methyl side groups. The same enzyme that cuts foreign DNA methylates the second strand of protected DNA. When that DNA separates and is duplicated during cell division, each of the progeny DNAs also carries protecting methyl groups.

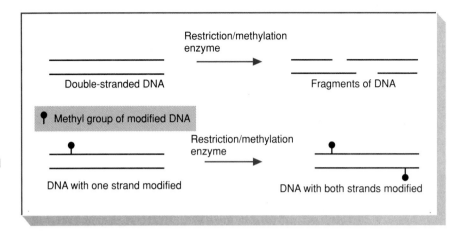

To cut intact DNA, restriction enzymes read information in the groove of the double helix rather than the base sequence of a single chain. With only four bases in DNA, the probability of finding a restriction sequence in any long piece of DNA is high. An organism having a restriction enzyme must protect its own DNA and methylation enzymes are used to provide this protection. These enzymes can modify the bases in double-stranded DNA by attaching a methyl group. If one strand of DNA is already methylated, these enzymes methylate the second strand. This chemical modification is done in a site along the groove rather than in a site used in base pairing (see Fig. 19.7).

To study genes, scientists cut the DNA with restriction enzymes. To determine the sequence of the bases in each restriction fragment, a series of successively shorter copies, all of which originate at the same end, is prepared. Copies can be either complimentary DNA fragments or RNA fragments, depending on the enzymes used to synthesize them. Many copies of each fragment must be made so there will be enough material to detect. Detection is simplified by starting each copy with a radioactive phosphorus atom in the first phosphate group.

Frederick Sanger and Walter Gilbert (b. 1932) shared a Nobel prize in 1980 for developing the methods used to determine the sequence of bases in nucleic acids (Sanger had received an earlier Nobel Prize for determining the sequence of amino acids in insulin). One approach is to build successively longer copies that all end at the same base, for example, A. To do this, a small amount of a modified base A′ that stops the chain from growing is included in the mixture. Then all copies start with the same base and end at sites encoded for A. Alternatively, complete copies of the whole restriction fragment can be made and then nicked with a small amount of a chemical that selectively cleaves the polymer at A. Using either process, one can prepare a mixture of copies, all with a common origin and all ending at sites coded for A (see Fig. 19.8).

These chains can be separated according to length using an electric field and a supporting polymeric gel. The electric force moving a fragment through the gel is proportional to its size, but the retarding frictional force increases rapidly with the increasing size. Hence, the smaller the fragment, the further it migrates. By using compounds labeled with a radioactive phosphorus isotope, scientists can locate the positions of the nucleic acid fragments in the gel. When separate samples ending in A, T, C, and G are run side by side, one can read the sequence of bases from the sequence of spots in the gel. Since the base pairing of genetic copying is known, the template sequence can be inferred.

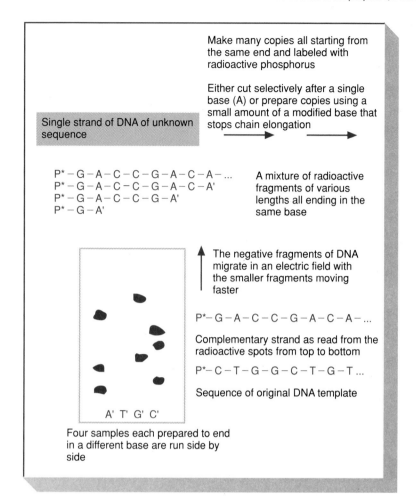

Single strand of DNA of unknown sequence

Make many copies all starting from the same end and labeled with radioactive phosphorus

Either cut selectively after a single base (A) or prepare copies using a small amount of a modified base that stops chain elongation

P*－G－A－C－C－G－A－C－A－...
P*－G－A－C－C－G－A－C－A'
P*－G－A－C－C－G－A'
P*－G－A'

A mixture of radioactive fragments of various lengths all ending in the same base

The negative fragments of DNA migrate in an electric field with the smaller fragments moving faster

P*－G－A－C－C－G－A－C－A－...

Complementary strand as read from the radioactive spots from top to bottom

P*－C－T－G－G－C－T－G－T...

Sequence of original DNA template

A' T' G' C'

Four samples each prepared to end in a different base are run side by side

Figure 19.8
Scientists use a mixture of enzyme-catalyzed reactions and chemical reactions to make complementary copies of an unknown fragment of DNA. They separate these copies and use a radioactive marker to detect them. Using these methods, it is possible to read the sequence of bases in very large fragments of DNA.

■ Pharmaceutical Drugs for the Treatment of Cancer and AIDS

Drugs used in chemotherapy of fast-growing cancers and in the treatment of certain viral diseases, including AIDS, resemble the chemical reagents used by scientists to determine the sequence of bases in nucleic acids. These drugs have structures similar to the monomers used as building blocks for the synthesis of DNA or RNA. They are taken up by the enzymes synthesizing nucleic acids, but they block the successful completion of a complimentary strand (see Fig. 19.9).

These drugs are also highly toxic. Because they act on the enzymes making DNA, they block cell division for both healthy and tumorous cells. Chemotherapy drugs can be used to kill cancer cells because the cancer cells are reproducing at a far greater rate than normal cells, and thus are necessarily making far more new DNA. (These same drugs cause the hair loss experienced by people undergoing chemotherapy, for they effectively shut down the synthesis of new hair cells as well.)

The task of finding antiviral drugs is far more difficult than finding drugs to combat infections by larger, more complex organisms. Viruses encode only a small number of proteins, and viral reproduction takes place entirely within a host cell. Antiviral drugs have a base (like those found in DNA or RNA) to promote recognition and binding, and a chemically-modified sugar group to block enzyme action.

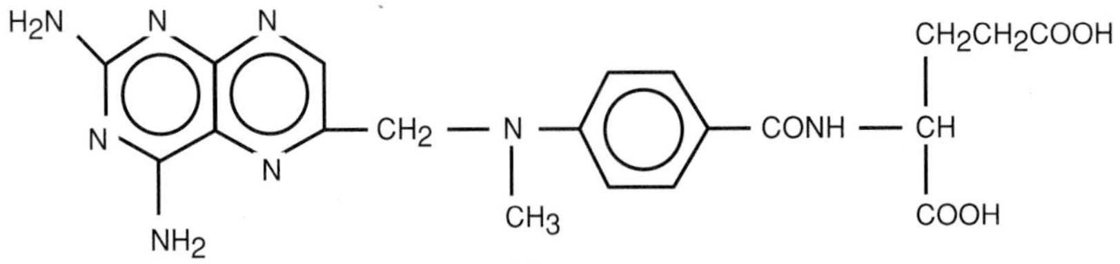

(16)

Methotrexate

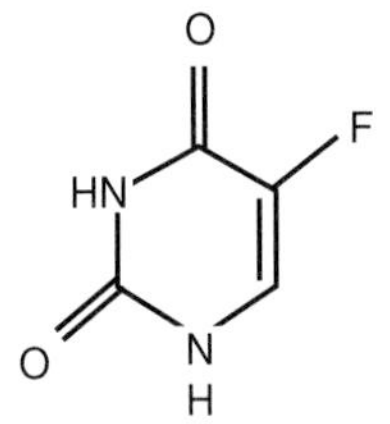

(15)

Fluorouracil

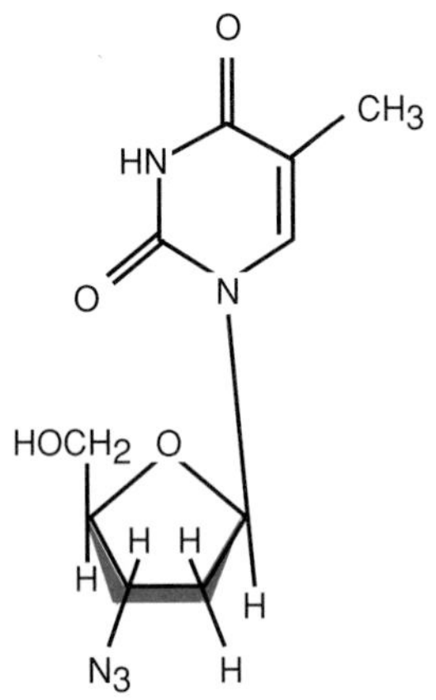

AZT

Figure 19.9
Structures of two compounds used in cancer chemotherapy and of AZT, a compound used in the treatment of AIDS.

The targets of the antiviral drugs now in use are viral-encoded enzymes that catalyze nucleic acid synthesis, so the chemist's task is made more difficult since the drugs cannot be too destructive of the host enzymes that also catalyze nucleic acid synthesis.

■ Informational Macromolecules—A Language Model

The development of techniques to determine the sequence of nucleic acids and proteins has permitted a close examination of their structures. These studies show that protein and nucleic acid structural sequences can be treated literally as messages. The messages are expressed in a language with words, phrases, and other grammatical structures. The messages can be copied, modified, and transmitted. Some of the most provocative studies have resulted from textual comparisons of a DNA, the mRNA transcribed on it, and the translated protein. The results of some of these studies, together with new questions raised by molecular biologists follow.

The first DNAs to be fully sequenced were those of small viruses. These viruses coded for only a few proteins. It came as a great surprise that the genetic information for some of these proteins was encoded in overlapping DNA segments.

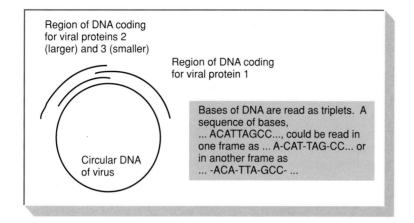

Region of DNA coding
for viral proteins 2
(larger) and 3 (smaller)

Region of DNA coding
for viral protein 1

Circular DNA
of virus

Bases of DNA are read as triplets. A
sequence of bases,
... ACATTAGCC..., could be read in
one frame as ... A-CAT-TAG-CC... or
in another frame as
... -ACA-TTA-GCC- ...

Figure 19.10
One of the first DNAs to be fully
sequenced was that of SV40, a small
virus. Scientists were surprised to
learn that the DNA encoding three
proteins found in the viral coat had
regions of overlap. The reading
frame for viral protein 1 differed
from that of the other two proteins.

DNA strand

| Exon | Intron | Exon | Intron | Exon |

Nuclear RNA transcription

Intron
transcript

Folding to bring ends of exons
together

Exon transcript

Excision and splicing

Messenger-RNA

Figure 19.11
Many genes come in pieces. DNA
expressed in proteins occurs in
segments (exons); these segments are
separated by intervening sequences
(introns). The strand of DNA is
copied completely; then it is edited
to remove transcripts of introns and
to complete the messenger RNA on
which the protein is assembled.

That a DNA phrase can be found in two protein sentences is evidence of complex punctuation. Alternate interpretations of a text must rely on a larger frame of reference than base triplets alone can furnish (see Fig. 19.10).

In contrast, when RNA transcripts of DNA genes from cell nuclei are examined, the results are striking. Far more RNA is synthesized on the DNA template than is necessary for the translation of the protein. The RNA in the nucleus differs from the messenger RNA found in the cytoplasm where protein is synthesized. The information to make most proteins is not encoded in a continuous segment of DNA. Instead, nuclear RNA is edited to form the mRNA at sites in the nuclear membrane. Genes are not continuous; they come in pieces. The DNA that is finally expressed in proteins occurs in segments known as exons that are separated by intervening sequences called introns.

The examination of the protein segments encoded by each exon has been determined for the genes that encode hemoglobin and lysozyme, proteins of known three-dimensional structure. The middle exon of the protein in hemoglobin encodes that portion of the protein that makes all contacts with the heme ring. Similarly, the active site cleft of the enzyme lysozyme is encoded by a single exon. These findings suggest that exons encode functional segments of proteins (see Fig. 19.11).

It has been suggested that introns facilitate evolution. According to this model, genes evolve when gene fragments evolve. A given piece of DNA may be repeated several times in a longer strand of DNA. Some of these repeated sequences, dormant in gene expression, may undergo mutation. If at a later time, a modified fragment is reinserted into the active gene pool, a favorable mutation may result.

Finally, for proteins secreted by a cell, it has been found that the newly synthesized protein inside the cell is larger than the protein found outside of the cell. The precursor protein synthesized on the mRNA has an additional short sequence of amino acids called a leader sequence. This leader sequence is excised as the protein crosses the cell membrane. It would appear that the language of gene expression includes shipping labels for export, perhaps even zip codes.

■ The Development of a Biotechnology Industry

The biotechnology industry seeks to develop new drugs for diagnosis and treatment by using the scientific knowledge of biochemistry and molecular genetics. In favorable cases, it is possible to insert genetic instructions for the synthesis of a protein into microorganisms, to have these organisms synthesize many copies of the desired material, and to harvest the desired protein. These techniques are being used to produce human growth hormone for the treatment of dwarfism and human insulin for the treatment of diabetes. The availability of human insulin benefits those patients who develop adverse reactions to the related-but-different insulin obtained from the pancreases of pigs.

Tools used to study the structures of genes have been adapted to develop tests for genetic defects using DNA from a very small number of cells. This DNA is cut by restriction enzymes, and the fragments are separated. A fragment carrying the gene of interest is cloned to make many copies. When these copies are cleaved, the pattern of fragments can indicate whether the DNA is that of the normal gene or one carrying a known mutation.

Genetic "fingerprinting" is used as a tool in criminal investigation and is presented as evidence in court cases. A laboratory investigator may recover a sample of DNA from a blood stain and then use a variety of restriction reagents to cleave individual samples of the DNA. The separation of the individual fragments from each test provides a set of patterns that act as genetic fingerprints. Differences in patterns can be used to prove that two samples are from different individuals. The absence of differences can be used to show a high probability that the samples are from the same individual.

Scientists are using molecular biology in an attempt to increase agricultural productivity. They are attempting to insert genes that carry specific disease resistance into the DNA of susceptible crops. In still another, more ambitious effort, they are investigating whether the genes that enable legumes to develop nodules and act as hosts to nitrogen-fixing bacteria can be transferred to the genetic material of crops that require large amounts of nitrogen fertilizers.

■ Questions—*Chapter 19*

1. The United States did not sign the biodiversity treaty presented at the 1992 Earth Summit meeting in Brazil. The treaty called for developed and undeveloped nations to share revenues from medicines developed from protected species. In what ways is such an arrangement fair? In what ways is it unfair?

2. Is cellulose an informational macromolecule? Why or why not?

3. Two different dipeptides can be formed using alanine and glycine. Draw the structures for each.

4. How many different dipeptides can be formed from the twenty amino acids found in proteins? (Note that the amino acids in dipeptide may be alike or different.)

5. How many different tripeptides can be hydrolyzed to give alanine, serine, and lysine?

6. How many different tripeptides that can be hydrolyzed to give alanine, serine, and lysine react with Sanger's reagent to give an amine-labeled serine after hydrolysis?

7. If a fresh sample of the tripeptide described in question six is partially digested to give a dipeptide containing the amino acids serine and lysine, what is the amino acid sequence in the original tripeptide?

8. The open-chain structure of the sugar ribose found in RNA is shown below. Draw the open-chain structure of 2-deoxyribose, the sugar found in DNA.

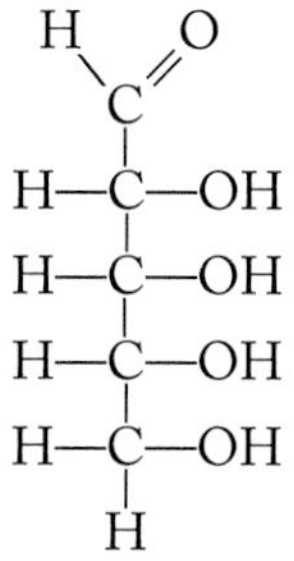

$$
\begin{array}{c}
\text{H} \quad \text{O} \\
\diagdown \ \diagup \\
\text{C} \\
| \\
\text{H—C—OH} \\
| \\
\text{H—C—OH} \\
| \\
\text{H—C—OH} \\
| \\
\text{H—C—OH} \\
| \\
\text{H}
\end{array}
$$

9. What is the complementary base sequence to AAGCT?

10. RNA is synthesized on a DNA template. What base in DNA codes for the base U in RNA?

11. For which of the following base-pairs is the hydrogen-bonding attractions greatest, AT or GC?

12. When heated, DNA unwinds in a phase change similar to that occurring in the melting of a crystalline solid. Use your answer to question eleven to explain why DNA containing more than 55% GC pairs is stable at temperatures at which DNA containing less than 45% GC pairs is denatured.

13. Why is a minimum of three bases needed to encode each of the twenty amino acids found in proteins?

14. Why would redundancy in the three base code for an amino acid help prevent the incorporation of incorrect amino acids in a protein?

15. Some large planar molecules are potent mutagens. Because they fit between parallel bases in a strand of DNA, they can cause errors in the synthesis of a complementary strand. How would these changes, called a "frame shift" mutation, differ from the change in a point mutation, where an incorrect base incorporated into an otherwise correct complementary strand?

Measurements and Calculations

C hemists commonly measure quantities of chemicals in one of three ways. They measure the weight of solids and liquids using a balance. When working with gases, they measure volume, temperature, and pressure. The volume of a given gas at a fixed temperature and pressure can then be converted to a weight. In many applications, chemists work with solutions and they frequently measure volumes of solutions with known concentration. From the volume and concentration measurements, the weight of a substance in solution can be calculated.

■ Chemical Equations Relate Moles of Products and Reactants

Because atoms are not measured directly, an interpretation of balanced equations in measured quantities is desirable. A balanced equation is a statement about the moles of elements and compounds in chemical reactions. Because reactions are often run using weights of reactants and products, it is useful to interconvert moles and weights. Using the mole to express quantities in chemical reactions, a variety of relationships between reactants and products can be calculated.

Sample Calculation

What weight of CO_2 is formed from 1000 g of C_8H_{18}?

$$2\ C_8H_{18} + 25\ O_2 \rightarrow 16\ CO_2 + 18\ H_2O$$

The balanced equation shows that 8 mole of CO_2 are formed for each mole of isooctane burned. To solve the problem, the weight of C_8H_{18} is converted into moles, the balanced equation is used to relate moles of product to moles of reactant, and, finally, the number of moles of CO_2 is converted into a weight:

$$\text{Molecular weight of } C_8H_{18} = 8 \text{ mole C } \frac{12.01\ g}{1 \text{ mole C}} + 18 \text{ mole H } \frac{1.008\ g}{1 \text{ mole H}}$$

$$= 114.2\ g$$

$$\text{MW of } CO_2 = 1 \text{ mole C } \frac{12.01\ g}{1 \text{ mole C}} + 2 \text{ mole O } \frac{16.00\ g}{1 \text{ mole O}} = 44.01\ g$$

$$\text{Weight } CO_2 = \frac{1000 \text{ g isooctane}}{114.2 \text{ g/mole}} \ \frac{8 \text{ mole } CO_2}{1 \text{ mole isooctane}} \ \frac{44.01\ g}{1 \text{ mole } CO_2}$$

$$= 3089\ g$$

■ Gas Calculations

Because gases expand or contract much more than liquids or solids, it is necessary to consider the ways by which the quantities of gases might be measured. Pressure, volume, and temperature are measured more often than weight.

Boyle's Law

For a sample of a gas at constant temperature, the product of the pressure times the volume is found to be a constant. This relation is known as Boyle's law.

$$P\,V = \text{constant}$$

Pressure can be expressed using a variety of units. An atmosphere of pressure is equal to 760 torr, the pressure exerted by a column of mercury 760 mm high. A barometer reading of 720 torr is equal to 720/760 = 0.95 atm.

When the pressure of a gas changes at constant temperature, the volume of the gas at the new pressure can be related to the original volume and pressure by applying Boyle's law. At constant temperature, the product $P_1 V_1$ at an initial set of conditions has the same value as the product $P_2 V_2$ under new conditions. (The subscript 2 refers to the new conditions, and the subscript 1 refers to the initial conditions.)

$$P_1 V_1 = P_2 V_2$$

To solve for V_2, both sides of the equation are divided by P_2:

$$V_2 = \frac{P_1}{P_2}\, V_1$$

Note that a similar equation relates the new pressure of a gas to the initial pressure when the volume changes at constant temperature.

$$P_2 = \frac{V_1}{V_2}\, P_1$$

Sample Calculation

What volume would 5000 L of helium in a balloon occupy if the pressure were decreased from 1.000 to 0.800 atm at constant temperature?

$$V_2 = V_1\, \frac{P_1}{P_2} = 5000\ \text{L}\ \frac{1.000\ \text{atm}}{0.800\ \text{atm}}$$

$$V_2 = 6250\,\text{L}$$

Charles' Law

At constant pressure, a gas expands when heated. A graph for gas volume versus temperature at constant pressure is linear. Extrapolation of the plotted lines to very low temperatures gives the value $T = -273°\,\text{C}$ for the intercept at $V = 0$. This relationship, known as Charles' law, is summarized in the following equation:

$$V = (\text{constant})\, [T\,(°\text{C}) + 273]$$

If a new temperature scale is chosen, the equation for Charles' law can be simplified. On the kelvin scale, $T(K) = T(°C) + 273$. Using the kelvin scale for temperature, Charles' law becomes:

$$V = (\text{constant})\,[T(K)]$$

$$\frac{V}{T} = \text{constant}$$

For a change of temperature at constant pressure, Charles' law can be used to relate the new volume to the initial volume and temperature, V_1 and T_1, and to the final temperature T_2:

$$\frac{V_2}{T_2} = \frac{V_1}{T_1}$$

$$V_2 = \frac{T_2}{T_1}\,V_1$$

Sample Calculation

What volume would be occupied by 1.000 L of a gas if it were warmed from 20 to 30° C (293 to 303 K)?

$$V_2 = V_1\,\frac{T_2}{T_1} = 1.000\ \text{L}\ \frac{303\ \text{K}}{293\ \text{K}}$$

$$= 1.034\ \text{L}$$

Note that Charles' law and Boyle's law are idealizations of the behavior of gases. All gases condense to liquids if cooled sufficiently at a high enough pressure.

A Combined Gas Law

A general equation that relates the volume of a gas at a new temperature and pressure to its initial volume, temperature, and pressure can be derived:

$$\frac{P_1 V_1}{T_1} = \frac{P_2 V_2}{T_2}$$

(At constant temperature, this formula reduces to Boyle's law; at constant pressure, it reduces to Charles' law.) By multiplying both sides of this equation by T_2/P_2, an expression for the new volume is obtained:

$$V_2 = \frac{P_1}{P_2}\,\frac{T_2}{T_1}\,V_1$$

Rather than try to remember subscripts in this equation, it is more useful to apply physical reasoning. If a gas is cooled, it contracts. In this case the ratio of initial and final temperatures should give a fraction smaller than 1. A gas also contracts if it is compressed. Again, the ratio of initial and final pressures should give a number smaller than 1. If a gas is heated, it expands. The initial volume should be multiplied by a ratio of temperatures greater than 1 to find the new volume. If the pressure on a gas is lowered, it expands. A ratio of pressures greater than 1 corresponds to expansion.

Sample Calculation

A 400 mL sample of O_2 gas is collected at 22° C and 0.95 atm of pressure. What volume would the gas occupy at 0° C and 1.00 atm?

$$T(K) = T(°C) + 273$$

$$T = 22°C + 273 = 295 \text{ K}$$

$$V_2 = V_1 \frac{P_1}{P_2} \frac{T_2}{T_1}$$

$$= 400 \text{ mL} \frac{0.95 \text{ atm}}{1.00 \text{ atm}} \frac{273 \text{ K}}{295 \text{ K}}$$

$$= 352 \text{ mL}$$

The Molar Volume of Gases

At 1 atm and 273 K, the volume occupied by a mole of nitrogen, oxygen, or most other gases is 22.4 L (small differences exist). This value can be used to calculate the number of moles of a gas from P, T, and V measurements or to calculate the volume of a given weight of a gas under standard conditions.

Sample Calculation

What is the volume of 11 g of CO_2 gas at 273 K and 1.00 atm?

$$V = \text{moles } CO_2 \frac{22.4 \text{ L}}{\text{mole}}$$

$$= \frac{11 \text{ g } CO_2}{44 \text{ g/mole}} \frac{22.4 \text{ L}}{\text{mole}}$$

$$= 0.25 \text{ mole} \frac{22.4 \text{ L}}{\text{mole}}$$

$$= 5.6 \text{ L}$$

■ Solutions

It is frequently convenient to measure quantities of chemicals by measuring volumes of solutions because it is faster and easier to measure volumes of solutions than to weigh substances. To express the concentration of solutions, chemists often use units of molarity (moles/L, M). For example, a 1.00 M solution of NaCl contains 1.00 mole of NaCl/L of solution. To prepare this solution, a chemist would dissolve 1.00 mole of NaCl in water and then dilute the solution to a final volume of 1.00 L. If the concentration of a solution is known, the measured volume can be used to calculate the number of moles of the dissolved substance.

Sample Calculation

What is the concentration of a solution of NaOH (FW = 40.0) if 10.0 g of NaOH is dissolved in water and diluted to 500 mL?

$$C = \frac{mole}{L} = \frac{10.0 \text{ g NaOH}/40.0 \text{ g/mole NaOH}}{0.500 \text{ L}}$$

$$= \frac{0.250 \text{ mole NaOH}}{0.500 \text{ L}}$$

$$= 0.50 \text{ } M$$

Sample Calculation

What weight of NaOH is required to make 2.0 L of a 0.20 M solution?

$$C = \frac{mole}{volume}$$

$$\text{Moles} = C \, V$$

$$\text{Moles}_{\text{NaOH}} = \frac{0.20 \text{ mole}}{L} \, 2.0 \text{ L} = 0.40$$

$$\text{Weight}_{\text{NaOH}} = (\text{moles}_{\text{NaOH}}) \, (\text{FW}_{\text{NaOH}})$$

$$= 0.40 \text{ mole } \frac{40 \text{ g}}{1 \text{ mole}} = 16 \text{ g}$$

Answers to Selected Questions

Chapter 1

1. There are only a small number of planets against a background of a very large number of stars.

3. Gases that are very soluble in water cannot be collected over water.

5. Water cannot be an element.

7. 44 g

9. Limestone has been decomposed into simpler (lighter) substances.

11. Samples A and B have the same composition by weight.

13. The properties of compounds differ from the properties of the elements combined.

15. The ratio of the masses of oxygen combined with 70 g of iron in the two compounds is $\dfrac{30 \text{ g}}{20 \text{ g}} = \dfrac{3}{2}$.

17. With the same fixed weight of hydrogen, the ratio of the weights of carbon in the two compounds is $\dfrac{4.89}{1.63} = \dfrac{3}{1}$.

19. No, Dalton did not say that an atom of an element had the same properties as the element.

21. If two compounds are formed from a pair of elements and if atoms combine in small whole number ratios, then with a fixed number of atoms of one element, the numbers of the second element would be a small, whole-number ratio. Since all atoms of an element have the same weight, the same fixed number of atoms of the first element would have the same weight in the two compounds. The ratio of the weights of the second element would be in the same small, whole-number ratio as the ratio of atoms.

Chapter 2

1. HO_2

3. UF_3, UF_6

5. (a) 5 (b) 6, 3 (c) 2

7. 50

9. Correct formulas are required to determine atomic weights.

11. 65.4, Zn

13. (a) 64.1 (b) 80.1 (c) 56.1 (d) 136.1

15. families-similar properties and formulas of compounds; rows-increasing atomic weights (unless family assignment indicated a reversal of atomic weight or the presence of a gap for a yet undiscovered element)

17. (a) AsH_3 (b) GeH_4 (c) SeH_2 (d) HI

19. Eka boron is scandium (Sc), and eka aluminum is gallium (Ga).

Chapter 3

1. $Ba^{+2} + 2 e^- \rightarrow Ba$

3. No, the radiation observed was the beta decay of radioactive "daughters" of U-238.

5. Some atoms are not "indestructible."

7. Radioactivity is associated with a particular chemical element. Radioactivity was associated with the nucleus after the development of a nuclear model for atomic structure, for it was the nucleus, not the electrons, that was unique to a given element.

9. With only a small number of drops, the largest common divisor might not correspond to the charge on a single electron. For example, all the drops in a small sample might have an even number of electrons, giving a value twice as large as the accepted value.

11. The fraction of alpha particles passing directly through a foil would be similar since the nucleus of all atoms is very small compared to the distance between nuclei. However, an alpha particle would be deflected less by the lighter, less positive nucleus of an element having a lower atomic number than gold.

15. $^{238}_{92}U$ has 146 neutrons; $^{14}_{6}C$ has 6 protons and 8 neutrons; $^{35}_{17}Cl$

17. In Cl^-, the same positive nucleus is surrounded by one more electron than in an atom of Cl. The increased repulsion between negative electrons causes the anion to be larger than the neutral atom.

19. (a) $^{222}_{86}Rn$, (b) $^{4}_{2}He$, (c) $^{218}_{84}Po$, (d) $^{214}_{82}Pb$

21. $^{30}_{15}P \rightarrow ^{30}_{14}Si + ^{0}_{+1}e$

Chapter 4

1. longest, blue; shortest, red
3. neon with eight outer electrons
5. As minor constituents of air, the physical properties of the noble gases were masked by the properties of nitrogen and oxygen. Unlike other elements, they were not found combined in solid compounds that could be isolated and studied.
7. The value of $(\frac{1}{2^2} - \frac{1}{n^2})$ increases and approaches $\frac{1}{4}$ as n gets larger.
9. The smallest value of $(\frac{1}{2^2} - \frac{1}{n^2})$ is $\frac{5}{36}$ when n = 3.

   As n becomes larger, the value in parentheses approaches $\frac{1}{4}$.
11. 50
15. An electron behaves as a wave in the modern description of the atom.
17. (a) 1 (b) 3 (c) 5 (d) 7 (e) 9
19. Complete the following table.

atomic number	element	protons	inner electrons	kernel charge	outer electrons
4	Be	4	2	+2	2
12	Mg	12	10	+2	2
20	Ca	20	18	+2	2
8	O	8	6	+6	6
16	S	16	10	+6	6
34	Se	34	18	+6	6

21. O 2p, Ca 4s, Fe 3d, Li 2s, Ag 4d, Hg 5d, U 5f

Chapter 5

1. −3, −1, +2, +3
3. (a) LiI; (b) K_2O; (c) CaS; (d) $BaCl_2$; (e) AlF_3
5. lattice with Mg^{+2} and O^{-2} ions
7. $Ba^{+2} > Sr^{+2} > Ca^{+2} > Mg^{+2}$
9. CsI has the largest cation and the largest anion so the attractive forces between the +1 cation and the −1 anion should be weaker, and the compound lower-melting.
11. (a) KNO_3; (b) Li_2CO_3; (c) $MgSO_4$; (d) NH_4Br
13. (a) SiF_4; (b) CS_2; (c) NCl_3; (d) P_2O_3 or P_2O_5
15.
$$\overset{\displaystyle H}{\underset{\displaystyle H}{H-\overset{|}{\underset{|}{Si}}-H}} \qquad H-\overset{\cdot\cdot}{\underset{\displaystyle H}{\underset{|}{P}}}-H \qquad H-\overset{\cdot\cdot}{\underset{\cdot\cdot}{\underset{|}{S}}}: \qquad H-\overset{\cdot\cdot}{\underset{\cdot\cdot}{Cl}}:$$
17. (a) triangular pyramid; (b) bent; (c) tetrahedral; (d) linear

Chapter 6

1. KI, BaS and AlF_3 are ionic.
3. AgF is ionically bonded because the difference in electronegativity between Ag and F is greater than the differences between Ag and the other halogens. Ionic bonding is favored when there is a large difference in electronegativity.

5. BF_3 is planar with boron at the center of an equilateral triangle formed by the fluorines. The dipole moments of the three B—F bonds cancel.
7. Nitrogen gas consists of diatomic molecules with weak intermolecular forces. Diamond is a giant covalent molecule.
9. Ammonia molecules form strong hydrogen bonds to water molecules; methane molecules do not.
11. If a water molecule was linear, it would not have a dipole moment and would not solvate ions well.
13. HF is a very polar molecule, and C_8H_{18} is not. Polar molecules can solvate ions to help overcome lattice forces.
15. Unless there is a large hydrophobic group present in a soap molecule, it cannot form stable micelles capable of carrying off grease.

Chapter 7

1. Water can damage electrical equipment causing short circuits. It also damages grain, promoting spoilage.
3. The shorter the lifetime that a molecule has in the atmosphere, the less likely it is to diffuse into the upper atmosphere where it can react to damage the ozone layer.
5. Some small fraction of CFCs released at ground level are destroyed or removed before these compounds interfere with the ozone layer. In contrast, the release of these compounds in the stratosphere delivers a high local concentration of ozone-depleting chemicals. However, the release of these compounds in outer space has no effect on the ozone layer.
7. The widespread use of technologies that capture and recycle CFC substitutes would minimize their impact on global warming.
9. The volume of CFCs used for the delivery of medicines is only a small fraction of their overall use.
11. Some volcanic eruptions put large quantities of sulfur oxides and dust high in the atmosphere. Droplets containing sulfur compounds and dust particles reflect some of the incoming solar radiation, shading and cooling the surface of the earth.

Chapter 8

1.
$$H-Cl + H-\overset{\displaystyle H}{\underset{\displaystyle H}{\underset{|}{O}}} \rightarrow Cl^- + H-\overset{\displaystyle H}{\underset{\displaystyle H}{\underset{|}{O^+}}}$$
3. basic, a weak base
5. (a) 1.0 mole/L; (b) 1.0×10^{-5} mole/L; (c) 1.0×10^{-7} mole/L; (d) 1.0×10^{-12} mole/L
7. 0.060 mole/L
9. $2\,OH^- + CO_2 \rightarrow CO_3^{-2} + H_2O$
11. (a) and (c) $H_3O + NH_3 \rightarrow H_2O + NH_4^+$; (b) and (c) $OH^- + NH_4^+ \rightarrow H_2O + NH_3$; An aqueous solution of ammonia and ammonium chloride is a buffer.
13. (a) PH_3; (b) PH_3; (c) OH^-
15. $HAc + HCO_3^- \rightarrow Ac^- + H_2O + CO_2$

Chapter 9

1. (a) +5, +5, +3 (b) +6, +4, +2, −2 (c) 0, +2, +4, +7 (d) −1, −1, +1, +7
3. Species being oxidized: (a) CH_4 (b) Fe^{2+} (c) Fe^{2+} (d) I^- (e) Cl_2; species being reduced: (a) O_2 (b) Cl_2 (c) MnO_4^- (d) Br_2 (e) Cl_2
5. Ease of oxidation is in the order Cu > Ag > Au.
7. (a) +3; (b) nitric acid; (c) Cl^-
9. The outer electron in I^- is more easily removed (oxidation), because it is farther from a +7 kernel than the outer electron in Cl^-.
11. Cl^-, NH_4^+
13. decreases, decreases, increases, decreases
15. $2\,SO_2 + O_2 \rightarrow 2\,SO_3$; $H_2O + SO_3 \rightarrow H_2SO_4$
17. (a) $NaCl + H_2SO_4$; (b) electrolysis; (c) electrolysis

Chapter 10

1. 10 minutes; 20 minutes
3. To find the fraction of a half-life that has elapsed, it is necessary to know both the original, and the final, amount of U-238 present. Pb-206 is a measure of the U-238 that has decayed.
5. After three half-lives, the rate of decay is one-eighth the original rate. The object is about 1,750 years old.
7. The fraction of carbon 14 in the atmosphere is sensitive to the rate of neutron production initiated by incoming cosmic rays.
9. Even though collisions between molecules was occurring at the same frequency, the reaction would be faster at the higher temperature because a greater fraction of the collisions would have sufficient energy for the molecules to react.
11. The addition reaction occurs at the surface of the catalyst. Platinum prepared by the reduction of powdered platinum oxide has a very large surface area. Grinding the metal would produce a smoother surface with lower area.
13. (a) increase rate; (b) increase rate; (c) increase rate
15. Hydronium ion is a reactant in the overall reaction. In the first step, hydronium ion reacts with methanol in an acid-base reaction. In the second step, the conjugate acid of methanol reacts with a chloride ion to form methyl chloride and water.
17. Binding interactions include both hydrogen bonds involving hydrogen atoms of N—H bonds of sulfanilamide, and weak interactions between the non-polar ring and non-polar groups on the enzyme.

Chapter 11

1. Only part of the energy in food is converted into work.
3. 3.74 calories
5. more than; Hydrogen bonded water is more ordered in the liquid state.

7. Molecules of H_2O in the vapor state are moving independently of one another. In contrast, molecules of H_2O in the liquid state are associated and highly oriented by strong hydrogen bonds.
9. The half-life of an isotope reflects the average behavior of a large number of identical nuclei. With a small sample, fluctuations in the rate of decay and the presence of background decay events make it more difficult to distinguish the idealized average behavior.
11. To extract and concentrate low-grade iron ore requires more energy than the extraction of high-grade iron ore. The added energy costs increase the cost of the iron produced.
13. (a) decreases; (b) increases
15. By reducing the amount of oxygen present at combustion temperatures, the equilibrium between the reactants N_2 and O_2 and the products NO and NO_2 is shifted to reduce the amounts of nitrogen oxides.

Chapter 12

3. Natural gas contains smaller amounts of sulfur and nitrogen compounds than most petroleums. Coal contains the greatest quantity of contaminants.
5. Sulfuric acid from the oxidation of sulfur-containing compounds in the oil would attack the metal of the engine.
7. Marble reacts with strong acids:

$$CaCO_3 + H_3O^+ \rightarrow Ca^{2+} + HCO_3^- + H_2O$$

17. These tailings contain radioactive daughters of uranium-238. For example, radioactive radon gas would escape from this loose material.

Chapter 13

1. −6; triangular with three bridging oxygens
3. $CaCO_3 + H_3O^+ \rightarrow Ca^{+2} + HCO_3^- + H_2O$
5. (a) The glass would be more like sodium silicate. It would be too low-melting and liquid to be readily shaped. (b) The glass would be more like sand; it would be too high-melting and viscous to be readily shaped.
7. Cement gains strength over a prolonged period of time as hydrated crystals form and grow.
9. (a), (c), and (d)
11. Many metals lose strength at elevated temperatures. Others are readily oxidized when hot. Cement does not have strength at high temperatures when water is driven from its hydrated crystals.
13. In a dry clay, the associated cations are closer, on average, to the negative aluminosilicate plates than in a moist clay. Hence, the force of attraction between positive and negative charges is greater.

15. Soils differ greatly in the numbers and kinds of ion exchange sites that can bind mineral ions.

17. Negative aluminosilicate plates repel one another. They can only come together when their negative charge is balanced by the positive charge of cations such as Mg^{2+} and K^+ present in seawater.

Chapter 14

1. (d)

3. Fe^{+2} is harder to reduce than Cu^{+2}; Ag is harder to oxidize than Cu.

5. $CuS + O_2 \rightarrow Cu + SO_2$; $SnO_2 + C \rightarrow Sn + CO_2$

7. Reactions between two solids are slow due to the limited area of contact between them.

9. It is important to know what is present in a mixture of metals in order to know what properties the alloy prepared from that mixture would have.

11. A coating of tough, adhering, impenetrable MgO protects the surface of magnesium.

13. Most of the aluminum in scrap aluminum is already reduced, and a major cost in producing aluminum is the reduction of aluminum oxide.

15. As in the case of Al^{+3}, the high positive charge on Fe^{+3} pulls electrons from the surrounding water molecules, causing them to be more acidic.

17. Impurities that are both easier to oxidize and harder to oxidize are removed in the electrolytic purification of copper because the process includes both an oxidation step and a reduction step. Because the impurities present in aluminum ores (particularly iron compounds) contain metals easier to reduce than aluminum, these impurities would be reduced along with the aluminum and would weaken the protective oxide coating on the aluminum.

19. The energy required to make a pound of glass from raw materials is far less than the energy required to produce a pound of aluminum from ore, yet the transportation costs in collecting each would be nearly the same.

Chapter 15

1. The alcohol boils higher than the isomeric ether. More energy is required to overcome hydrogen bonds between alcohol molecules than is required to separate polar ether molecules.

3. (a)
(b) $CH_2=CH-CH_3$

(c)
(d) $CH_3-CH_2-O-CH_3$

5. (a)-(1); (b)-(3); (c)-(2)

7. The formation of hydrogen bonds between the —OH group of alcohols and solvent water molecules increases the solubility of alcohols in water.

9.

11.

13. (a)
(b)

(c) (d) $CH_3CH_2NH_2$

(e)

15.

17. The high kernel charge of fluorine pulls electron density away from the —OH, weakening the bond to hydrogen and making the oxygen atom more positive. This shift in electron density makes it easier for the acid to donate H^+ to a water molecule or any other base.

19. (a) butanoic acid (butyric acid) (b) propanal (propionaldehyde) (c) 3-pentanone (d) ethyl methanoate (ethyl formate) (e) ethanamide (acetamide)

21. (a)
$+ H_2O$

(b)
$+ HO-CH_3$

23.

Electrons are pulled toward the oxygen making nitrogen more positive and less able to share its lone pair of electrons.

Chapter 16

1. More refining steps are required to produce high-octane compounds, branched alkanes, and aromatic hydrocarbons than to produce low-octane, linear alkanes. Because some energy is required for each refining step, there are losses in each step.

5. Plants to manufacture ammonia are located near oil refineries, because the hydrogen used in the manufacture of ammonia is a by-product of the refining of petroleum.

7. Natural gas consists largely of methane. Ethylene is a by-product of petroleum refining.

9. Unless there is a large hydrophobic group present in a soap or a detergent, it would not form stable micelles capable of carrying off grease.

11. Much of the NO_2 produced at ground level, or at low levels in the atmosphere, dissolves in water and returns to earth before ever reaching the ozone layer.

13. When light is absorbed, an electron is lifted from a lower energy orbital to a higher energy orbital. This process weakens chemical bonds, making weak bonds more susceptible to cleavage.

Chapter 17

1. 16,400; 4

3.

5. (c)

7. Linear polyethylene chains stack better than the chains in branched polyethylene, and so high-density polyethylene has stronger crystalline regions.

9. $C_4H_9\text{-N=N-}C_4H_9 \rightarrow 2\ C_4H_9^{\bullet} + N\equiv N$
This compound decomposes to form free radicals that initiate polymerization.

11. Di-*n*-octyl phthalate is a plasticizer that causes polyvinyl chloride to be more flexible.

13. more; harder

15. The chains are more nearly parallel in a textile fiber.

17.

19. (a) A polyester made with a long-chain diol would have too little crystallinity and strength. (b) Reaction of a diacid with glycerol gives a rigid, three-dimensional thermosetting resin that cannot be spun to form fibers. (c) A linear polyester made using phthalic acid would not stack well to form parallel chains, so it would lack the strength to form a good fiber.

21. anions

Chapter 18

1.

3. $C_6H_{12}O_6$, $C_6H_{12}O_6$

5. Enzymes that cleave β-glycoside linkages break down cellulose.

7. Cod and tuna are active at cold temperatures. Lipids containing unsaturated fatty acids are more liquid and less rigid at these temperatures.

9. A diet that combines corn and beans provides the balanced mixture of essential amino acids needed to synthesize new proteins.

11.

13.

15. The function of a protein, enzyme, or zymogen depends on its three-dimensional shape. If the zymogen had the same shape as the enzyme at the catalytic site, it would be a far more effective catalyst.

17. Membrane-bound proteins have non-polar groups at their surfaces in contact with membrane lipids. In contrast, water-soluble proteins have polar groups at their surfaces in contact with water.

Chapter 19

1.

5. $3 \times 2 \times 1 = 6$

7. H_2N-serine-lysine-alanine-OH

9. TTCGA would be the complementary sequence with the complimentary chain running in the opposite direction.

11. The GC pair with three hydrogen bonds is more strongly held than the AT pair with two.

13. If only two bases were required, $4 \times 4 = 16$ possible pairs are possible. This is not enough to code for all the amino acids found in proteins. With three bases, there are a total of $4 \times 4 \times 4 = 64$ possible combinations.

15. A point mutation gives rise to a single amino acid substitution. A frame-shift mutation causes the reading frame for all subsequent amino acids to be incorrect.

A

acid a substance that increases the hydronium ion concentration in aqueous solution; a proton donor

acid, Lewis a molecule or ion that can act as an electron-pair acceptor

acid-base reaction transfer of a proton from a proton donor to a proton acceptor

acid-base reaction, Lewis formation of a covalent bond by electron-pair donation

acid rain acidic precipitation resulting from the emission of sulfur dioxide and nitrogen oxides

acidic (aqueous) solution a solution that contains a greater concentration of H_3O^+ than OH^- ions; pH<7

activation energy the energy that reactants must have for reaction to occur

addition reactions addition of atoms or groups to a double or triple bond

alcohol ROH, where R is an alkyl group

aldehyde RCHO, where R is an alkyl or an aryl group

alkali metal a column I metal

alkaline earth metal a column II metal

alkanes hydrocarbons with only single bonds

alkenes hydrocarbons with double bonds

alkyl group group containing one less hydrogen than an alkane

alkyl halides RX, where R is an alkyl group and X is a halogen atom

alkynes hydrocarbons with triple bonds; acetylenes

alloy intimate mixture of two or more metals, or metals plus nonmetals, in a substance that has metallic properties

α decay (alpha decay) emission of an α particle by a radioactive nuclide

α particle (alpha particle) a dipositive helium ion, He^{2+}

amalgam alloy of mercury and another metal

amide $RCONH_2$, where R is an alkyl or an aryl group

amine RNH_2, R_2NH, R_3N, where R is an alkyl or an aryl group

anhydrous denoting the absence of water

anion negatively charged ion

anode electrode at which oxidation occurs

aqueous solution a solution in water

aromatic hydrocarbons benzenelike hydrocarbons

aryl group group having one less hydrogen than an aromatic hydrocarbon

asymmetric atom an atom bonded to four different groups

atmosphere all of the gases that surround the earth

atom smallest particle of an element that can participate in a chemical reaction

atomic mass unit one-twelfth the weight of one carbon-12 atom; the unit of atomic and molecular weights

atomic number the number of protons in the nucleus of each atom of an element

atomic orbital the space in which an electron with a specific energy is most likely to be found

atomic weight the average weight of the atoms of the naturally occurring element relative to one-twelfth the weight of an atom of carbon-12

Avogadro's law equal volumes of gases, measured at the same temperature and pressure, contain equal numbers of molecules

Avogadro's number the number of atoms in exactly 12 g of carbon-12; 6.022×10^{23}

B

base a substance that in aqueous solution increases the hydroxide ion concentration; a proton acceptor

base, Lewis a molecule or ion that can act as an electron-pair donor

basic (aqueous) solution a solution that contains a greater concentration of OH^- ions than H_3O^+ ions; pH > 7

battery two or more voltaic cells combined to provide electric energy

β decay (beta decay) electron emission by a radioactive nuclide

β particle (beta particle) high-speed electron produced by a nuclear reaction

binary compound compound containing only two elements

binding energy (nuclear) the energy required to decompose a nucleus into its component nucleons

boiling point temperature at which a liquid boils (bubbles of vapor form throughout the liquid)

bond length average distance from nucleus to nucleus in a stable compound

breeder reactor a nuclear reactor that produces more fissionable atoms than it consumes

buffer solution a solution that resists changes in pH when small amounts of strong acid or strong base are added

C

calorimeter a device for measuring the heat given off or absorbed by a chemical reaction

carbohydrate a polyhydroxy aldehyde, a polyhydroxy ketone, or a compound derived from these subunits

carbonyl group C=O

carboxylate acid RCOOH where R is an alkyl or an aryl group

carboxylate ion $RCOO^-$, where R is an alkyl or an aryl group

catalyst a substance that increases the rate of a reaction but may be recovered from the reaction unchanged

cathode electrode at which reduction occurs

cathode rays streams of electrons flowing from the cathode to the anode in a gas discharge tube

cation a positively charged ion

ceramic a material made from clay that has been hardened by firing at high temperature

chain reaction a reaction taking place by a series of repeated reaction steps

changes of state interconversions between the solid, liquid, and gaseous states

chemical bond a force that acts strongly enough between two atoms to hold them together

chemical equation symbols and formulas representing the total chemical change that occurs in a chemical reaction

chemical equilibrium a dynamic equilibrium in which the amounts of the species present do not change with time

chemical formula the symbols for the elements combined in a compound and the subscripts which indicate how many atoms of each element are present

chemical kinetics the study of reaction rates and reaction mechanisms

chemical properties properties that can only be observed in chemical reactions

chemical reaction process in which at least one substance is changed in composition and identity

chemistry the study of matter and its transformations

chirality the property of having nonsuperimposable mirror images

cis-trans isomers compounds with identical groups arranged in different ways on either side of a double bond or rigid ring

collision theory a reaction that occurs when particles collide, only a very small portion of these collisions result in reaction

combustion a process or instance of burning; a high-temperature reaction with oxygen that releases heat and light

compound a substance of definite composition in which two or more elements are chemically combined

condensation movement of molecules from the gaseous phase to the liquid phase

continuous spectrum radiation at all wavelengths

covalent bond bond based upon electron-pair sharing; the attraction between two atoms that share electrons

critical mass smallest mass that will support a self-sustaining nuclear chain reaction

cross-linking bonding between adjacent chains in a polymer

crystalline solid substance in which the atoms, molecules, or ions have a characteristic, regular, and repetitive three-dimensional arrangement

D

deionized water water with almost all ions removed

density mass per unit volume

diatomic molecule a molecule made of two atoms

diffraction scattering of light by regular array of lines or points

dipole moment measure of the polarity of a chemical bond or molecule

doping addition of impurities to semiconducting elements

double bond two electron pairs shared between the same two atoms

E

electrochemistry study of oxidation-reduction reactions that either produce or utilize electric energy

electrode a conductor through which electric current enters or leaves a conducting medium

electron negatively charged subatomic particle

electronegativity a measure of the ability of an atom to attract electrons to itself

element a substance composed of only one kind of atom

emission spectrum the spectrum of radiation emitted by a substance

enantiomers mirror-image isomers

endpoint the point at which chemically equivalent amounts of reactants in a titration have been combined

entropy a measure of the randomness or disorder of a system

enzyme a protein that catalyzes a biological reaction

equilibrium state of balance between opposing changes

ester RCOOR, where R's are alkyl or aryl groups

eutrophication process in which a lake grows rich in nutrients and fills with organic sediment and aquatic plants

excited state state of energy reached by electrons when an atom or molecule has absorbed sufficient extra energy

exothermic process a process that releases heat

F

family the elements in a single vertical column in the periodic table; also called a group

faraday the amount of electricity represented by 1 mole of electrons; 96,500 C

flotation concentration of metal-bearing mineral in a froth of bubbles that can be skimmed off

fossil fuels natural gas, coal, and petroleum

free radicals reactive species that contain an odd number of electrons

frequency the number of wavelengths passing a given point in unit time

functional group a group that contributes a characteristic chemical behavior to the molecule

G

gas-discharge tube a glass tube that can be evacuated and in which there are sealed electrodes

Gay-Lussac's law of combining volumes a law that in a chemical reaction, the ratios of the volumes of the gases involved, measured at the same temperature and pressure, are small, whole numbers

gene a portion of DNA that codes for a specific protein

glass an amorphous solid formed when a mixture of silica and other compounds is melted and then cooled rapidly

greenhouse effect the effect of warming by absorption and re-emission of radiation

ground state lowest-energy state of an electron in an atom

H

half-life the time it takes for one-half of the nuclei in a sample of a radioactive isotope to decay

halides binary compounds of halogen atoms with other elements

halogen a column VII element

hard water water that contains dipositive ions that form precipitates with soap, or upon boiling

heat capacity the amount of heat required to raise the temperature of a given amount of material by $1°\,C$

heterogeneous reaction a reaction between substances in different phases

homogeneous reaction a reaction between substances in the same gaseous or liquid phase

hydrocarbon a compound composed of carbon and hydrogen

hydrogenation addition of hydrogen to a molecule

hydrogen bond unusually strong dipole-dipole attraction between a group in which hydrogen is bonded to N, O, or F and a lone pair of electrons on another atom or ion

hydronium ion H_3O^+

hydroxide ion OH^-

hydroxyl group —OH

I

indicator an organic acid or base that changes color when it reacts with hydronium ion or hydroxide ion

inhibitors substances that slow down a catalyzed reaction

initiation production of the first reactive intermediate in a chain reaction

inorganic chemistry the chemistry of all the elements and their compounds, with the exception of hydrocarbons and hydrocarbon derivatives

intermolecular forces the forces of attraction or repulsion between individual molecules

ion an atom or group of atoms that has a net positive or negative charge

ion exchange replacement of one ion by another

ionic bonding bonding based on the attraction between positive and negative ions

isomers compounds that differ in structure, but have the same molecular formula

isotopes atoms with different mass numbers, but the same atomic number

K

kernel the nucleus and the inner shell electrons of an atom

ketone $R_2C=O$

L

lattice energy energy liberated when gaseous ions combine to give a crystalline, ionic substance

law statement of an empirical relation between phenomena that is always the same under the same conditions

law of conservation of mass the law that in chemical reactions, matter is neither created nor destroyed

law of definite proportions the law that in a pure compound, elements are combined in definite proportion by weight

law of multiple proportions the law that if two elements combine to form more than one compound, a fixed weight of element A will combine with two or more different weights of element B so that the different weights of B are in the ratio of small whole numbers

leaching a process of extracting a soluble compound from insoluble material

Le Châtelier's principle the precept that if a system in equilibrium is subjected to a stress, a change that will offset the stress will occur in the system

line spectrum radiation at only certain wavelengths

lone pair pair of outer electrons not involved in bonding

M

mass an intrinsic property; the quantity of matter in a body

mass number the sum of the number of protons and the number of neutrons in a nucleus

melting point temperature at which the solid and liquid phases of a substance are in equilibrium

metallic bonding the attraction between positive metal ions and surrounding freely mobile electrons

metallurgy all aspects of the science and technology of metals

mineral a naturally occurring substance with a characteristic range of chemical composition

mixture any combination of two or more substances in which the substances combined retain their identity

molarity moles of dissolved substance per liter of solution

mole an Avogadro's number of a chemical compound or species

molecular weight sum of the atomic weights of the total number of atoms in the formula of a chemical compound

molecule smallest particle of a pure substance that has the composition of that substance and is capable of independent existence

N

natural radioactivity decay of radioactive isotopes found in nature

net ionic equation an equation that shows only the species involved in the chemical change

neutral (aqueous) solution a solution that contains equal numbers of H_3O^+ and OH^- ions

neutralization the reaction of an acid with a base

neutron fundamental, subatomic particle that carries no charge and has a mass almost the same as the mass of the proton

nitrogen fixation the formation of compounds from molecular nitrogen

noble gas a column VIII element

non-polar bond a bond in which electrons are shared equally between two atoms

normal boiling point boiling point at 1 atm pressure

normal hydrocarbon chain an unbranched chain

_n_-type semiconductor negative electrons are the majority of the current carriers; contains a donor impurity

nuclear-binding energy energy that would be released in the combination of nucleons to form a nucleus

nuclear fission splitting of a heavy nucleus into two lighter nuclei of intermediate mass number

nuclear fusion combination of two light nuclei to give a heavier nucleus

nuclear reactions reactions that result in changes in atomic number, mass number, or energy state of nuclei

nuclear reactor equipment in which nuclear fission is carried out at a controlled rate

nucleon a proton or a neutron

nucleus a central region in the atom that is very small by comparison with the total size of the atom, and in which all of the mass and positive charge of the atom are concentrated

O

octet rule atoms tend to combine by gain, loss, or sharing of electrons so that each atom holds or shares eight outer electrons

olefins hydrocarbons with double bonds; alkenes

optical isomerism occurrence of pairs of molecules of the same molecular formula that rotate plane-polarized light equally in opposite directions

optically active compound that which rotates the plane of vibration of plane-polarized light

ore a mixture of minerals from which a particular metal or several metals can profitably be extracted

organic chemistry the chemistry of carbon compounds and their derivatives

osmosis the passage of solvent molecules through a semipermeable membrane and from a more dilute solution into a more concentrated solution

oxidation any process in which the oxidation number increases

oxidation number a number that represents the positive or negative character of atoms in compounds found by a set of rules

oxidation-reduction reactions reactions in which oxidation and reduction occur together; also called redox reactions

oxides binary compounds of oxygen

oxidizing agent an atom, molecule, or ion that can cause another substance to undergo an increase in oxidation number

P

peptide bond (linkage) the amide linkage joining two amino acids

period a horizontal row of elements in the periodic table

periodic law the physical and chemical properties of the elements are periodic functions of their atomic numbers

periodic table a chart showing all the elements arranged in columns with similar chemical properties

petroleum cracking a process of breaking large molecules into small molecules, usually in the gasoline range

petroleum isomerization conversion of straight-chain alkanes into branched-chain alkanes

petroleum reforming conversion of noncyclic hydrocarbons to aromatic compounds

pH $-\log [H_3O^+]$

photoelectric effect the giving off of electrons by certain metals when light shines on them

photon a single quantum of radiant energy (1 hv)

photosynthesis reaction of carbon dioxide and water in green plants to give carbohydrates

physical properties properties that can be exhibited, measured, or observed without changing the composition and identity of a substance

polar bond a covalent bond in which electrons are shared unequally

pollutant an undesirable substance added to the environment, usually by the activities of earth's human inhabitants

polyatomic ions ions that incorporate more than one atom

polymer a large molecule made of many units of the same structure linked together

polypeptide a polymer formed from amino acids joined by peptide linkages

positron particle identical to an electron in all properties except charge, which is +1

precipitate solid that forms during a reaction in solution

pressure force per unit area

products new substances produced in a chemical reaction

propagation production of the species that initiate further steps in a chain reaction

proton fundamental, subatomic particle with a positive charge equal in magnitude to the negative charge of the electron

_p_-type semiconductor a semiconductor in which positive holes are the majority of the current carriers; contains an acceptor impurity

pure substance a form of matter that has identical physical and chemical properties regardless of its source

Q

quantized measurements measurements restricted to quantities that are multiples of the basic unit, or quantum, for a particular system

quantum numbers numbers that specify the amounts of energy of a system, such as an electron in an atom

R

radiation energy traveling through space

radioactive isotopes isotopes that decay spontaneously

radioactivity spontaneous emission by unstable nuclei of particles, or of electromagnetic radiation, or of both

rate constant the proportionality constant between the rate and the reactant concentrations

rate-determining step the slowest step in a reaction mechanism

rate equation an equation that gives the relationship between the reaction rate and the concentration of the reactants

reactants substances that are changed in a chemical reaction

reaction mechanism the pathway from reactants to products

reaction rate speed with which products are produced or reactants are consumed in a particular reaction

reagent any chemical used to bring about a desired chemical reaction

redox reactions oxidation-reduction reactions

reducing agent an atom, molecule, or ion that can cause another substance to undergo a decrease in oxidation number

reduction any process in which oxidation number decreases

resonance hybrid actual molecular structure of a molecule or ion which is not adequately described by a single Lewis structure

reversible reaction a chemical reaction that can proceed in either direction

S

salts neutral ionic compounds composed of the cations of bases and the anions of acids

saponification hydrolysis of an ester by a base

saturated hydrocarbons hydrocarbons containing only single covalent bonds

silica SiO_2

silicates compounds containing metal cations and silicon-oxygen groups

single bond a bond in which two atoms are held together by the sharing of two electrons

slag molten mixture of minerals

smelting a process for producing a free metal from its ore

solubility the amount of solute that can dissolve in a given amount of solvent

solution homogeneous mixture of the molecules, atoms, or ions of two or more substances; a single-phase mixture

solvation interaction of a solute molecule or ion with solvent molecules

solvent the component of a solution usually present in the larger amount

stable isotopes isotopes that do not decay spontaneously

states of matter the gaseous state, the liquid state, and the solid state

steel alloys of iron that contain carbon (up to 1.5%) and usually other metals as well

strong acid an acid that is virtually 100% ionized in dilute aqueous solution to produce hydronium ions

strong base a base that is completely dissociated in dilute aqueous solution to produce hydroxide ions

subatomic particle a particle smaller than the smallest atom

T

theory unifying model or set of assumptions that explains phenomena and laws based on observation

thermodynamics the study of energy and its transformations

titration measurement of the volume of a solution of one reactant that is required to react completely with a measured amount of another reactant

triple bond three electron pairs between the same two atoms

U

unsaturated hydrocarbons hydrocarbons with double or triple covalent bonds

unsaturated solution a solution that can still dissolve more solute under the given conditions

V

voltaic cell a cell that generates electric energy from a spontaneous redox reaction

W

wavelength the distance between two peaks or troughs on adjacent waves

weak acid a compound or ion that reacts with water only to a small extent to produce hydronium ions

weak base a compound or ion that reacts with water only to a small extent to produce hydroxide ions

weight the force exerted on a body by gravity

TABLE OF THE ELEMENTS

Name	Symbol	Atomic number	Name	Symbol	Atomic number
Actinium	Ac	89	Molybdenum	Mo	42
Aluminum	Al	13	Neodymium	Nd	60
Americium	Am	95	Neon	Ne	10
Antimony	Sb	51	Neptunium	Np	93
Argon	Ar	18	Nickel	Ni	28
Arsenic	As	33	Niobium	Nb	41
Astatine	At	85	Nitrogen	N	7
Barium	Ba	56	Nobelium	No	102
Berkelium	Bk	97	Osmium	Os	76
Beryllium	Be	4	Oxygen	O	8
Bismuth	Bi	83	Palladium	Pd	46
Boron	B	5	Phosphorus	P	15
Bromine	Br	35	Platinum	Pt	78
Cadmium	Cd	48	Plutonium	Pu	94
Calcium	Ca	20	Polonium	Po	84
Californium	Cf	98	Potassium	K	19
Carbon	C	6	Praseodymium	Pr	59
Cerium	Ce	58	Promethium	Pm	61
Cesium	Cs	55	Protactinium	Pa	91
Chlorine	Cl	17	Radium	Ra	88
Chromium	Cr	24	Radon	Rn	86
Cobalt	Co	27	Rhenium	Re	75
Copper	Cu	29	Rhodium	Rh	45
Curium	Cm	96	Rubidium	Rb	37
Dysprosium	Dy	66	Ruthenium	Ru	44
Einsteinium	Es	99	Samarium	Sm	62
Erbium	Er	68	Scandium	Sc	21
Europium	Eu	63	Selenium	Se	34
Fermium	Fm	100	Silicon	Si	14
Fluorine	F	9	Silver	Ag	47
Francium	Fr	87	Sodium	Na	11
Gadolinium	Gd	64	Strontium	Sr	38
Gallium	Ga	31	Sulfur	S	16
Germanium	Ge	32	Tantalum	Ta	73
Gold	Au	79	Technetium	Tc	43
Hafnium	Hf	72	Tellurium	Te	52
Helium	He	2	Terbium	Tb	65
Holmium	Ho	67	Thallium	Tl	81
Hydrogen	H	1	Thorium	Th	90
Indium	In	49	Thulium	Tm	69
Iodine	I	53	Tin	Sn	50
Iridium	Ir	77	Titanium	Ti	22
Iron	Fe	26	Tungsten	W	74
Krypton	Kr	36	Unnilhexium	Unh	106
Lanthanum	La	57	Unnilpentium	Unp	105
Lawrencium	Lr	103	Unnilquadium	Unq	104
Lead	Pb	82	Uranium	U	92
Lithium	Li	3	Vanadium	V	23
Lutetium	Lu	71	Xenon	Xe	54
Magnesium	Mg	12	Ytterbium	Yb	70
Manganese	Mn	25	Yttrium	Y	39
Mendelevium	Md	101	Zinc	Zn	30
Mercury	Hg	80	Zirconium	Zr	40